Innovation, Employment and Growth Policy Issues in the EU and the US

Paul J.J. Welfens · John T. Addison
Editors

Innovation, Employment and Growth Policy Issues in the EU and the US

Springer

Editors
Prof. Dr. Paul J.J. Welfens
EIIW - European Institute for International Economic Relations
University of Wuppertal
Rainer-Gruenter-Str. 21
42119 Wuppertal
Germany
welfens@uni-wuppertal.de

Prof. John T. Addison
University of South Carolina
Moore School of Business
Dept. Economics
1705 College Street
Columbia SC 29208
USA
ecceaddi@darla.badm.sc.edu

ISBN 978-3-642-00630-2 e-ISBN 978-3-642-00631-9
DOI 10.1007/978-3-642-00631-9
Springer Dordrecht Heidelberg London New York

Library of Congress Control Number: 2009926957

Cover design: WMX Design GmbH, Heidelberg

Printed on acid-free paper

Springer is part of Springer Science+Business Media (www.springer.com)

Preface

Economic development has cyclical dynamics and long-term dynamics – the latter are typically related to demographical changes, innovation and long-term institutional changes in open economies. Financial markets – that means mainly capital markets – and labor markets are affected in OECD countries both by innovations and institutional reforms. As regards demographics ageing is a typical challenge on both sides of the Atlantic, and pension reforms in industrialized countries have placed greater emphasis on capital markets than in previous decades. Innovation dynamics certainly are also quite important for all high wage OECD countries. The Lisbon Agenda has put particular emphasis on more growth, higher innovation dynamics and better exploitation of the advantages of a digitally networked society. Traditionally, the US has a lead in global innovations, and the US policy certainly has contributed to the American technological leadership. There still is a per capita income gap in favor of the US and the US labor market situation also looks relatively favorable, but in the five years since 2001 employment growth in the euro area was higher than that of the US. The euro area is, however, a rather heterogeneous set of countries which differ in terms of institutions, attitudes and reform progress – and everywhere governments are aware that there have to be reforms, not least in the context of globalization which bring a more complex and dynamic spatial structure of value-added. In short, policymakers in Europe are well aware of some of the key challenges and the Lisbon Agenda of the EU is one of the answers which Europe has formulated in its search to become more competitive. While the EU (and Japan) are ageing relatively fast, the US still is facing favorable demographic dynamics, but in the long-term ageing will affect all OECD countries, and this in turn raises new issues with respect to factor market adjustments, pension reforms, innovation dynamics and – in a broader global perspective – energy policies. What progress has been made on both sides of the Atlantic, what reforms are promising, and what benchmark is useful for those willing to learn? The analytical contributions and the policy papers presented in this book suggest key answers to these topics and the relevant issues.

The economic relations between the EU and the US are a key element of international dynamics where trade, capital flows and cooperation between policymakers are crucial fields of joint interest. In a transatlantic perspective, one may emphasize both the competition among firms in the world markets (often are multinational companies) which have split up the value-added chain across many countries. Trade and foreign direct investment thus go together and since the mid-1980s foreign investment indeed has grown faster than trade. Portfolio capital flows and the internationalization of banking in both the US and the EU are also key elements of globalization dynamics on both sides of the Atlantic. The US has been the economic leader since the early 1990s as its growth rate consistently outpaced that of real GDP in the EU and eurozone, respectively.

It was not before 2006 that real GDP in the eurozone increased as fast as in the US, in 2007: the eurozone's growth rate reached 2.6% and thus exceeded that of the US by 0.4% points (figures are based on preliminary calculations of the Deutsche Bundesbank). The fact that the eurozone achieved 2.3% annual growth in the period 2005–2007 and thus had nearly had closed the gap vis-á-vis the US (2.7% growth p.a. in the same period) is strongly influenced by the improvement of Germany's economic growth which improved from a meager 0.3% p.a. in 2002–2004 to 2% in the period 2004–2007. Part of US economic growth might reflect an upward bias through high capital inflows, including foreign direct investment inflows. Such inflows to some extent reflected the favorable expectations stemming from technology dynamics, including the high progress rate in the field of information and communication technology – easy financing conditions supported by (partly doubtful) financial engineering of investment banks in Wall Street also have contributed to the favorable US performance which seemed to outpace the growth dynamics in Europe. Several EU countries had high net capital outflows in the decade after 1995, while the US has moved towards high net capital inflows. Part of these inflows came from Asia and China, respectively.

Both the EU and the US face a long-term challenge from China as the country has launched an impressive economic and technological catching-up process, and this in turn will reinforce leading OECD countries' quest for enhancing innovation and growth – and such goals often can be achieved only if policymakers adopt adequate reforms in the domestic economy and effectively exploit the opportunities of modern globalization. Capital market dynamics as well as labor market reforms and innovation policies naturally become important in this context.

These facts and developments need to be explained, and investment dynamics as well as innovation performance plus factor market developments might have contributed to these changes. Changes in companies' performance, policy reforms as well as institutional modernization might have contributed to the transatlantic convergence in growth rates. Taking into account the different population growth rates in the US and the eurozone – about 1% in the US vs. about 0.5% p.a. in the eurozone in 1996–2007 – one may point out that real net domestic product per capita in 2005–2007 achieved 1.8% annual growth on both sides of the Atlantic while the eurozone stood at a modest 0.4 percentage points in 2002–2004 (compared to 0.4% in the period 2002–2004; 1996–2001 witnessed about 2% p.a. for both the eurozone and the US). In the medium term the EU eastern enlargement seems to favor the position of the eurozone and the EU, respectively: accession countries have achieved relatively high growth and the catching-up process with growth rates somewhat above the EU15 may be expected to continue for many years. The relatively high unemployment rates of the EU27 – with some EU countries actually having achieved full employment at turn of the century – has gradually come down after 2001, but it is unclear to which extent cyclical or structural influences have been decisive.

With EU eastern enlargement the European Union has become bigger and also more diverse in terms institutional settings, at the same time the EU27 is eager to catch up with the US. Beyond this supranational policy goal which naturally is linked to the EU innovation policy (and that of its member countries) there are key challenges at the national level in EU member countries: labor market reforms as well as reforms in social security are among the most difficult problems in some of the large EU countries. Ageing in European societies is proceeding faster than in the US and this raises major issues for policy makers. Capital markets dynamics will be affected in the course of pension reforms in the EU and elsewhere; this also will have effect on the current account position in the long-term. As economic globalization is unfolding – with a rising role of China and India in the production of industrial goods and the provision of digital services, respectively – the high- wage OECD countries will seek to specialize more in technology-intensive production of goods. Here innovation policy is a key challenge for both the US and the EU, and it is quite interesting to consider the different policy approaches pursued and to focus on key sectors. The knowledge society is a useful concept which is unfolding on both sides of the Atlantic. It is, however, clear that the internet age brings more truly global market places and more worldwide competition than before so that digital modernization has many global aspects.

Both the EU and the US have a common interest in energy security and in energy policies which support sustainable development and economic modernization in combination with ecological progress. Cooperation and competition both play a role in the field of energy markets and innovation in energy-intensive industries and activities, respectively. However, the US and Europe have pursued rather different policy approaches in core fields of energy policy over many years; the issues concerning the Kyoto Protocol and global warming are difficult to reconcile with US ambitions to maintain high economic growth in a framework with maximum policy autonomy. US economic growth – strong in the 15 years after 1991 – is facing new problems in the context of a financial market crisis and a US banking crisis which could not only bring about several years of slower growth in the US but also critical international spillovers for EU countries and other partner countries.

Policy concepts in the US and Europe are similar in many fields, including capital market liberalization and pension reforms, but there are also distinct areas where the policy approaches differ markedly. This partly reflects different weighting of values and differences in terms of global political ambitions; both the US and the EU will jointly face the challenge of China and other new players in the world economy which at the beginning of the 21st century is less shaped by OECD countries than in the second half of the last century. Economic globalization also is a topic of common concern, and outsourcing and off-shoring are key elements here; both in Europe and the US there have been heated discussion about the gross effects and net effects of increasing economic fragmentation, and sometimes the economic debate is rather confusing here (not to mention the widespread notion in the media that international outsourcing/offshoring imply net losses in terms of jobs and welfare for the US and the EU, respectively). Thus the issue of international economic fragmentation and rising international outsourcing and off-shoring, respectively, are of special interest – and the analyses presented here come up with new findings and suggest that the "bazaar hypothesis" is not convincing for Germany.

From a US perspective transpacific economic relations have gained importance since the 1990s when transpacific trade started to exceed transatlantic trade. However, the transatlantic trade and investment links – with foreign direct investment strong in both directions – rest on strong pillars as one finds much trade and FDI in the field of technology-intensive products and production, respectively. For the US the EU has posed a new challenge, namely in the form of the euro zone, and this holds despite the fact that it is only a sub-group of the EU. The euro zone has not only become a rather stable and dynamic core of the EU: solid growth of trade and foreign direct investment within the eurozone has been observed, with Germany remaining the leading exporting country of the Community.

In their opening chapter, Welfens and Borbély offer a wide-ranging review of structural change, innovation, and growth in the European Union. Their focus is on industrial structural change – its problems and its prospects – in the expanded EU. Their starting point is trade and growth theory. An analysis of structural change along its main dimensions (relative goods and factor prices, shifts in sectoral output and employment shares, and the respective contributions of process and product innovation) is first presented. Next, capital mobility is introduced as well as Sinn's controversial characterization of the large German trade surplus against the backdrop of the increase in international outsourcing. The authors then flesh out the model to show that growth, at least in the medium term, hinges on both demand and supply-side dynamics, with the structure of output and the intensity of trade contributing to growth. Finally, in this exegesis on structural change, innovation, and growth, they offer some dynamic Schumpeterian considerations. The bottom line is that the ability of firms from EU15 countries to rely on imported intermediate products from EU accession countries is the basis for gaining competitiveness in both the global economy and vis-à-vis the United States. It enables them to become more price competitive while restructuring domestic outsourcing in the EU15, making it more focused on producing technologically advanced intermediate products than heretofore.

A detailed set of empirical regularities are investigated along two main dimensions: innovation traits and structural change, and Sinn's bazaar effect. International competitiveness is evaluated on the basis of revealed comparative advantage indicators (RCAs) and export unit values (EUVs). Indices by industry are presented for Germany, Hungary, and Italy, each of which countries have vibrant exporting sectors. But, as the authors caution, a flourishing export market does not necessarily translate into competitiveness or success in coping with structural change. What counts here is the ratio of domestic to foreign value added in production, as well as the ratio of domestic to foreign intermediated inputs in production. For this reason Welfens and Borbély turn their attention to the empirical importance of intermediate imports. Again enter Sinn's so-called bazaar effect, which is that a declining part of the final product's value added is generated domestically via international outsourcing and offshoring.

Cautioning that a declining share of domestic value added in production does not necessarily imply a problem for the domestic economy (as long as the sum of domestic value added share plus domestic intermediate inputs' share in total production does not decline), the authors consider the bazaar effect for six key industries in seven countries. While there is evidence of such an effect (e.g. chemicals) there are some big differences between industries, and often a rising share of domestic intermediates.

Further, all of this obviously applies to the "gross" bazaar effect. A country's imported intermediates may themselves contain its exported intermediates The authors make a stab at calculating the net bazaar effect for the German motor vehicle industry vis-à-vis the rest of EU15 and observe a further reduction of the Sinn effect.

Finally, by way of a summing up of their diverse findings, the authors offer some policy interpretations and conclusions. These offer comfort to some countries but not others (e.g. Germany and Italy). Overall, they conclude that the EU as a whole should benefit from enlargement and globalization. The devil is as always in the detail, residing in a country's labor market institutions, its R&D programs, and its ICT and educational policies, inter al. The Lisbon strategy also receives a provisional nod and a wink.

In the second contribution to this volume, Joachim Möller investigates the vexed question of wage compression in Germany. As the author notes, Germany is widely characterized as having a "deformed" wage distribution, with compression from below stemming from collective bargaining and the social wage. With insufficient variation in low skill wages – Sinn's accordion effect – employment does the accommodating! That is, there is inadequate employment in low-wage service industries and correspondingly high and persistent unskilled worker unemployment. One remedy that has therefore commended itself to German observers is the abandonment of institutional and other regulations.

Recently this interpretation has come under attack from a variety of sources. Möller's careful paper is in this revisionist tradition. His approach is to examaine the distribution of wages in Germany using the German IAB-Beschäftigten-stichprobe (IAB-BST) and, since his analysis turns on a difference-in-differences approach using the United States as the counterfactual distribution, the Current Population Survey/Outgoing Rotation Group (CPS-ORG) for the United States. His empirical analysis is restricted to full-time workers using harmonized data. Results are presented for the aggregate wage distribution(s) and by gender and broad skill category.

Abstracting here from the particular US results, Möller finds scant evidence of wage compression from below, or an accordion effect. In other words, for all low-skilled workers inter-quantile distances are higher below, not above the median. This also holds for females in general. The accordion effect, where observed, is found only among the upper reaches of the male worker distribution (i.e. among skilled workers. His explanation is that German unions home in on Facharbeiter as the standard. But all of this has to do with within-group wage dispersion. As Möller notes, an alternative scenario is one in which between-group wage differentials are too narrow, not that one-sided pressure obtains. The possibility of this type of deformation is no less pressing a research concern than asymmetric deformation.

In his comparative treatment, John Addison examines the links between worker representation, employee involvement/innovative work practices (EI/HPWPs), and various indicators of firm performance. Having reviewed the closely linked theoretical arguments favoring worker representation in its two principal guises and EI/HPWPs, he proceeds to review the empirical evidence on each and on their interaction for the United States and Germany. For the United States, union effects on performance are most typically adverse; where positive, they are likely of small magnitude. The independent influence of EI/HPWPs is frankly mixed, and it is as well to recognize up front the frailty of our understanding on what practices (or bundles of practices) work. Moreover, their positive effects on productivity – where observed – have to be assessed against the backdrop of their cost implications. That said, there is some indication of a positive interaction with unionism.

Despite its position in theory as the exemplar of voice, initial research into the effect of works councils on firm performance reached almost universally pessimistic conclusions. Pace the US evidence, however, more recent studies using large data sets have been more upbeat. Too upbeat in the author's view: excessive praise of the institution is just as misguided as excessive revulsion against it. On balance, however, the German evidence on both worker representation and EI/HPWPs, and their interaction, is more positive than for the United States. But a number of caveats attach to this statement. First, and most obviously, works councils are not unions, and we are only now seeing union effects in Germany, beginning with the union premium and extending to consider the union-works council nexus. Second, there is the vexed question of the endogeneity of the works council institution and EI/HPWPs themselves. Unfortunately, progress in analyzing one dimension is not matched along the other. Third, there is the issue of whether works councils and HPWPs are substitutes or complements. Fourth, beneficial aspects of each obtain devolve on productivity, with the result that cost considerations continue to cast a long shadow. Finally, the stability of relationships in the face of changes in the law and indeed the haemorrhageing of formal collective bargaining is a real concern.

Partly by way of addressing this latter issue, Addison concludes with an addendum summarizing the British evidence. That evidence is both instructive and challenging. Instructive in the sense of charting major changes in union effects attendant upon changes in the law undercutting union bargaining power. And challenging both because of the instability of underlying relationships (not just that between unionism and firm performance) but also because of more fundamental ambiguity in the relation between workplace representation and EI/HPWPs reported for that country.

The paper by Joachim Winter aims at discussing the macroeconomic and capital market effect of population ageing and of fundamental pension reforms. As an analytical starting point Winter takes the overlapping generations (OLG) model which is well suited for simulations. The OLG approach then is combined with demographic dynamics – exogenous in the simulations – and the OLG approach combined with the public pension system. The OLG model has standard features such as a neoclassical production function, convex capital adjustment costs and age-dependent labor productivity as well as standard household maximization behavior. As the approach presented considers an open economy with capital flows, modelling of the international capital market is crucial – including the degree of capital mobility which is changing in the various stages of simulations undertaken by Winter. The author's focus is on those EU countries that are most affected by population ageing: France, Germany and Italy are key countries in this respect. The author also points to some of the key insights from the US debate and argues that Germany's pension reform proposals (Rürup Commission) were adequate.

Three panellists looked at selected issues of capital market dynamics: Volker Clausen considers the macroeconomic perspective of ageing and summarizes key insights from recent debates which emphasize that ageing will both undermine saving – as life-cycle models of savings predict negative savings rates of those who retire – and investment where the latter is related to the decline of the labor force which changes relative factor prices and thus implies a lower rate of return on capital. The net effect of ageing on the savings-investment ratio is somewhat unclear and it is the net effect which determines international capital flows. As regards Germany, the author shows that the savings-GDP ratio has increased while the investment-GDP ratio has fallen with the result that the Federal Republic of Germany has become a major net capital exporter. The net foreign asset position has improved, both through accumulation effects and favorable valuation effects. Next Freddy Van den Spiegel discusses capital market perspectives and points out that the real interest rate has shown a long-term decline in both the EU and the US; the increase of the global savings rate could have contributed to this development. As regards ageing dynamics he points out that governments have several instruments that could influence the pace of ageing; lengthening of working life and immigration are two important elements. As the EU's ageing is occurring in the new set-up of economic globalization, the adjustment pressure will be less severe than without such internationalization. Finally, in his analysis of the US economic outlook Holger Schmieding puts the focus at first on some favorable economic globalization dynamics contributing to relatively high US corporate profit rates. Also, the role of low US central bank interest rates and high Asian savings for the long US upswing receiver are emphasis. The author points out the mechanics of the subprime crisis and the role of asset-backed securities in financing the real estate boom on the one hand, while on the other hand he observes that the aggressive interest rate cuts of the FED could help the US to move to economic recovery relatively soon. There is another favorable aspect as emphasized by Schmieding. Although the ratio of debt service of households to income has increased from 12% of disposable income in the late 1990s to 14% in 2007 asset incomes (interest and dividends) stood at 19% of disposable income in 2007, so that there are no signs that households in the US are overindebted. While the authors expect some cyclical adjustment problems he is optimistic about long-term global rebalancing.

EU countries have achieved a long-term economic catching-up vis-à-vis the US so that innovation has become a major driver of economic growth in Europe and the US. Lothar Funk and Axel Plünnecke thus focus on the innovation dynamics in selected OECD countries. Their analysis thus evinces a Schumpeterian perspective. The authors first take a broader look at the literature and emphasize some key findings from innovation research – including the ACEMOGLU conjecture that countries that have moved close to the technology frontier will have to rely more on radical innovations combined with fast adaptation to technological progress. The authors discuss the required conditions for innovation and present statistics on human resources and R&D-GDP expenditure ratios for various countries. Among the countries the authors focus on the growth stars, the catching-up countries and the laggards. Ultimately, the authors try to identify clusters of innovativeness and suggest key policy conclusions.

Charles Wessner takes a broader look at innovation dynamics and entrepreneurship in the US. His focus is on the Small Business Innovation Research (SBIR) program which has become a major pillar of US R&D policy. The author argues that new challenges from China and other countries – includung a more general concern about US competitiveness – have stimulated policymakers in to emphasize Schumpeterian dynamics of small and medium-sized firms. Wessner discusses funding problems of technology-oriented start-up companies in the US and highlights the strategic orientation of the Small Business Innovation Research program. Moreover, the key factors for success are analyzed and some new elements of R&D policies for Europe (and elsewhere) discussed.

A key field of modern Schumpeterian dynamics concern pharmaceutical innovation. Mario Villarreal and Elizabeth DuPre analyze the links between TRIPS and pharmaceutical innovation dynamics and access to medicines. The authors' analytical starting point is the economics of patents. Moreover, the process of pharmaceutical innovations and the costs and risks involved are discussed. Since patent protection is so important in the pharmaceutical industry the globalization process presents crucial challenges for this sector and the international protection of intellectual property rights. Villarreal and DuPre consider the key problems of WTO rounds and TRIPS, respectively – with compulsory licensing and parallel importing representing the two important policy fields. Finally, the authors focus on the perspectives of the Doha round and policy alternatives.

In a broader perspective, Schumpeterian dynamics concern the expansion of the digital knowledge society in the 21st century. The paper of Ernst Helmstädter confronts the very enthusiastic political expectations about the knowledge society with the austere analytical view of the economics discipline. The political strategies for establishing a competitive knowledge-based economy are numerous. Because such strategies seem to believe in scientific knowledge the climate for the sciences looks very favorable. In contrast, the contribution of the economics discipline to the analytical understanding of the use of knowledge in society is still very rudimentary. This contribution tries to assemble the considerable fragments from the history of economic thought and proposes out of these experiences how to proceed in developing an economic theory of the knowledge economy. The approach is based on New Institutional and Evolutionary Economics. Tentative political conclusions are drawn.

With EU eastern enlargement more than ten new member states have joined the Community and economic and technological catching-up are crucial elements in the modernization of Eastern Europe and the New Member Countries (NMS). The perspective of the new member countries is quite interesting since they have had to combine systemic transformation, economic opening up and modernization. This triple perspective is the context of the contribution of Andrea Szalavetz. The role of exploiting the experience of catching-up in Asia is discussed as is the role of capital deepening and investment in information and communication technology. The author highlights the structure of investment in selected countries and the dynamics of industry where capital-saving technological change in the manufacturing sector is emphasized. In-depth analysis of NMS' technology accumulation experience compared to advanced economies reveals that, beyond the quantity differences related to their actual development status, important quality differences also have to be taken into account.

Innovation performance can be measured in various ways. Benchmarking is useful for policymakers with the aim of finding out the best way to harness Schumpeterian dynamics. Berhard Iking highlights the findings from EU comparisons – across countries – and from regional comparisons in Germany (Bavaria, Baden-Württemberg and North-Rhine Westfalia). The European Innovation Scoreboard is a well-known concept which facilitates measuring various dimensions of innovativeness. The descriptive statistics not only show a strong variance of innovation indicators across countries but also considerable differences between the regional technology powerhouse Baden-Württemberg – followed by Bavaria – and the largest West German state (in terms of population), namely, North-Rhine Westfalia. Human resource building and knowledge creation are identified as key challenges for policymakers.

A crucial field of EU technology policy concerns the energy sector. Global climate change and issues related to the security of energy supply – mainly concerning Russian gas supplies – as well as high oil prices motivates EU energy technology policy. Jochen Hierl and Peter Palinkas discuss the Community's policy and emphasize the relevance of the EU's Green Paper "Secure, Competitive and Sustainable Energy for Europe" which highlights the key topics and issues. The debate about peak oil, the increasing dependency of the EU countries in the field of energy imports, the post-Kyoto approach to climate policy and the EU's single energy market – starting in July 1, 2007 – are the crucial aspects considered. A European strategic energy technology plan will be adopted and this which underscored the relevance of Schumpeterian dynamics and of EU innovation policy in the energy sector. The fact that energy policies all have a long-term perspective and the EU's strategy to focus both on technology push and market pull forces is emphasized by the authors. Among the crucial policy fields discussed are renewables, carbon capture and storage and energy efficiency.

The US has not only a national energy policy but there are also important regional activities organized by a group of New England and Mid-Atlantic states which agreed in 2005 to create a CO_2 cap and trade system for electricity known as the Regional Greenhouse Gas Initiative. Other US states also have launched corresponding initiatives. Moreover, there are several potential fields for transatlantic cooperation.

The paper of Fred Sissine summarizes policies in the US that aim to reduce greenhouse gas emissions (GHG) that contributing to global climate change. The status and direction of activities in that country are also described. This provides a framework that may be used to compare and contrast with policies underway in the European Union (EU) to curb emissions

Caution is required when assessing NMS' technological upgrading experience in an appropriate technology context for at least two reasons. One is that in the NMS, the factors underlying technology choice are still different from the textbook factors. The other reason is that simplistic classifications of technology level may yield erroneous conclusions in assessing the quality of structural upgrading as well as in assessing the human capital intensity of value adding activities or the local direction of technical change. This supports the claim that, besides the well-definable measures of convergence, soft factors also have to be considered when assessing catching-up performance.

Political science analyses are quite important in a transatlantic perspective. Martin List takes a closer look at recent debates and explains both the realist paradigm in international (transatlantic) relations and the institutionalist and constructivist perspectives. The author favors a "critical perspective" as a kind of shared transatlantic approach and argues that more cooperation could be useful for both sides. Common values also are considered as important pillars for joint strategies.

Jackson Janes highlights in his contribution – an extended dinner speech – the historical challenges for the transatlantic relations and takes a closer look at foreign policy issues and security policy aspects. The different role of economic interdependency and political interdependency are discussed and the crucial impact of September 11 on the US public and policy, respectively. In the view of Janes the opportunities for transatlantic cooperation have increased; with special lessons to be drawn from the cooperation between the US and Germany. Jointly part of the global agenda and finding new solutions to old problems could be a key point for future transatlantic cooperation.

All of the above contributions are revisions of papers presented at two workshops organized by the EIIW in Wuppertal – on EU-US economic and technological relations. Transatlantic economic dynamics and policy issues have been on the research agenda of the European Institute for International Economic Relations for many years. With the creation of the Schumpeter School for Business and Economics at the University of Wuppertal (in October 2008) the EIIW considers Wuppertal as becoming an ideal place for conducting research on innovation dynamics. Institutional changes, technological progress and new policy approaches will be recurrent topics for comparative transatlantic and global research.

We greatly appreciate the technical support of Martin Keim and Deniz Erdem (EIIW) as well as the editorial assistance of Michael Agner (University of Odense). This book is offered in hard copy and – in this age of digital "integration" – as an e-book as well to provide an accessible catalyst promoting wider discussion. In terms of editing, the project is another transatlantic joint venture. In this connection,

we appreciate the support of the Haniel Foundation, Duisburg, and the Wolfgang Ritter Foundation, Bremen. We hope that our joint intellectual bridge will stimulate debate between the academic community and policy-makers.
Wuppertal, Paris, Belfast and Columbia (USA), September 2008

Wuppertal, Germany Paul J.J. Welfens
Columbia, USA John T. Addison

Contents

List of Figures

List of Tables

Contributors

John T. Addison Moore School of Business, University of South Carolina, Columbia, SC, USA, ecceaddi@moore.sc.edu

Dora Borbély EIIW at the University of Wuppertal, Germany

Volker Clausen Department of Economics, University of Duisburg-Essen, Germany

Elizabeth DuPre American Enterprise Institute, Washington, USA

Lothar Funk University of Applied Sciences Düsseldorf, Germany, lothar.funk@fhduesseldorf.de

Ernst Helmstädter Institut für Arbeit und Technik (IAT), Gelsenkirchen, Germany

Jochen Hierl European Investment Bank, Luxembourg

Bernhard Iking ZENIT - Zentrum für Innovation und Technik mbH, Mülheim/Ruhr, Germany

Jackson Janes AICGS/The Johns Hopkins University, Washington, USA

Martin List FernUniversität at Hagen, Germany

Joachim Möller Institute of Economics, University of Regensburg, Germany

Peter Palinkas European Parliament, DG IV, Luxembourg

Axel Plünnecke Institut der deutschen Wirtschaft, Köln, Germany, pluennecke@iwkoeln.de

Holger Schmieding Bank of America, London, UK

Fred J. Sissine Congressional Research Service, Washington, DC, USA

Andrea Szalavetz Institute for World Economics, Budapest, Hungary

Freddy Van den Spiegel Fortis Bank, Brussels, Belgium

Mario Villarreal American Enterprise Institute, Washington, DC, USA

Paul J.J. Welfens EIIW (European Institute for International Economic Relations), University of Wuppertal, welfens@eiiw.uni-wuppertal.de

Charles Wessner National Academies, Washington, DC, USA

Joachim Winter University of Munich, Munich, Germany

Chapter 1
Structural Change, Innovation and Growth in the Single EU Market

Paul J.J. Welfens and Dora Borbély

1.1 Introduction

In the 1990s US economic growth exceeded that of both the EU15 and Euro zone. It was only in 2006 that the EU growth – much influenced by Germany's economic recovery – gained speed and in 2007/2008 output growth of the Euroarea and the EU27 is expected to exceed US growth. As the EU's growth rate is rising relative to the US one may expect that the US current account deficit-GDP ratio will improve; one should, however, note that the EU's surplus vis-à-vis the US is rather modest, the main surplus countries in a bilateral perspective are Japan and China. With China's exchange rate more or less fixed to the dollar (and similar settings in Asian countries), the bilateral exchange rate movements in Asia will hardly help the US to strongly reduce the current account deficit in the medium term. To the extent that the high US current account deficit should continue for many more years to come one might have to cope with a sharp real depreciation of the dollar in the long run. One may argue that the US banking (subprime) crisis and the associated fall of consumption automatically will reduce imports and stimulate – along with a real depreciation of the US dollar of 2004–2007 – exports in the medium term. However, the enormous bilateral trade deficit with China continues to grow strongly, while the net export position vis-à-vis the euro zone will improve only gradually.

As regards EU–US economic relationships the Merkel initiative of 2007 to reinforce transatlantic economic cooperation could help the US to improve its current account position, not least since effective liberalization of services markets on both sides of the Atlantic will generate more trade in services, and the US as the world's No. 1 services exporter stands to benefit from the policy initiative adopted in 2007. The envisaged Transatlantic Trade Council will regularly analyze the progress achieved in major fields, such as financial market services, intellectual property rights, pharmaceutical testing procedures and patenting. Joint initiatives

P.J.J. Welfens (✉)
EIIW (European Institute for International Economic Relations), University of Wuppertal
e-mail: welfens@eiiw.uni-wuppertal.de

P.J.J. Welfens, J.T. Addison (eds.), *Innovation, Employment and Growth Policy Issues in the EU and the US*, DOI 10.1007/978-3-642-00631-9_1,

in raising security in international transportation – read: anti-terror activities – also are expected to play a certain role. The field of energy policy is a rather thorny field where one finds joint interests in a strategic perspective but for various reasons will have considerable problems on the way to parallel actions, e.g. in the field of a new Kyoto protocol for the period after 2012. Telecommunications could become a more promising field, not least since internet-based telephony is growing and in new global telecommunications market one may find considerable benefits to be derived from a consistent set of regulatory rules on both sides of the Atlantic.

As regards macroeconomic issues the cooperation between the US and the EU is rather modest, and this problem hardly will be much affected by the Transatlantic Trade Council and the related activities. The phenomenon of a high current account deficit-GDP ratio and the potential for a hard landing of the dollar has not been discussed between the EU and the US, except for multilateral talks at the OECD or the G-8 meetings. On the one hand, US politicians might not be too concerned about large changes in the dollar exchange rate, on the other hand the institutional setting in Brussels is opaque: A US treasury secretary might consider discussing the potential of a sharp change in the dollar exchange rate with the head of DG II (Macroeconomics) or with the head of the Euro group plus the ministers of Finance of the UK – plus the respective ministers in Denmark, Sweden and EU accession countries. Moreover, the EU which argues that it should have a larger impact on international economic relations due to its increased number of member countries and the increase in population and real income after the EU eastern enlargement is rather weak as it fails to speak with one voice in international issues and in global institutions. The idea that it would be a blessing in each member country were to have a national seat at the IMF is self-deceptive, as the combined weight of national representation at the IMF (and in other international organizations) effectively is likely to be lower than a single EU/Euroland seat; provided that such a single seat also goes along with the creation of an effective Euroland Treasury in Brussels.

As regards the debate about the US current account deficit it obviously is important to consider the price responsiveness of exports and imports of goods and services. Here the empirical analysis is rather unclear if one considers the results of Chin (2006) who shows that standard modeling of the US export function gives a good empirical fit while the import function is much more difficult from an econometric perspective. Similar problems are also known to play a role in Germany, and these problems underline the potentially important role of international outsourcing and offshoring – the latter involves foreign direct investment and trade in intermediate products with a multinational company. As regards the US current account dynamics one should not only consider outsourcing and offshoring which can affect the price elasticity of imports – for an import demand function with a good empirical fit for the US see IMF (2007) – but also growth dynamics in the EU and the US. The US investment-GDP ratio has increased only modestly after 1993, at the same time the US savings rate has dropped, and this, along with the US budget deficit could also explain the US current account deficit. However, the long-term transatlantic growth differentials also are an important aspect.

In the literature we find considerable evidence that information & communication technology (ICT) plays an important role for the growth differential US vs. EU15: Jorgensen and Stiroh (2000), Colecchia and Schreyer (2002), Oliner and Sichel (2002), Stiroh (2003), Inklaar, O'Mahony, and Timmer (2003) and Van Ark and Piatkowski (2004) have argued that ICT production and the use of ICT – that is ICT investment – are important drivers of productivity growth. Comparing the periods 1995–2000 to 1979–1995 the Inklaar et al. analysis of labor productivity growth in the US and EU-4 finds a rise of 1.25 percentage points in the US and a reduction of 0.27 points in the EU. The growth accounting estimates show that labor quality changes have reduced in both the US and the EU-4 labor productivity. The employment reallocation effect in the US was good for +0.05 points, but in the EU-4 the figure was –0.06 points. ICT producing industries generated similar impacts on productivity growth in the US and the EU, namely 0.04 and 0.03 percentage points respectively. As regards the impact of ICT using industries the EU did not reach even half the increase of the US which was 0.29 points – the main effect stemming from financial services (0.17 in the US; 0.02 in EU-4). Non-ICT capital deepening contributed to 0.08 points in the US and –0.45 points in the EU. Total factor productivity contributed 0.79 points in the US, but only 0.13 points in the EU-4. The impact from ICT producing industries was rather similar on both sides of the Atlantic (0.36 in the US vs. 0.24 in the EU), but in ICT using industries there were much bigger differences, in particular wholesale trade, retail trade and financial services seem to be problem areas for Western Europe. Weak EU-15 productivity increases and slow growth are all the more unsatisfactory since Germany, France, Italy and Spain suffer from high unemployment rates and since slow growth in 2000–2005 seems to indicate that the ambitious goals of the EU Lisbon Agenda – aiming at higher growth and employment by 2010 – will not be achieved. In the 1990s the investment-GDP ratio in the Euro zone was below that of the US and it also seems clear that the degree of factor market flexibility is lower in the Euro zone than in the US. The creation of the Euro zone was expected to contribute to output growth, however, growth has not accelerated; moreover, EU eastern enlargement is expected to stimulate growth in the EU15 and the accession countries, namely through trade creation – and the associated specialization gains – and foreign direct investment creation. EU accession countries have shown economic catching up but Germany and Italy face slow growth; both countries and France have stubborn high unemployment rates.

An EU study on the Lisbon Process (Denis, McMorrow, Röger, & Veugelers, 2005, p. 4) summarizes its findings as follows: "The structural nature of the EU's productivity downturn is confirmed by the analysis. . ., with the bulk of the deterioration emanating from an outdated and inflexible structure which has been slow to adapt to the intensifying pressures of globalization and rapid technological change. The EU's productivity problems are driven by the combined effect of an excessive focus on low and medium-technology industries (with declining productivity growth rates and a globalization-induced contraction in investment levels); an inability to seriously challenge the US's dominance in large areas of the ICT industry, as reflected in the relatively small size of its ICT in a range of ICT-using industries, although measurement issues severely complicate an assessment of the gains from ICT production and diffusion. The post-1995 differences in EU–US productivity patterns are fundamentally driven by the US's superiority in terms of its capacity to produce and absorb new technologies, most notably in the case of ICT. Healthy knowledge production and absorption processes are mutually supportive elements of any successful long run productivity strategy. Evidence is presented which suggests that the US's overall innovation system is superior to that of the EU's, both in terms of the quality and funding of its knowledge sector and the more favorable framework conditions prevailing. The repeated ability of the US system to direct resources towards the newer, high technology (and often high productivity growth), industries is a reflection of the quality of the interrelationships between the different actors in its innovation system and of an economic and regulatory framework which has the capacity to transform excellence in knowledge creation into globally competitive industrial structure. The systematic inadequacies of the EU's innovation system are highlighted by the experience of the ICT industry, with the history of this industry suggesting that a 'national champions' strategy in high technology industries is highly problematic."

Our analysis will not focus much on the ICT – contributing about 1/3 to US productivity growth in the 1990s – since many ICT issues have been largely explored elsewhere (Audretsch & Welfens, 2003; Barfield, Heiduk, & Welfens, 2004; Welfens et al., 2004). We also will not look into the intra-US differences in productivity growth which are considerable and show that the West – the 12th district of the Federal Reserve System – has recorded a much higher productivity growth in the second half of the 1990s with 3.5% p.a. than the rest (2.6%) of the US (Daly, 2002). Rather we want to focus on problems of and prospects for industrial structural change in EU25: The end of the Soviet Union and the transformation of Eastern Europe has opened up more than two dozens post-socialist economies for trade and foreign direct investment; eight relatively poor eastern European countries joined the EU on May 1, 2005. Taking into account growth theory and trade theory, it is clear one should expect a medium-term catching-up process and considerable trade creation in the context of EU eastern enlargement. In 2007 Bulgaria and Romania also joined the Community, which will cause further structural adjustment in the enlarged single market with additional options for outsourcing, trade in final products and foreign direct investment. Further expansion plans will be difficult to realize since the negative referenda in France and the Netherlands had largely tilted the scales to the No-side due to strong popular opposition to the Turkish EU enlargement project envisaged by the European Council and the European Commission.

The combination of economic globalization – the rise of trade and FDI (partly related to the opening up of China) in combination with the digitization of the world economy – and EU eastern enlargement have generated considerable pressure for structural change in the EU25. In the eastern European new member countries – post-socialist transition economies – economic opening up, systemic transformation and the rise in real per capita income have brought strong shifts in relative prices and hence structural change. With the Europe Treaties of the early 1990s, opening up the EU15 markets for prospective EU accession countries external impulses overlapped the early transformation process which included restructuring and privatization of firms. Moreover, foreign direct investment inflows have considerably contributed to the modernization of the supply side, technology transfer and growth in most east European countries. In 2004, the year of full EU membership for eight eastern European countries, per capita GDP stood at 45% (at PPP figures) of EU15. Growth rates in those countries have exceeded those in Western Europe in the 15 years after the start of transition in 1990, which was marked by transformational recession; the EU implemented an asymmetric trade liberalization strategy in the 1990s when the so-called Europe Treaties with potential access countries from Eastern Europe offered relatively generous access to the EU market. In the period 1990–2005, there has been some economic catching-up in Eastern Europe where Poland, Hungary, the Czech Republic and the Slovak Republic were rather successful in attracting FDI inflows. The latter came mainly from Western Europe and the US; the sectoral focus was not only on industry but it included the services sector – in particular banking and finance – as well. Restructuring and economic modernization in Eastern Europe's low wage economies (at the beginning of the 20th century, wage rates were about 1/5 of those in EU15) generated growth and stimulated trade with EU15 where many firms realized outsourcing to or off-shoring in accession countries.

The following analysis looks first at the theory of structural change and selected approaches on innovation and growth (Sect. 1.2) before we take a closer look at empirical aspects of economic dynamics in Europe (Sect. 1.3); we are interested in describing the dynamics of structural change and the developments of revealed comparative advantages and other trade indicators – this includes aspects of the role of imported intermediates in exports and of exports in imported goods on the one hand, on the other hand the question as to whether negative RCAs of EU15 countries vis-à-vis accession countries are positively correlated with positive RCA positions of EU15 vis-à-vis the US. We also look at some key aspects of the "bazaar effect" which emphasize the problem of hollowing out in the sense that exports contain an increasing share of imports. The final section suggest on the one hand various policy options for both EU15 and the accession countries, on the other hand we present some conclusions for EU–US economic political relations.

1.2 Theory of Structural Change, Innovation and Growth

1.2.1 Basic Dimensions of Structural Change

Economic globalization implies that there will be considerable changes in the relative price of tradables which in turn will cause relative factor price adjustment which – in a neoclassical perspective –should largely follow the logic of the Heckscher-Ohlin Samuelson (HOS) model: Countries which are relatively richly endowed with unskilled labor, skilled labor, capital and technology will specialize in those goods which use the relatively abundant respective input factor intensively. However, in the existing world economy there are some critical deviations from basic assumptions of the HOS model; we have economies of scale, network effects (endogenous growth of demand) in the ICT sector, technology spillovers – hence positive externalities – and foreign direct investment, which are all not in the standard HOS model. Take for instance the Stolper-Samuelson theorem which sees factor immobility and perfect competition in a neoclassical world with free trade,: If the relative price of good i is rising the relative factor reward to that factor will increase which is used relatively intensively in the production of good i. However, with scale economies and network effects a certain modification is necessary: If p_i falls the factor reward of the intensively used factor will rise if there is a dominant network effect or scale effect. Take as an application the following case: At first glance the long-term relative fall of ICT prices imply a relative fall of remuneration of software engineers, but network effects and scale effects bring about a relative rise of the remuneration of such engineers.

With tradable prices adjusting across countries and factor prices reacting to output prices there will be real income effects and effects in factor markets; this can include unemployment to the extent that wages are rigid downwards for workers for which firms' relative demand is falling. As modern globalization includes the opening up and industrialization of China, it is obvious that the relative prices of labor intensive goods will fall which in turn will reduce the relative wage rate of unskilled labor.

Structural change and growth go together since structural change in a competitive economy should relocate resources from low-productivity sectors towards high-productivity sectors on the one hand; and from low stages of value-added (production of simple intermediate products) to more advanced stages (semi-finished goods) and finally to the production of finished goods on the other. An upward move in terms of quality or technological refinement will typically be accompanied with rising – relative – unit export values so that the marginal value product of labor in the respective sector rises. An internationally improving competitiveness in the respective sector can also be measured by the RCA, the revealed comparative advantage.

The main dimensions can be summarized as follows, namely changes in

- relative prices of goods: this concerns in a broad perspective the ratio of nontradables to tradables $\varphi =: [P^N] / [P^T]$ which is expected to increase in parallel with relative per capita income. This relative price change typically goes along with a real appreciation of the currency (Balassa-Samuelson effect): The ratio P/[eP*] will rise over time as either the nominal exchange rate e – with given price levels at home and abroad P and P*, respectively – will fall or the domestic price level P will rise (at given e and P*, respectively); one should note that the overall price level $P = :[P^N]^a[P^T]^{(1-a)}$.which implies that $P = \varphi^a P^T$ (a is a parameter indicating the share of nontradables in overall consumption). As regards tradables prices one may anticipate that countries catching-up will record a growing share of intra-industrial trade; this holds for eastern European accession countries (Borbély, 2004, 2005). This change in the composition oftrade will go along with

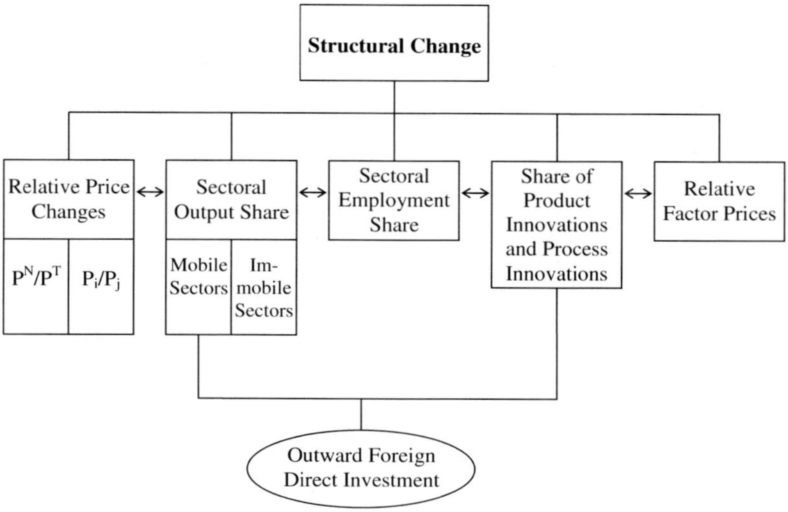

Fig. 1.1 Structural change and its five dimensions

a rise in the average export unit value reflecting a shift towards a greater share of high quality goods and other goods fetching a premium in world markets – the latter can include product innovations which allow to get a higher price in the market (Welfens, 2007).

- relative prices of input factors: the factor used relative intensively in the good whose output is rising will benefit in terms of relative factor rewards (Stolper-Samuelson theorem); e.g. if the production and export of (unskilled) labor intensive goods in China increases, the wage rate of unskilled workers is raised – note that this leaves open the role of the price of nontradables which may be expected to affect the reservation wage rates and international factor price convergence. If unskilled labor intensive production is increasing in China it is rather unlikely that unskilled labor intensive production in Eastern Europe will also increase unless transportation costs amount to effective regional market demarcations. Thus one may expect that eastern European countries will specialize partly in goods using unskilled labor intensively, but also on goods using skilled labor intensively. By contrast leading EU15 countries – following the logic of Heckscher-Ohlin – will specialize increasingly in goods which are technology intensive or knowledge intensive and thus particularly require skilled labor. This does not rule out some employment of unskilled workers, but it will mainly be in the nontradables sector, in particular in the services sector. From this perspective, it is not surprising that Germany's specific unemployment rate of unskilled workers increased strongly in the 1990s; at the beginning of the 21st century it was twice as high than the average unemployment rate.

- shifts in sectoral output shares; and this will include shifts in the share of intermediate imports. Leading OECD countries may be expected to specialize not only more on technology and knowledge intensive production, but they also are likely to increasingly outsource production both nationally and internationally; the opening up of Eastern Europe which already started in the early 1990s in the context of the Europe Treaties of the EU has stimulated international outsourcing towards eastern Europe.
- shifts in sectoral employment shares: Following the Stolper-Samuelson theorem the globally increasing relative demand for knowledge-intensive and technology-intensive goods will raise the relative wage of skilled worker which in turn improves opportunities for expansion of the education system and training/retraining activities; countries with a rather flexible supply side in the education system thus stand to particularly gain (e.g., the US, the UK, and the Netherlands).
- share of process innovation vs. product innovations: Given the growing role of software in all sectors and all countries, one may assume that process innovations will dominate in the digital world economy. Moreover, we may assume that modern software development amounts to capital-saving technological progress so that effectively capital is relatively more abundant in the early 21st century than in the 20th century, which could lead to a fall in the relative price of capital and a decline of the capital rental rate relative to the wage index (a composite index for skilled and unskilled labor). The Rybczynski theorem says: An exogenous increase in the endowment of production factor j – given relative goods prices – will lead to higher output of that good which is using the more abundant factor (j) relatively intensively. Therefore we expect a global growth of software intensive – more generally of ICT intensive – goods production; skilled labor which is largely complementary to ICT stands to benefit from this development. Ignoring the early transition period with its many distortions (including variable political risk premium) the Rybczynski theorem might also be applied in the context of FDI flowing to Eastern Europe where the yield on investment should decline relative to the wage ratio. Taking a look at Asia one may apply a similar reasoning to China. However, the Rybczynski theorem overlooks the importance of network effects and positive regional technology spillovers which obviously are relevant if many multinationals invest in a particular region. Such network effects and positive spillover effects are bound to at least transitorily raise the rate of return of foreign investors.

Given the opening up of China in the 1980s and Eastern Europe and Russia in the 1990s on the one hand and the expansion of ICT – facilitating the international organization of production abroad (e.g., through digital reporting and networked management activities) – on the other, there are new opportunities for foreign direct investment. Moreover, the global innovation race has accelerated as is visible in rising R&D-GDP expenditure ratios in OECD countries so that ownership specific advantages – here technology advantages – of firms play a greater role. Following the OLI approach of Dunning (1977) who explains FDI in terms of a combination of ownership specific advantages, locational advantages and internationalization advantages (savings on transactions costs through firm-internal transactions), we may expect that FDI growth will exceed that of global output. FDI clearly is a non-neoclassical element in modeling of open economies, and it is unclear whether standard neoclassical theorems hold without modifications. As FDI is often combined with innovation dynamics we also have to consider Schumpeterian dynamics and hence deviations from the neoclassical world of perfect competition. As a macroeconomic implication we should carefully distinguish between GDP and GNP where a simple asymmetric model of an open economy with FDI would have to consider that in the case of identical GDPs (Y in country I which is the host country, Y* in the source country) – produced with a Cobb-Douglas technology according to $Y = K^{ß}L^{1-ß}$ and abroad $Y = K^{*ß}L^{*1-ß}$ – the GNP in the domestic host country is $Z = Y - ßY$ (where ßY is profit accruing to foreign investors assumed to own the total capital stock K) while GNP in the source country is Y*+ßY (Welfens, 1997): If $ß = 1/3$ and population L in both countries is the same, the relative per capita income position y/y* (y = :Y/L) is 2:1 in favor of the source country so that there is no real convergence even with free trade and free capital flows. Both the US and in west European countries witnessed a lively debate about outsourcing and off-shoring in the late 1990s.

1.2.2 Capital Mobility and Other Non-Neoclassical Imperfections

In reality we have a world economy with both trade and factor mobility, in particular mobility of capital (FDI). How is the familiar Vinerian analysis of trade creation and trade diversion affected by the existence of FDI? We assume a three country world in which country A and country B form a customs union. Country C is assumed to have a comparative advantage in the production of the imported good which is a capital good. With a tariff-ridden import price of $k'^{**}(1+T)$ the quantity imported is J_o – coming from country C; J_o is assumed to be identical to depreciations. The creation of a customs union raises imports to J_1 – now coming from country B; but we also have trade diversion in the sense that country C is no longer exporting to country A. There is trade creation as J_1 exceeds Jo; the difference between J_1 and J_o is Greenfield investment. By implication GDP in country A will increase which also implies a rise in GNP and hence a rightward shift of all demand functions, including the import demand function (JJ). Hence we have an outward shift of the JJ_0 curve to JJ_1. Moreover, the presence of foreign investors will stimulate product innovations and this will bring about an upward rotation of the import demand curve so that JJ_2 is the relevant curve; this implies additional trade creation so that from a theoretical perspective J_{01} (point E_2) is the relevant import quantity. We clearly can see both in the right-hand panel and in the left-hand panel a considerable welfare gain induced by FDI.

As regards mobility of capital one should distinguish between mobile Schumpeter (real: technology-intensive) sectors and immobile Schumpeter industries, where the latter refer to an effective inability to separate research and development (R&D) from the production process (Klodt, 1993); relevant industries are the air and space industry and the special machinery industry so that international wage cost differentials for labor are irrelevant.

One should note that the existence of immobile Schumpeterian industries can impair international factor price equalization. Another impediment – from a theoretical perspective – refers to the existence of nontradables. As regards the role of the latter, the digitalization of the world economy has reduced the share of nontradables in overall output. Both the internet and digital networks have enhanced tradability of many services since it has effectively rendered the supply side or the user more mobile across countries.

Structural dynamics in a digital world economy bring about shifts in the structure of output and can go along with long-term relative factor price changes; changes in goods prices can also play a role. From the perspective of a small open economy in catching-up countries one may assume that the prices of certain goods are subject to an exogenous long-term decline – as it is observed with ICT goods – or a long-term increase as it is the case with sectors with small process innovation rates plus a global income elasticity of demand exceeding unity.

From a radical macroeconomic point of view one might argue that the composition of output is rather irrelevant; in a pure neoclassical perspective with Harrod-neutral progress the level of the growth path of output per efficiency unit of labor, and hence labor productivity, is determined by the savings rate, the growth rate of the population, the rate of capital depreciation and the exogenous growth rate of

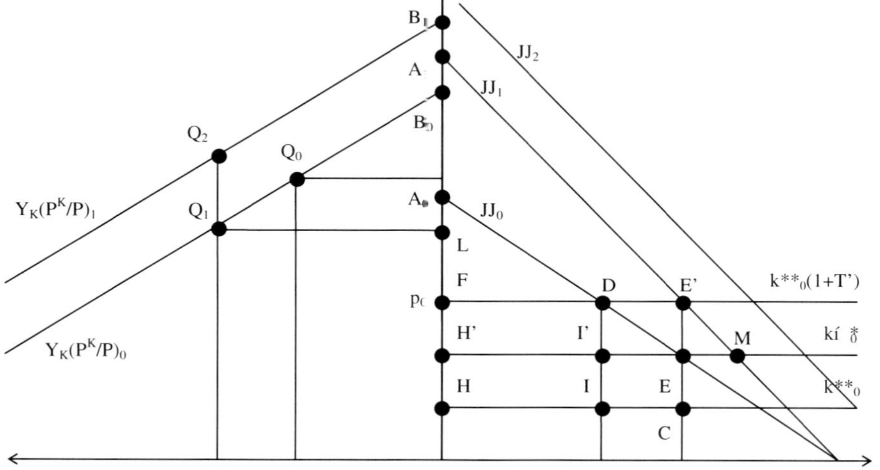

Fig. 1.2 Trade creation and FDI-induced product innovation
Source: Welfens (2007).

technological progress; if production is determined by a Cobb-Douglas production function $Y = K^{ß}(AL)^{1-ß}$ – where A is the level of labor-saving technology – we can additionally state that the parameter ß influences the steady state situation. The growth rate of progress itself is exogenous. In endogenous growth theory one may focus on several ingredients which explain the growth rate of progress, including such factors as:

- human capital accumulation (Lucas, 1988)
- positive external effects from capital accumulation (Romer, 1990)
- R&D expenditures (Grossman & Helpman, 1991)
- intermediate products which allow for the production of a greater variety of final products which stimulates demand; and this in turn stimulates output growth (Romer, 1990; Grossman & Helpman, 1991; Bretschger, 1998)

A major aspect of structural change is the increase in international outsourcing which leads to the problem of the bazaar economy – to use the term dubbed by Sinn (2005). Sinn has argued that Germany's large trade balance surplus is not really impressive, rather Germany's economy would be facing a major problem in international competitiveness as increasingly German exports consist of exported goods containing rising shares of imported intermediates.

1.2.3 Demand-Side Perspective of Structural Change

Some aspects and problems of the concept of the bazaar economy hypothesis are easily understood if one takes a look at the equilibrium condition for the goods market in an economy with unemployment: Here the aggregate demand C+I+G+X-J determines output Y (C is consumption, I investment, G government consumption, X exports, J imports).

$$Y = C + I + G + X - q^{*}J \tag{1.1}$$

As we assume that $C = cY$, $I = b'Y$ and $G = \gamma Y$, $X = x(q^{*},Y^{*})Y$; $J = j(q^{*})Y$ we can write in growth rates (denoting $x = X/Y$; $j = J/Y$, $q^{*} = eP^{*}/P$ where e is the nominal exchange rate and P the price level; * stands for foreign variables, g for growth rates):

$$g_Y = cgv + bgv + \gamma gv + xg_x - jp_J \tag{1.2}$$

For simplicity we have set q^{*} equal to unity. The growth contribution of exports ix xg_X which suggests for the case of $x = 0.3$ and $g_X = 0.10$ that the contribution from exports is 0.3 percentage points. However, imported intermediates stand for 1/3 of value-added in exports the true growth contribution of exports is $x'g_X = 0.2$ times $0.10 = 0.2$ percentage points. From this perspective it is clear that a rising share of intermediate exports reduces the growth impact of a given export growth.

However, the expression "given export growth" is a problem here since a profitable dynamic international division of labor which leads to a rising share of intermediate exports should raise the growth rates of exports; this indeed is the case if we assume an export function

$$X = x\left(q^{=}, Y^{*}\right) \&e'^{\,\lambda^{v}} \tag{1.3}$$

where e' is the Euler number and v the share of differentiated products in exports and λ a positive parameter. Hence we have assumed that with a rising share of intermediate imports – leading to a rising share of product varieties – the growth rate of exports will increase. Moreover, we might even assume that a rise in v will cause a real appreciation, and the net effect on X of this rise in v could still be positive. Moreover, if there is trade in intermediate products we should focus on the net bazaar effect: In country I's imports from country II one will find intermediate products which country II had imported from country I (country I's exported intermediate products). Hence the negative growth impact of imports is smaller than a look at jg_J suggests as the true j' – representing value-added abroad in country II relative to GDP in country I – is smaller than j. Finally, one may point out that in the case that the macroeconomic production function does not only include factor inputs K and L but also the degree of high-technology specialization (Jungmittag, 2004) a rising share of intermediate imports in exports – more generally more traded intermediate inputs in global output – implies a positive growth effect if trade in intermediate products stimulates high technology specialization. Such specialization effects is likely to be highly asymmetric, namely in favor of countries with a high ratio of R&D expenditures relative to GDP and a particular focus on high technology as proxied by the share of high-tech patents in overall patents.

1.2.4 A Demand-Side and Supply-Side Growth Perspective

While short term analysis might argue that aggregate demand determines output, long run growth analysis typically argues that the production function – and the respective factor inputs – determines output growth. However, in a medium term perspective one may argue that both the demand side and the supply side determine output growth (Welfens, 2005). One may argue that a meaningful medium-term policy perspective comes up with rather different results which point to considerable needs for more empirical analysis. The approach suggested also allows one to take into account both demand side impulses (Y^d is aggregate demand: the sum of planned consumption and investment plus exogenous government demand plus net exports) and long run supply side effects (the production potential Y^{pot} as proxied by a simple production function). The analytical description can be found in the Annex 1.

Medium term output growth thus not only depends positively on the growth rate of capital accumulation g_K, the growth rate of labor input g_L, the rate of (Harrod-neutral) technological progress g_A, and the growth rate of foreign output g_{Y*} (the foreign growth rate will affect g_Y the more, the higher foreign output Y^* relative to Y is), but also on the import-GDP ratio j and the export-GDP ratio; this is in line with many empirical studies finding a significant impact of trade intensity on growth. Moreover, growth depends negatively on the term s'; this being said does not, of course, rule out that the savings rate $s = 1–c$ has a positive impact on the level of the long run growth path which is in accordance with long run growth theory. One should note that from a theoretical perspective the growth rate of labor saving technological progress might depend on the trade intensity x+j (or any suitable index reflecting the relative intensity of exports and imports); a more refined view might introduce specific weights for the impact of low, medium and high-technology trade intensity; one also should note in this context the empirical findings of Jungmittag (2004) who finds that the degree of high-technology specialization Ω' has a significant positive impact on economic growth of EU15 countries. If we assume that the growth rate of technological progress depends on the trend innovation input ratio (r'), namely weighted past R&D-expenditures relative to Y, and on the trend degree of high technology specialization (Ω') we can – using positive parameters f' and f'', respectively – replace g_A by $f'r'+f''\Omega'$. In an economy with FDI inflows one would have to make additional modifications; in particular one may assume that $g_A = f'r'+f''\Omega' + f'''F$ where F is the share of the capital stock owned by foreign investors.

The approach presented suggests that both the supply-side dynamics and the demand side dynamics are important for medium term growth. Moreover, the structure of output and the intensity of trade can contribute to output growth.

1.2.5 Dynamic Schumpeterian Perspectives

If one assumes that Schumpeterian rents are a positive function of technology intensity and also rise with the degree of outsourcing – the highest profit rate being in the production of finished goods since at that stage of production the firm has a high leverage over suppliers – a promising strategy for economic catching up is to combine technological upgrading and a long-term focus on the expansion of finished goods.

This perspective is not an argument for naïve industrial policy but foremost for a reasonable policy strategy:

- strong emphasis on competition and free trade policies
- clear focus on human capital formation and training
- adequate emphasis on inward FDI promotion where over time policymakers could try to encourage production of finished products and high value-added services
- adequate R&D support by government.

It is unclear whether economic and technological catching-up really could rely on leap-frogging. Typically, firms in a given sector will have to achieve a certain

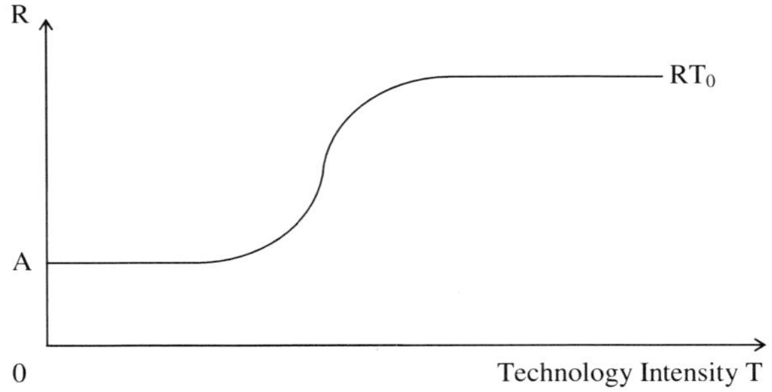

Fig. 1.3 Structure of production and profit rate R ("Schumpeter Ladder")

technological graduation over time. In an early stage of development and economic catching up, diffusion of new technologies will be more important than innovation; only after a certain per capita income and a critical level of R&D-GDP expenditures have been reached may one expect there to be strong and sustained innovation dynamics – and only in the long run can high technology dynamics play a considerable role in OECD countries which have achieved a high per capita income. From this perspective EU accession countries will gradually climb up the technology ladder where those able to attract high FDI inflows relative to GDP should have particularly favorable perspectives of economic and technological catching-up. As accession countries move up the technology ladder the EU15 countries will have to increasingly specialize on high technology goods and high technology services which implies enormous challenges for the education system. In a long run perspective this also holds for accession countries since catching-up in terms of per capita income will raise the relative price of nontradables which in turn is likely to stimulate demands for higher real wage increase such that firms in the tradable sector will increasingly be able to survive only if they start more comprehensive outsourcing to other countries which still have low wages, e.g. the Ukraine or Russia or many Asian countries.

A more complex perspective is obtained in a three country perspective. We may state the following hypothesis: In a triangular perspective of international

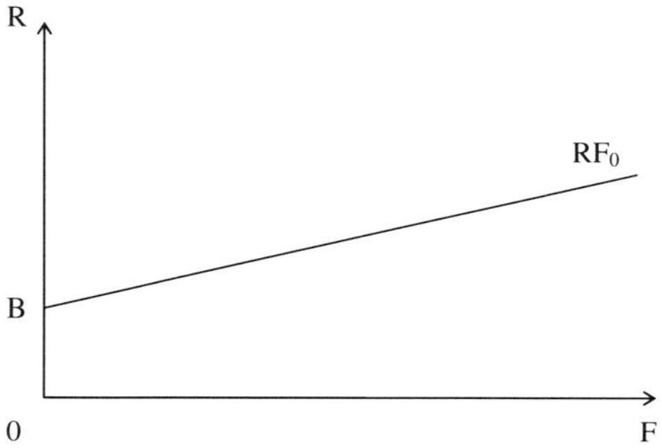

Fig. 1.4 Position in chain of value-added and profit rate R; (F =: Final Assembly)

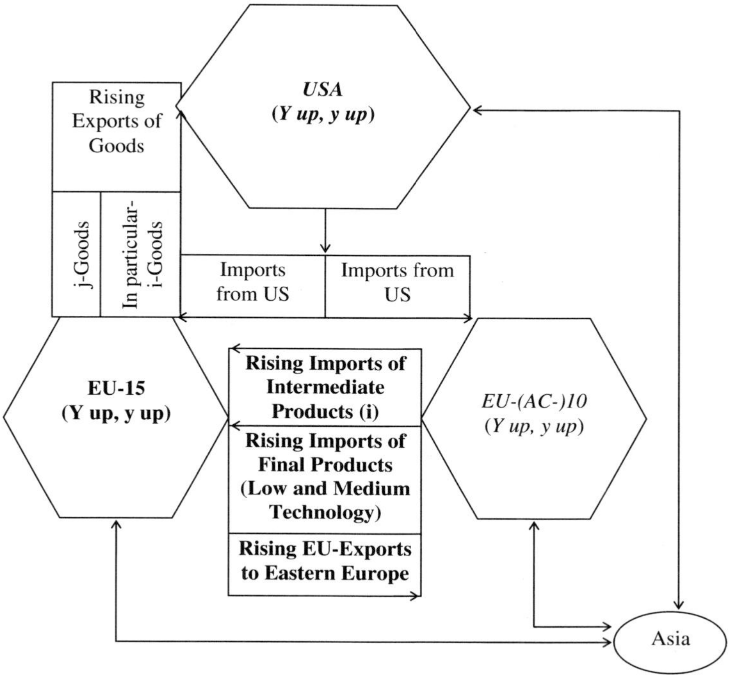

Fig. 1.5 Triangular perspective on trade, structural change and efficiency gains

specialization one may expect that EU15 countries positive RCA in exports vis-à-vis the US be positively correlated with those import fields in which EU15 has a negative RCA vis-à-vis Eastern Europe. This means that the new international division of labor after EU eastern enlargement is such that at least part of the EU15 sectoral import structure within EU25 is reinforcing external competitiveness as measured by the relative sectoral net export position in the USA. Similarly, Japan's net export position in the US could benefit from its import structure vis-à-vis Asian trading partners.

In this triangular perspective the ability of firms from EU15 countries to rely on imported intermediate products from EU accession countries is the basis for gaining competitiveness in both the global economy and the US. EU15 firms' outsourcing to Eastern Europe not only allows them to gain in terms of price competitiveness as cheaper imported intermediate products replace domestic intermediate products (or there is even true new outsourcing with cost advantages). It also is attractive to increase outsourcing to eastern Europe as a means to restructure domestic outsourcing in EU15, namely in a way that domestic outsourcing is more focused on producing technologically advanced intermediate products than previously. Following Romer (1990), Grossman and Helpman (1991, Chap. 3) and Bretschger (1998, Chap. 7), one may argue that a complex "high-technology" product Y^h is produced according to the following CES function using intermediate products x_i (i = 1,2... n).

$$Y^h = \left[\sum (x)_i^{\beta'}\right]^{1/\beta'} \tag{1.4}$$

If all intermediate products have the same cost structure we will have identical quantities of each intermediate product produced such that x1 = x2 = ... xn. Thus we get the typically symmetrical equilibrium

$$Y^h = xn^{1/\beta} \tag{1.5}$$

Assuming – to make sure that positive profit rates exist – that $0<\beta'<1$ we finally get (with X denoting the sum of all inputs)

$$Y^h / X = n^{(1-\beta')/\beta'} \tag{1.6}$$

Hence factor productivity is a positive function of the number of intermediate products. To the extent that economic opening up and regional integration lead to an increase in n, we can expect a positive productivity effect. This should translate into a higher level of the growth path.

Given slow development in EU societies, it would be wise for Europe to stimulate technological progress, productivity growth and output growth. Accelerated productivity growth and stronger innovation dynamics would be all the more important for EU countries since in the long run Europe faces problems from its ageing societies. The simulations of McMorrow and Roeger (2004) suggest that long run growth will reduce in the EU15 countries where the ageing phenomenon in the first half of the 21st century will be stronger than in the US.

Facing many theoretical issues in the context of structural change, innovativeness and growth, it is quite important to take a closer look at empirical analysis.

1.3 Empirical Analysis

The following section will concentrate on the analysis of selected innovation traits, structural change and the bazaar effect mainly through descriptive statistics. We will focus on selected EU-15 countries as well as some new EU member states and partially the USA as well.

1.3.1 Selected Innovation Traits and Structural Change

Product innovations allow for the increase in product prices in world markets and hence the earning of high incomes (wages and profit). Process innovations are equivalent to cost reductions and allow firms to fetch higher market shares and high incomes, in particular if price elasticity is larger than unity or if increased market share also allows for the exploitation of dynamic scale economies (e.g., learning by doing effects). Innovation dynamics can be assessed in different ways:

- Innovation expenditures, usually scaled by sales ("R&D intensity"); this in an R&D input indicator
- Patents per capita (R&D output indicator)
- Product innovation rate (new products to the market in % of sales, survey data, innovation output indicator)
- Diffusion rate (new-to-the-firm products, figures are from surveys)

Taking a closer look at selected EU countries, one finds that Sweden, Germany and Finland were leading in R&D intensity in manufacturing (6.4, 4.7 and 3.9, respectively, in 2003; EU average 3.45; see Table 1.1). France and the Netherlands achieved 3.1, the UK 3.0. Germany's R&D intensity in the services sector was much weaker, namely 1.6 compared to the EU average of 1.8. Sweden was a clear leader in this field. France and the UK recorded 1.6 and 1.4, respectively. It is interesting to observe that in the field of product innovations in manufacturing, Germany was below the EU average despite its leading position in R&D intensity. Finland, Sweden and France were leading countries in the field of product innovations. This suggests that the German innovation system might have considerable efficiency problems. A similar picture is found in production innovation in the services market. As regards diffusion indicators, Germany is a leading EU country. Moreover, Sweden and Germany recorded a high ratio of New-to-firm to New-to-market in the manufacturing industry, which points to relatively fast diffusion (this could reflect strong competition).

Against such apparent innovation weakness, one might consider it surprising that Germany has such a high current account surplus, e.g. 5% of GDP in 2002. However, 90 billion net exports recorded in 2002 would quickly melt away if full employ-

Table 1.1 European Innovation Scoreboard, 2003

European Innovation Scoreboard 2003 - selected mamber states

	EU15	DE	FR	NL	AT	FI	SE	UK
Innov exp manuf	3.45	4.71	3.08	3.07	2.83	3.91	6.42	2.96
Innov exp serv	1.83	1.64	1.57	0.78	0.92	0.96	19.11	1.39
New-to-mark prods manuf	10.5	7.1	9.5	–	8.4	27.2	3.5	9.5
New-to-mark prods serv	7.4	3.7	5.5	–	4.3	12.2	9.3	–
New-to-firm prods manuf	28.6	40.3	17.5	23.8	23.1	31.1	32.1	–
New-to-firm prods serv	18.8	16.4	17.1	13.9	12.8	18.8	23.7	–
New-to-firm/New-to-mark prods manuf	2.7	5.7	1.8	–	2.8	1.1	9.2	–
New-to-firm/New-to mark prods serv	2.5	4.4	3.1	–	3.0	1.5	2.5	–

Source: European Commission (2003, p. 27), Brussels and own calculations.

ment could be restored; investment would increase by about 10% or by about Euro 20 bill., consumption also by about 5% or 60 bill., which would leave net exports down at Euro 10 bill. The assumption here is that consumption is a positive function of disposable income and a negative function of the expected unemployment rate. Investment is assumed to depend negatively on the real interest rate and the expected unemployment rate. To put it differently, a high net export position of a country with a high unemployment rate cannot simply be considered an indicator of high international competitiveness. Rather, it largely reflects weak domestic demand. The reduction of net exports in the case of rising employment and hence a falling expected and actual unemployment rate will hold even if one takes into account the expansionary impact of higher employment on the supply side. This perspective is, of course, not to deny that in a situation of high net exports (and also in the case of net imports: see the US in the 1990s), certain sectors are positively successfully-specialized in production and export of technology intensive or innovative products.

International competitiveness in specific sectors can be assessed on the basis of revealed comparative advantage indicators (RCA: sectoral relative export share in country *i* as compared to the same industries relative export share on the EU15 single market, with an indicator above unity indicating a sectoral competitive advantage) or with respect to export unit values. A sectoral increase in the weighted export unit value indicates an improved competitiveness in the EU single market as higher prices can be fetched in a very competitive market. (There might, however, be cases for which changes in market power or government intervention also affect the export unit value).

According to the Heckscher-Ohlin theory, the RCA should depend on relative factor endowments. Higher RCAs and higher export unit values in certain sectors are likely to contribute quite strongly to output growth in the long run. Scale intensive sectors and science intensive sectors are obviously two potentially relevant sectors. In a high wage economy, emphasis on science-based products can strengthen competitiveness through product innovations which will temporarily lead to rising export unit values and hence higher profitability. This is a Schumpeterian perspective which leads away from perfect competition. Scale intensive products also imply that the perfect competition model does not hold. In some cases, scale intensive products exhibit both static and dynamic scale economies so that high production volumes could be combined with first mover advantages.

Interestingly, the US has achieved a higher export unit value in all fields where it has enjoyed a positive comparative advantage. This suggests a positive feedback mechanism in the sense that a higher export unit value goes along with increased profitability, which in turn reinforces investment and hence should contribute to an improving RCA.

The US has achieved a strong increase in the GDP weighted export unit value in NACE 30, 32, 33 and 35, respectively: manufacture of office machinery and computers; manufacture of radio, television and communication equipment and apparatus; manufacture of medical, precision and optical instruments, watches and clocks; manufacture of other transport equipment (e.g. airplanes). US companies apparently are well positioned to fetch higher prices in those sectors which represent a relatively large share of the economy. In the fields of NACE 32 and 35, the improvements in export unit values also represent a large share of US exports. The rise in the export unit value was quite impressive in NACE 32 and NACE 35, for which the respective value doubled and nearly quadrupled, respectively. In NACE 35 the US export value is five times as high as in the case of Germany, ten times as high as in the case of Italy and about fifty times as high as in the case of Hungary.

As regards export unit values and the change of export unit values over time, one should also take a look at weighted export unit values so that the relative economic significance of certain sectors can be understood. As regards Germany, it is well-known that the country has a positive RCA – read RCA above unity – in both the automotive industry and in other transport equipment (NACE 34 and 35). Taking a closer look at German industry, one can see that specialization in terms of RCA changed slightly in the decade after 1993. Germany has one important loser industry (see by contrast Italy), namely NACE 19 which not only represents

Table 1.2 USA – RCA, EUV, EUV weighted with the sectoral export shares of manufacturing and of GDP

	RCA			EUV 2001	EUV 1993	dEUV	EUV 2001	EUV 1993	dEUV
NACE rev.1 (2-digit)	2000/ 2001	EUV 2001	EUV 1993	Weighted (export share)	Weighted (export share)	Weighted (export share)	Weighted (GDP share)	Weighted (GDP share)	Weighted (GDP share)
15	0.24	0.40	0.26	0.01	0.01	0.00	0.16	0.11	0.06
16	0.07	2.04	1.64	0.00	0.00	0.00	0.02	0.01	0.00
17	0.28	6.85	5.28	0.04	0.06	–0.02	0.95	0.67	0.28
18	0.11	28.16	17.75	0.06	0.11	–0.05	1.25	1.23	0.02
19	0.16	9.17	11.17	0.02	0.04	–0.03	0.39	0.48	–0.09
20	0.79	1.37	0.82	0.01	0.01	0.00	0.21	0.14	0.08
21	0.50	0.84	0.5	0.01	0.01	0.00	0.29	0.12	0.16
22	**1.1**	**14.21**	**9.48**	**0.14**	**0.11**	**0.03**	**3.07**	**1.20**	**1.87**
23	0.29	0.11	0.09	0.00	0.00	0.00	0.02	0.01	0.00
24	**0.91**	**3.95**	**2.25**	**0.52**	**0.32**	**0.2**	**11.56**	**3.50**	**8.06**
25	0.57	8.00	6.13	0.14	0.13	0.00	3.09	1.49	1.61
26	0.49	3.91	2.66	0.03	0.02	0.01	0.65	0.24	0.41
27	0.53	7.35	4.33	0.21	0.14	0.07	4.75	1.54	3.21
28	0.56	12.57	8.64	0.18	0.13	0.05	3.90	1.43	2.48
29	**1.24**	**20.61**	**14.22**	**2.16**	**1.57**	**0.59**	**48.12**	**17.43**	**30.70**
30	**1.4**	**144.84**	**117.88**	**22.11**	**20.29**	**1.82**	**492.25**	**225.17**	**267.08**
31	**1.31**	**35.9**	**25.52**	**1.6**	**1.01**	**0.58**	**35.52**	**11.22**	**24.29**
32	**1.93**	**252.79**	**125.42**	**27.94**	**8.35**	**19.59**	**622.02**	**92.67**	**529.35**
33	**3.64**	**150.75**	**84.41**	**13.41**	**7.53**	**5.88**	**298.58**	**83.58**	**214.99**
34	0.20	9.73	6.96	0.32	0.22	0.10	7.06	2.43	4.63
35	**4.73**	**299.91**	**76.1**	**53.81**	**10.36**	**43.45**	**1197.95**	**114.99**	**1082.96**
36	0.97	22.27	12.45	0.44	0.23	0.22	9.89	2.50	7.39

Note: Fields of positive RCAs are bold typed; strong improvement in GDP-weighted export unit value is underlined; fields of declining export unit value are in italics.

a negative RCA but also signifies declining export unit values: tanning and dressing of leather, manufacture of luggage, handbags, saddlery, harness and footwear; NACE 17 and 18 – they are classified as labor intensive sectors – also show weak international competitiveness (RCA<1); sectors 28 and 36 which also are labor-intensive show at least an improvement of export unit values. There is a high RCA in the manufacturing of fabricated metal products (NACE 28, not including machinery and equipment). It is also noteworthy that the export unit value has increased over time for this product group. In the field of office machinery and computers (NACE 30) – a sector which (together with NACE 32: telecommunications equipment) is considered highly relevant for productivity growth – Germany has a negative RCA. Worse yet, the export unit value in this sector has declined. NACE 32 has improved over time. The overall picture with respect to the long-term development of export unit values in German industrial export reveals that export unit values – average revenue per quantity unit (e.g., kilogram of steel, etc.) – showed few changes over the period from 1993 to 2001. Which sectors are most important for economic dynamics: In a narrow sense those sectors which show a positive RCA and a high weighted export unit export value; this at least is the concept presented here. As regards the economic significance of export unit values, it is indeed useful to take a closer look at weighted unit values where sectoral shares in overall manufacturing exports are taken as weights: considering only weighted indicators reaching at least 0.75 (hence export unit value must be high or the share of the respective sector in overall export of manufacturing) – see the bold figures in the respective tables – we see that 29, 30, 32, 33, 34 and 35 are crucial sectors for Germany.

Note that the change in the weighted export unit value of 32, 33 and 34 was positive in Germany over the period 1992–2001; and this should translate into relatively increasing wages for skilled workers as we may assume that these sectors are using skilled labor intensively. As regards Germany, 29, 33, 34 and 35 reveal an economically significant positive RCA; as regards Hungary we find 18, 30, 31, 32, 34 as positive RCA: 34 is an overlap with Germany. The fact that Hungary could improve the weighted export unit value strongly in 34, the automotive sector, points to a strong catching-up process in the Hungarian automotive sector. To the extent that this finding is representative of accession countries in Eastern Europe, Germany's automotive firms acting in the lower quality segments of the market might face profitability problems in their German plants. The new international division of labor in Europe suggests that mass production of standard cars will be largely relocated to Eastern Europe's low wage countries. Hence the respective regions will face serious labor reallocation challenges in the early 21st century.

NACE 30, 31 and 32 indicate successful Hungarian specialization. However, note that 31 and 32 – differentiated goods (this also includes 29) – illustrate relative footloose industries: the manufacturing of office machinery and computers (30) and of electrical machinery and apparatus n.e.c. (31) could internationally be relocated relatively quickly.

One should point out that RCAs typically follow relative sectoral patent positions. A rising share in global patents in the respective sector translates into an

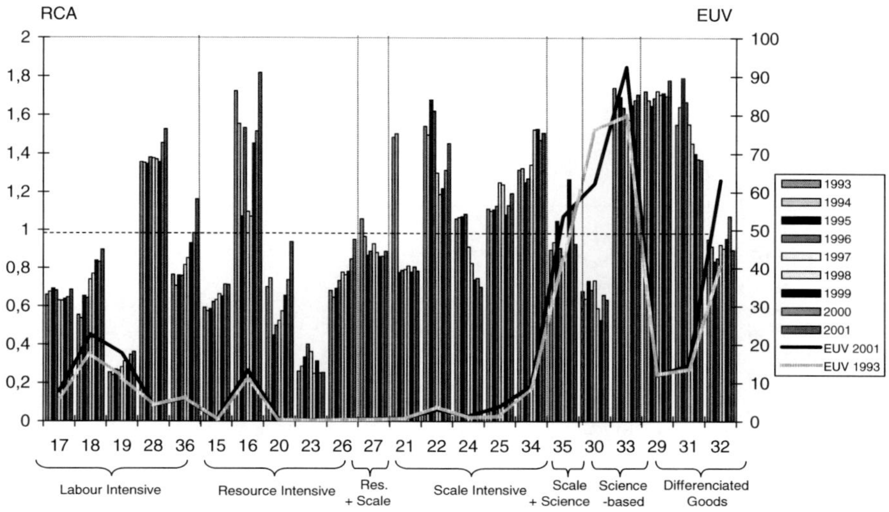

Fig. 1.6 Germany – RCA and export unit values

improved sectoral RCA with a time lag of 3–4 years. Hence expenditures on research & development and innovation policies are important.

Compared to the apparently stable German industrial specialization pattern, Hungary has launched a rather impressive catching-up process since reinforcing the RCAs in some technology intensive sectors and was also able to fetch higher export unit values – a proxy for its ability to extract high prices in competitive EU market – in EU-15 markets. Hungary has many fields which have shown a rise in the export unit value.

Germany's industry shows some clear fields of comparative advantage as does Hungary, an interesting case of new economic dynamics in an EU accession country. It is quite noteworthy that Hungary achieved higher export unit values in several sectors. The Table 1.4 shows that weighted improvements of export unit values were strong in 30, 32 and 34, essentially electronic products which represent scale-intensive goods, science-based goods and differentiated goods.

Italy has suffered in a traditional field of comparative advantage from a fall in the export unit value, namely in NACE 18 (manufacture of wearing apparel; dressing and dyeing of fur) which indicates stronger global price competition for an important sector of the Italian economy. There also was a strong fall in the export unit value in NACE 30 which is the crucial manufacturing of office machinery and computers, but in this group Italy also demonstrates a revealed comparative weakness as the RCA (Revealed Comparative Advantage) is much below unity. More encouraging news can be seen which respect to 35, which is close to an RCA exceeding unity and where the export unit value has improved. Very encouraging is also NACE 19 – with a high RCA and improved export unit value – which is a traditional strength of the Italian economy: tanning and dressing of leather, manufacture of

Table 1.3 Germany – RCA, EUV, EUV weighted with the sectoral export shares of manufacturing

	RCA			EUV 2001	EUV 1993	dEUV	EUV 2001	EUV 1993	dEUV
NACE rev.1 (2-digit)	2000/2001	EUV 2001	EUV 1993	Weighted (export share)	Weighted (export share)	Weighted (export share)	Weighted (GDP share)	Weighted (GDP share)	Weighted (GDP share)
15	0.71	0.62	0.52	0.03	0.03	0.01	8.55	2.93	5.62
16	**1.67**	**13.25**	**10.82**	**0.07**	**0.05**	**0.01**	**8.37**	**4.00**	**4.37**
17	**0.67**	**7.82**	**6.12**	**0.13**	**0.10**	**0.03**	**20.67**	**11.22**	**9.44**
18	**0.86**	**22.60**	**17.51**	**0.33**	**0.26**	**0.07**	**25.94**	**13.86**	**12.08**
19	0.35	17.65	11.39	0.08	0.05	0.03	0.89	4.96	–4.07
20	0.84	0.38	0.40	0.00	0.00	0.00	0.43	0.15	0.28
21	0.80	0.85	0.73	0.02	0.02	0.00	3.58	1.56	2.02
22	**1.38**	**3.11**	**3.83**	**0.03**	**0.04**	**–0.01**	**5.21**	**2.81**	**2.40**
23	0.25	0.26	0.18	0.00	0.00	0.00	0.64	0.10	0.55
24	0.72	1.43	1.01	0.16	0.11	0.05	28.49	11.78	16.71
25	**1.16**	**3.92**	**1.38**	**0.13**	**0.05**	**0.09**	**21.49**	**9.50**	**12.00**
26	0.90	0.43	0.44	0.01	0.01	0.00	1.17	0.60	0.57
27	0.88	0.66	0.51	0.04	0.03	0.01	6.50	2.10	4.40
28	**1.49**	**4.22**	**4.18**	**0.14**	**0.14**	**0.00**	**21.27**	**9.46**	**11.81**
29	**1.74**	**12.02**	**12.20**	**1.50**	**1.52**	**–0.02**	**197.64**	**96.45**	**101.19**
30	**0.65**	**62.26**	**76.05**	**4.26**	**5.24**	**–0.95**	**799.70**	**208.07**	**591.63**
31	1.37	14.70	13.64	0.69	0.64	0.05	101.16	35.49	65.68
32	**0.99**	**63.06**	**40.44**	**3.54**	**2.27**	**1.27**	**561.74**	**113.08**	**448.66**
33	**0.69**	**92.49**	**80.01**	**3.16**	**2.73**	**0.43**	**217.73**	**114.93**	**102.80**
34	**1.49**	**9.27**	**8.80**	**1.94**	**1.84**	**0.10**	**307.56**	**107.29**	**200.27**
35	**1.03**	**53.74**	**42.32**	**2.53**	**2.00**	**0.54**	**241.70**	**163.22**	**178.48**
36	**1.07**	**5.92**	**6.28**	**0.12**	**0.12**	**–0.01**	**19.83**	**8.25**	**11.58**

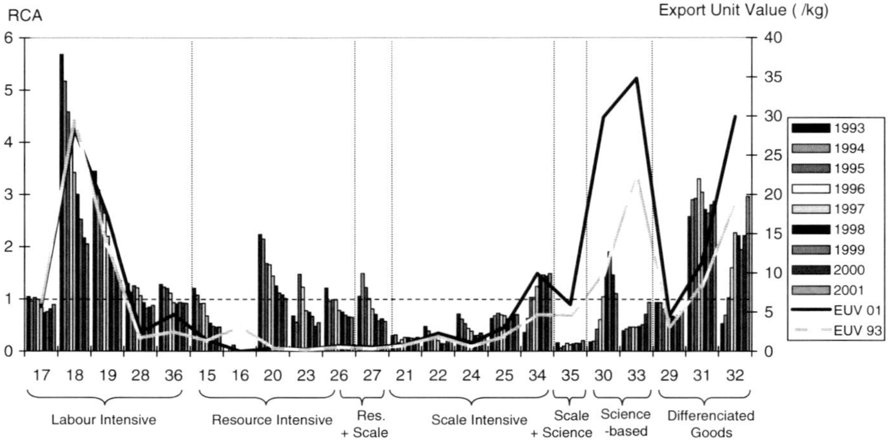

Fig. 1.7 Hungary – RCA and export unit values

Table 1.4 Hungary – RCA, EUV, EUV weighted with the sectoral export shares of manufacturing or respective sectoral shares in GDP

	RCA			EUV 2001	EUV 1993	dEUV	EUV 2001	EUV 1993	dEUV
NACE rev.1 (2-digit)	2000/2001	EUV 2001	EUV 1993	Weighted (export share)	Weighted (export share)	Weighted (export share)	Weighted (GDP share)	Weighted (GDP share)	Weighted (GDP share)
15	0.46	1.45	1.27	0.05	0.04	0.01	19.29	16.37	2.92
16	0.00	0.00	3.19	0.00	0.00	0.00	0.00	0.00	0.00
17	**0.85**	**5.72**	**6.29**	**0.11**	**0.12**	**–0.01**	**48.85**	**26.15**	**22.70**
18	**2.11**	**28.39**	**29.41**	**1.05**	**1.09**	**–0.04**	**494.83**	**438.19**	**56.65**
19	**1.42**	**17.21**	**13.31**	**0.28**	**0.21**	**0.06**	**118.38**	**70.66**	**47.72**
20	1.05	0.37	0.38	0.00	0.00	0.00	1.62	0.63	0.99
21	0.25	0.84	0.75	0.01	0.01	0.00	2.72	0.50	2.22
22	1.19	2.32	1.89	0.00	0.00	0.00	1.69	0.83	0.87
23	0.51	0.27	0.18	0.00	0.00	0.00	1.59	0.29	1.30
24	0.31	1.09	0.56	0.04	0.02	0.02	22.93	5.09	17.85
25	**0.70**	**3.04**	**1.93**	**0.07**	**0.04**	**0.02**	**27.75**	**4.43**	**23.32**
26	0.65	0.73	0.53	0.01	0.01	0.00	2.89	1.46	1.44
27	0.60	0.56	0.29	0.02	0.01	0.01	8.56	1.76	6.80
28	0.87	2.40	1.69	0.05	0.04	0.02	21.86	6.41	15.45
29	**0.73**	**4.51**	**3.16**	**0.29**	**0.20**	**0.09**	**113.75**	**26.54**	**87.21**
30	**1.28**	**29.81**	**9.74**	**3.59**	**1.17**	**2.42**	**2503.98**	**9.06**	**2494.92**
31	**2.83**	**11.36**	**8.32**	**1.10**	**0.81**	**0.29**	**489.95**	**74.79**	**415.16**
32	**2.59**	**29.91**	**18.76**	**5.06**	**3.17**	**1.89**	**1624.24**	**35.17**	**1589.07**
33	**0.82**	**34.83**	**22.37**	**0.80**	**0.51**	**0.29**	**198.11**	**20.40**	**177.71**
34	**1.46**	**9.93**	**4.68**	**2.35**	**1.11**	**1.24**	**941.11**	**20.94**	**920.17**
35	0.18	5.96	4.54	0.05	0.03	0.01	6.85	2.63	4.23
36	0.92	4.69	2.50	0.09	0.05	0.04	37.21	8.30	28.91

luggage, handbags, saddlery, harness and footwear. A successful adjustment is also found in NACE 28 and 29, respectively: Manufacture of fabricated metal products, except machinery and equipment (28) and manufacture of machinery and equipment n.e.c. (29). From this perspective Italy could benefit considerably from EU eastern enlargement both through rising exports of sophisticated consumption goods and of industrial goods.

So far we have analyzed the relative export position of selected countries as suppliers on the EU-15 market. However, having a flourishing export market does not necessarily mean that a country is competitive and is dealing well with the challenges of structural change. What counts is the ratio of domestic to foreign value added in production, as well as the ratio of domestic to foreign intermediated inputs in production. This will have an impact on how well factor markets can cope with structural change. Thus we will now turn to the analysis of the importance of intermediate imports, which brings us to the so-called bazaar effect.

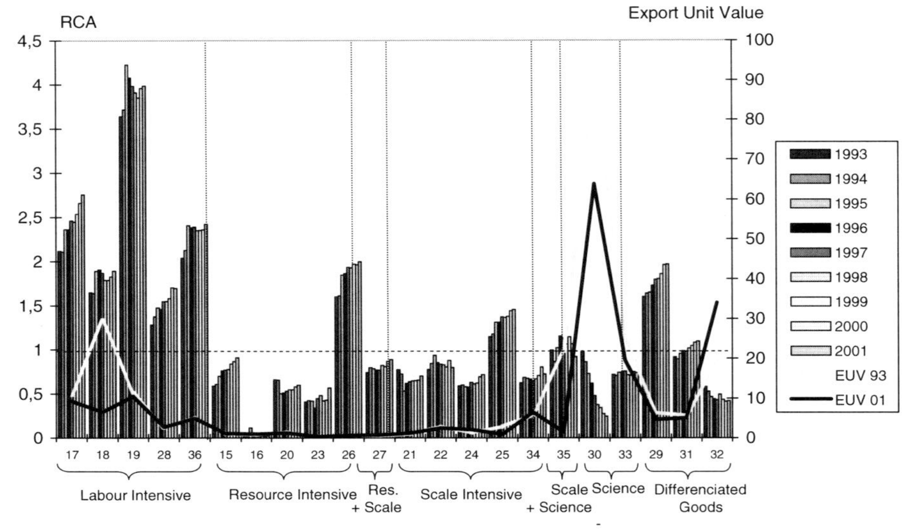

Fig. 1.8 Italy: RCA (revealed comparative advantage) and export unit value

1.3.2 The Bazaar-Effect

The bazaar-effect according to Sinn (2005) states that the share of domestic value added in total output of an industry *i* falls to the benefit of foreign countries. This means that a declining part of the final product's value added is generated domestically. There is a tendency toward outsourcing and offshoring, while the first implies the purchase of intermediate products from external firms and the latter indicates that a (mostly labor intensive) part of the value added production chain is relocated to a foreign country. In the extreme case, the economy would merely buy and sell products, just like on a bazaar. The question of outsourcing and offshoring is especially brigand within an EU25 perspective, since many western European companies offshore production to eastern Europe and also buy eastern European intermediate inputs.

Indeed, one can show that the share of domestic value added in production is declining in Germany, as well as in some other European countries. However, this does not necessarily imply a problem for the domestic economy. As long as the sum of domestic value added share plus domestic intermediate inputs' share to total production does not decline, there should not be a negative effect to the domestic factor markets. If, for instance, domestic value added in production declines by 2 percentage points, but at the same time the share of domestic intermediates in production rises by 2 percentage points, the share of domestic participation in production remains the same, causing no harm to the domestic economy. This is merely a relocation of economic activity among

Table 1.5 Italy – RCA, EUV, EUV weighted with the sectoral export shares of manufacturing and of GDP

	RCA			EUV 2001	EUV 1993	dEUV	EUV 2001	EUV 1993	dEUV
NACE rev.1 (2-digit)	2000/ 2001	EUV 2001	EUV 1993	Weighted (export share)	Weighted (export share)	Weighted (export share)	Weighted (GDP share)	Weighted (GDP share)	Weighted (GDP share)
15	0.84	1.07	1.04	0.07	0.07	0.00	7.93	4.70	3.23
16	0.01	0.70	0.75	0.00	0.00	0.00	0.01	0.00	0.00
17	2.79	9.53	10.83	0.59	0.96	–0.37	69.33	65.67	3.67
18	1.85	15.80	29.77	0.52	1.36	–0.83	61.52	92.67	–31.14
19	3.76	17.62	11.43	0.78	0.68	0.11	92.16	46.23	45.93
20	0.62	1.30	1.49	0.01	0.01	0.00	0.84	0.53	0.31
21	0.68	1.12	0.99	0.02	0.02	0.01	2.72	1.24	1.48
22	0.89	2.69	2.88	0.02	0.02	0.00	2.51	1.48	1.02
23	0.49	0.28	0.18	0.00	0.00	0.00	0.36	0.11	0.25
24	**0.65**	**1.79**	**1.30**	**0.17**	**0.10**	**0.07**	**19.81**	**7.00**	**12.82**
25	1.46	2.95	2.90	0.13	0.13	0.00	15.50	8.73	6.77
26	2.00	0.59	0.65	0.02	0.02	–0.01	2.10	1.70	0.39
27	0.92	0.71	0.57	0.04	0.02	0.01	4.23	1.71	2.52
28	**1.72**	**2.58**	**2.57**	**0.11**	**0.10**	**0.01**	**13.07**	**6.92**	**6.14**
29	**1.99**	**6.19**	**6.35**	**1.04**	**0.96**	**0.08**	**122.32**	**65.92**	**56.40**
30	0.29	56.58	89.47	1.81	4.88	–3.06	212.81	333.21	–120.39
31	1.10	6.54	5.67	0.24	0.19	0.05	28.73	13.04	15.69
32	**0.45**	**24.50**	**19.24**	**0.63**	**0.42**	**0.21**	**73.63**	**28.80**	**44.83**
33	0.72	20.28	24.72	0.36	0.43	–0.08	47.91	29.62	12.29
34	**0.77**	**6.32**	**5.81**	**0.78**	**0.30**	**0.30**	**91.19**	**32.51**	**58.68**
35	**0.95**	**24.99**	**21.79**	**0.90**	**0.09**	**0.09**	**105.96**	**55.71**	**50.25**
36	2.39	3.89	5.20	0.19	–0.10	22.38	22.38	19.76	2.62

domestic industrial sectors. For total manufacturing one can observe a decline in the share of value added plus the share of domestic intermediate inputs since the second half of the 1990s; this, however, is not valid for all industries. Therefore we will now turn the attention to the six most economically important German industries.

These are motor vehicles, machinery, chemicals, radio/TV/communication equipment, textiles/wearing apparel, and office machinery. These six industries make up to roughly 62% of German exports and 52% of German imports in the year 2004. The share of these six industries in the foreign trade position of the other countries considered in the analysis is also rather high (e.g., 60% of Hungarian exports, and 59% of Hungarian imports in 2002). We will analyze the extent of the bazaar effect for these six industries.

1.3.2.1 Motor Vehicles

The following figure shows the share of domestic value added, of domestic intermediate inputs and of foreign intermediate inputs in the total output of the motor vehicles industry in selected countries. These include three new EU member states, Slovenia, Hungary and Poland. Furthermore it includes Italy and Germany as the "sick men" in Europe; the Netherlands as a small open economy; and especially concerning the telecommunications industry, Finland has been included into the analysis.

The bazaar-effect as described by Sinn (2005) refers to a decline in the share of domestic value added in total output Indeed this is the case for motor vehicles in Germany, where the share of value added falls from 32.5% in 1995 to 21.8% in 2000. At the same time, however, the share of domestic intermediates rises from 51.1 to 60%. The total share of domestic formation of output therefore only slightly declines from 83.6 to 81.8%. This corresponds to relocation to foreign countries in the production of motor vehicles to the extent of 1.8% of output. However, still 81.8% of output is produced in the home country, either as value added in the motor vehicles industry or as intermediate production in other domestic industries. Thus the bazaar-effect is rather small in the German motor vehicles industry.

A similar rise in the share of foreign intermediates can be observed in the Netherlands, Finland and Italy, and to a greater extent in Poland and Slovenia. The Netherlands, Hungary and Slovenia, as being small open economies, are expected to have a higher overall share of foreign intermediates, which is shown in the Fig. 1.9. However, the share of foreign intermediates is considerably higher for the small new EU economies than for the Netherlands.

The values (+ or –) at the end of the rows indicate the tendency of Export Unit Values (EUV) since 1993. It seems that a problem might occur if a strong decline of the domestic share of production is accompanied by declining

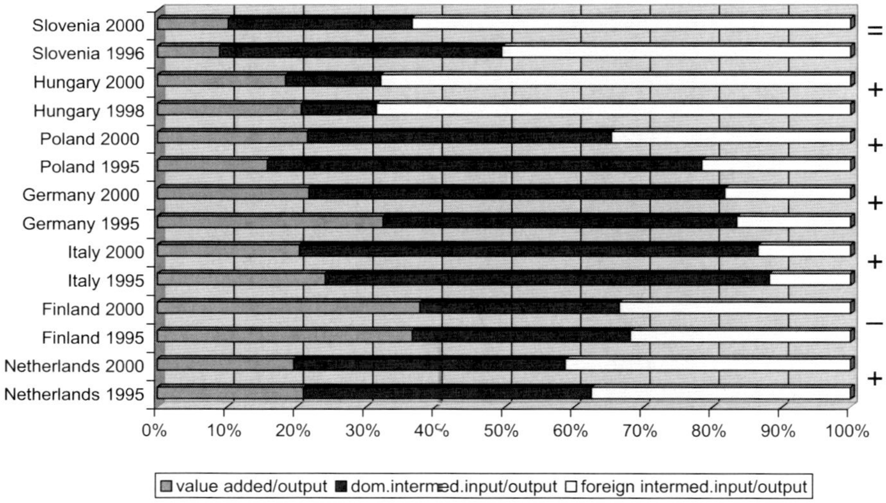

Fig. 1.9 Motor vehicles – bazaar effect

tendency of Export Unit Values. In the motor vehicles industry, only Finland shows a decline in EUVs; at the same time, however, that share of value added has increased.

1.3.2.2 Machinery and Equipment

Most countries show a considerably high share of domestic production in this area. The sum of the share of value added and domestic intermediates exceeds 70% in most cases. Only in some new EU member states can one find a little higher share of foreign intermediates: Slovenia and Hungary.

Except for Slovenia the share of foreign intermediate inputs has risen in the time period considered. A relatively strong rise occurred in Hungary with 11%, followed by Finland and Germany with roughly 4%. Although there is a negative tendency in the development of the EUVs in Germany, the situation is not alarming, since the share of foreign intermediates is still very low (16%). In most of the countries the EUVs tend to increase, except for Italy and the Netherlands, where they stagnate instead.

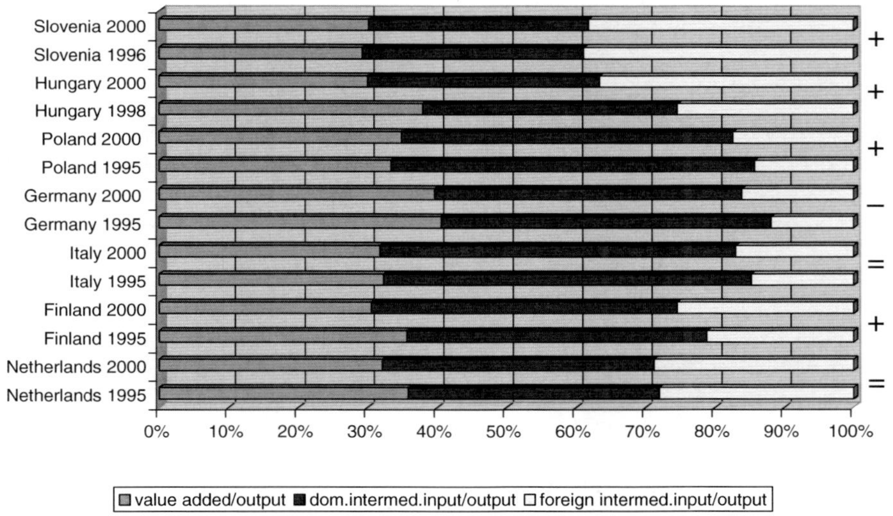

Fig. 1.10 Machinery and equipment – bazaar effect

All in all, the picture in the machinery industry is rather homogenous with a relatively high percentage of domestic activity in production.

1.3.2.3 Chemical Products

Homogeneity is also given in the chemicals production. In most countries the share of domestic value added makes up to 25–35% of the industry output. Except for

Italy, for which the share has remained constant, domestic value added in production is declining in western European counties, giving rise to the bazaar-effect. This is also the case for Hungary. However, both in Slovenia and in Poland the share is increasing considerably. At the same time, this rise cannot compensate for the loss in the share of domestic intermediates, thus the share of foreign intermediates in output has risen in both countries considerably (Slovenia 13%, Poland 14%). The only country, where the total domestic share in 2000 exceeded the previous value is Hungary. EUVs show a clear tendency to rise in all but one country, Slovenia.

1.3.2.4 Radio, Television and Communication Equipment

The Fig. 1.12 is not as harmonious any more in the radio, television and communication equipment industry.

The new EU member states stand out through relatively low shares of value added and domestic intermediates. This is valid for Slovenia and Hungary for both years considered and for Poland especially for the year 2000. Furthermore the share of foreign intermediates is extraordinarily high, reaching values of over 80% in Hungary. Also the very strong increase of the foreign intermediates is striking: 27% in Poland and 19% in Slovenia. As mentioned above, small countries are expected to have higher shares of foreign intermediates, since they are more dependent on foreign trade (especially foreign imports) than big open economies, but also compared to other western European small economies such as the Netherlands, these ratios are very high. Hungary with merely 13% of value added and 4% of domestic interme-

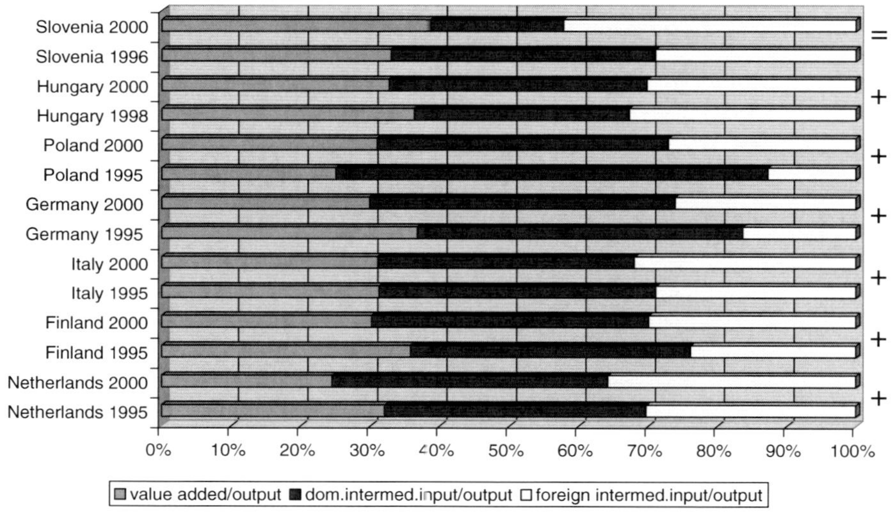

Fig. 1.11 Chemical product – bazaar effect

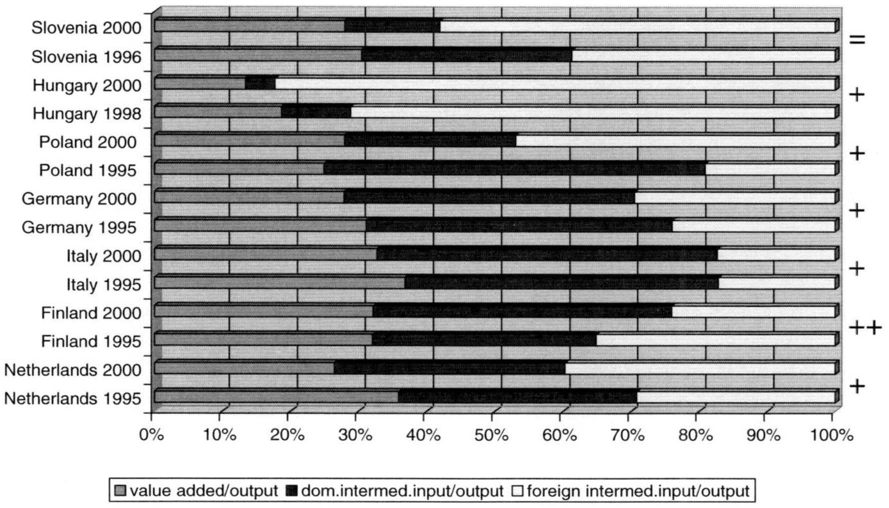

Fig. 1.12 Radio, television and communication equipment – bazaar effect

diates appears to be the real bazaar economy in the radio, television and communication equipment industry.

The only country with a relative stable share of foreign intermediates is Italy, and the only country with a decline in the share of foreign intermediates is Finland. The latter here is not surprising, and adding the outstanding rise in EUVs in this industry underlines the dominant position of Finland in the telecommunications market. However, it is worth noting that one can also observe a rise in EUVs in most other countries.

1.3.2.5 Textiles

In contrast, export unit values decline in most of the countries in the textiles industry. Only in Poland and Germany do the EUVs seem to stagnate, and there is weak increase in the Netherlands.

The Netherlands is also the only country, for which the share of foreign intermediate inputs has declined (by roughly 2%); in all other countries the share has increased. The strongest increase occurred in Hungary with 24%, followed by Poland with 17%, and Slovenia with roughly 7%. At the same time, value added rose by 2.5% in Slovenia and declined by almost 13% in Hungary. This does not underline the predictions of the traditional trade theory, namely that eastern European countries will mainly specialize in labor intensive production due to the abundance of labor. However, there are other labor intensive industries, such as wearing apparel and leather and fur, where some of the new EU member states have considerable comparative advantages.

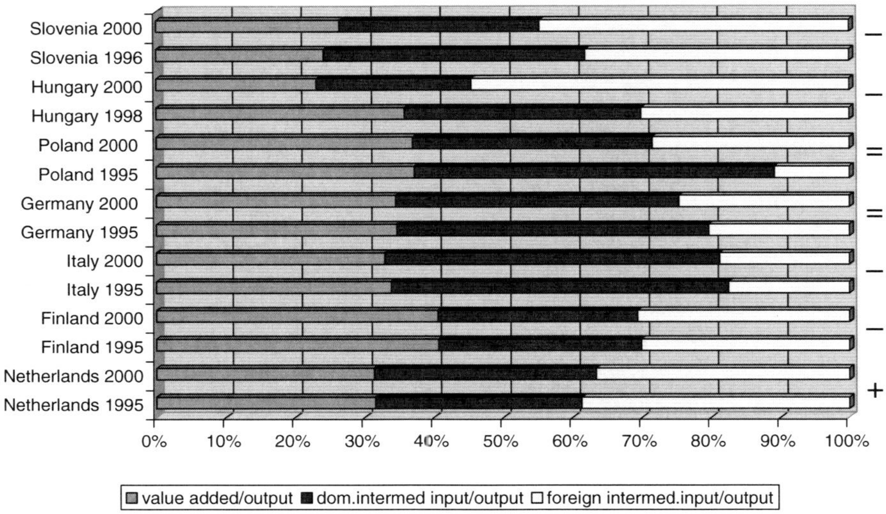

Fig. 1.13 Textiles – bazaar effect

1.3.2.6 Office Machinery

Finally, we take a look at the composition of the industrial production of office machinery. It reveals the most diversified picture of all industries.

In general the share of foreign intermediates is highest as compared to the other industries. Starting with 28% in Slovenia in 1995, the average share is roughly 40–60%, with extremely high shares of over 90% again in Hungary. But also in Finland and Italy, the share of foreign intermediates far exceeds 50% of output.

A clear bazaar effect can be identified in Finland, Italy, Germany and Hungary. In these countries the share of domestic value added decreased strongly. In all these countries the share was already rather low in 1995, but it almost disappeared in the year 2000, mostly in Hungary with 6% and in Finland with merely 2%. Especially Hungary, with a domestic intermediates share of 1.5%, seems to play only a bazaar function on the office machinery market.

However, we also find countries, within which the share of domestic value added increased in the second half of the 1990s. In the Netherlands it rose by 1.5%, in Poland by roughly 6%, and in Slovenia by 3%. In addition the share of domestic intermediates rose in the Netherlands by 4.6%, added up to an increase in the domestic share of production by almost 6%. This is the clear opposite effect to the bazaar. This effect is also visible in Poland with an increase in the domestic share of production by 1.6%.

At the same time, EUVs show a clear tendency to rise in Poland and fall in the Netherlands. The latter is the case for almost all other countries as well, which indicates that the bazaar-hypothesis is strongly relevant in the market for office machinery, particularly in Italy, Germany and Hungary.

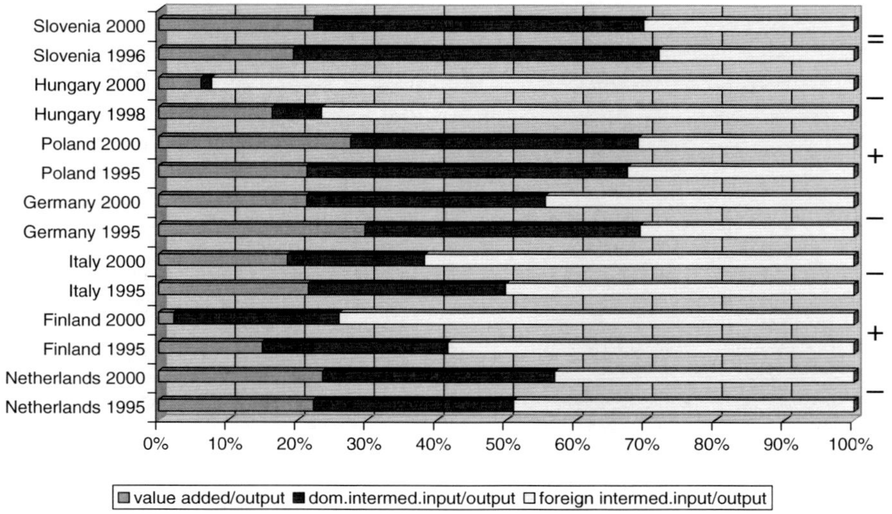

Fig. 1.14 Office machinery – bazaar effect

All in all, it is important to look at the sectoral level when analyzing the bazaar effect, since the differences among the industries are large. Also it is not the share of domestic value added that matters, but the sum of the shares of domestic value added and domestic intermediate inputs. Only if this sum declines can we see an increase in the foreign intermediates inputs, which might have negative implications on domestic factor markets such as the labor market.

1.3.3 Net Bazaar Effect

Finally one might not only look at the "gross" bazaar effect as described above, which controls for the intermediate imports in production and thus in exports in country I. One should also consider that imports of country I from country II also contain to some extent exported intermediates from country I to country II. Controlling for this would reduce the "gross" bazaar effect, which we call the "net-bazaar effect".

It is rather challenging to calculate the net bazaar effect for a country, because one would need statistical information on the share of intermediate products in exports in relation to the share of intermediate products in imports. We can, however, approximate by assuming that the import/export ratio in total foreign bilateral trade corresponds to the import/export ratio of bilateral trade in intermediates. Thus one can calculate the net bazaar effect on a bilateral basis. The effect is visible if one calculates the net bazaar effect towards the most important trading partners.

This is done in the following for the German – EU14 bilateral trade for the motor vehicle industry. First one has to identify the share of EU14 imports in German

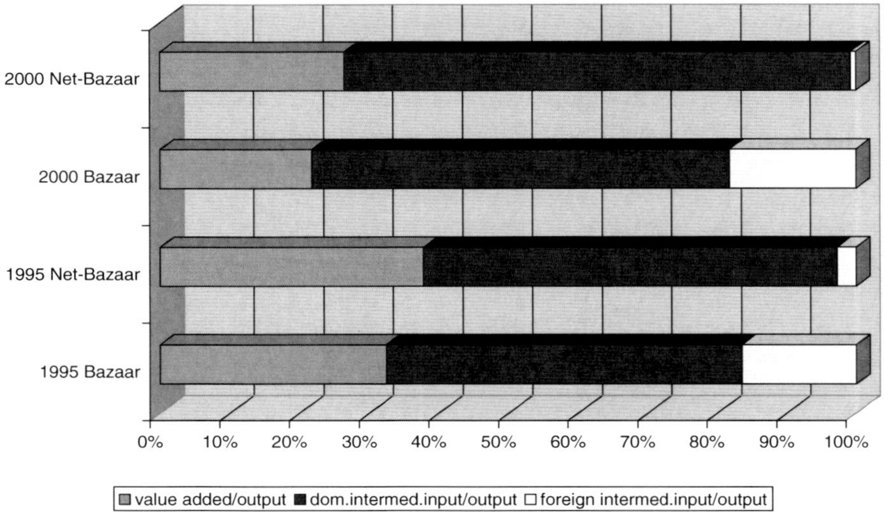

Fig. 1.15 The gross and the net bazaar-effect in the motor vehicles industry in Germany's trade towards EU14

total imports. This accounts for 60.4% in 1995 and for 54.8% in 2000. Furthermore, the import/export share of motor vehicles for the bilateral trade between Germany and EU14 in the year 1995 is 0.7, and in the year 2000 it is 0.56. Subtracting the intermediate exports contained in intermediate imports reduces the share of foreign intermediate imports accordingly.

According to the Fig. 1.15, the share of foreign intermediates declines in 1995 from 16.5 to 2.4%, and in 2000 from 18.2 to 0.7%. Thus the domestic share of production (measured as the sum of the share of value added plus the share of domestic intermediates) rises. The greater the economic importance of the trading partner, the greater the corresponding difference between the gross and net bazaar effect.

1.3.4 The Current Account and Intermediate Products Plus FDI

The current account can be specified on the basis of a simple export function, an import function and (net) profit transfers, which yields several new results compared to the traditional view of trade and the current account. To simplify we will consider inward foreign direct investment and thus an asymmetric two country perspective. Moreover, we will take into account the arguments of Froot and Stein (1991) who – emphasizing a world with imperfect capital markets – have argued that foreign investment will increase after a devaluation of the currency of the host country (country I): In international mergers and acquisitions, firms from country II will be more often successful after the appreciation of the currency of country II since firms then have higher equity capital – expressed in units of country I currency – so that a leveraged merger or acquisition will be easier than before.

Let us consider a set-up where country II is the only host country, country I is the only source country of FDI; and there are no portfolio capital flows. We will thus assume that FDI inflows in country II are proportionate to Y and a positive function of the real exchange rate $q^* = :eP^*/P$ (with e standing for the nominal exchange rate). The share of capital of country II owned by investors from country I is φ and the marginal product rule for factor rewards is assumed to hold in both countries. Production potential in the home country is denoted as Y' and in the foreign country as Y'^*. We assume that the higher the production potential in the country is, the higher the exports of country I will be, thereby leading to a supply-side impulse for exports as well as the standard foreign demand-side impulse; in a similar vein we make such an assumption on the import side so that real imports are a positive function of the foreign production potential and a positive function of GDP in country I. This implies at first sight an export function $X = \mathbf{Y}'^{\eta}q^{*\alpha}Y^*$ and an import function $J = \mathbf{Y}'^{*\eta*}q*^{-\alpha'}Y$; note that using both Y' and Y^* in the export function – and similarly Y'^* and Y in the import function – could also be understood as reflecting the fact that some sectors' exports are supply-side driven while others are demand-side driven. Such a view is consistent with the perspective developed by Welfens (2007), who argues that a hybrid macro model should consider a blending of a supply-side growth model and a Keynesian macro model. Incidentally, it is clear that this approach is consistent with the standard gravity equation. Hence we can express the effective export–import ratio X'' in real terms as real exports plus profit transfers received minus FDI flows abroad ($\varphi q^* ß^* Y^* – b'' Y^* q^{-ß'}$) relative to imports expressed in domestic goods units (the latter means we have to multiply J by q^*):

$$X'' = \left\{Y'^{\eta}q^{*\alpha}Y^* + \left[\varphi q^* ß^* Y^* - b''Y^*q^{-ß'}\right]\right\} \Big/ \left(q^* q^{*-\alpha'} Y Y'^{*\eta^*}\right) \tag{1.7}$$

In the standard textbook setting the square bracket is zero – plus the elasticities η and η^* are zero – so that the Marshall-Lerner condition would be $\mathrm{dln}X''/\mathrm{dln}q^* = \alpha + \alpha' - 1 > 0$, but here the situation is more complex; for the problem at hand, it is also useful to focus on exports plus profits transfers received minus FDI flows abroad minus imports. Moreover, if we want to consider exports of intermediate goods in the import function one may modify the import function as follows (using the notation x'' as a proxy for the intensity of international outsourcing/off-shoring and making the assumption that the potential for international outsourcing and off-shoring with a focus on intermediate products is proportionate to the foreign capital stock K^*):

$$J = q^{*-\alpha'}Y + x'K^*q^{*\alpha'} \tag{1.8}$$

The term $x''K^*q^{*\alpha''}$ indicates that "round trip exporting of intermediates" is more attractive if country I's currency has depreciated. Additionally, one should note that imports should not depend on real GDP but on national income, which is Y + $q^*\varphi ß^*Y^*$where φ is the share which country I investors holds in the capital stock abroad. Hence an adequate specified import function reads

$$J = q^{*-\alpha'}\left[Y + q^{*}\varphi\beta^{*}Y^{*}\right] + x'K^{*}q^{*\alpha'} \tag{1.9}$$

In order to have a rather simple expression we rewrite real imports as

$$J = q^{*-\alpha'}\left[Y + q^{*}\varphi\beta^{*}Y^{*}\right] - x''K^{*}q^{*-\alpha'} \tag{1.10}$$

For the special case of $\alpha' = \alpha''$ this expression could now be written as $q^{*-\alpha'}\{[Y + q^*\varphi ß^*Y^*] - x''K^*\}$. However, we do not want to focus on this special case and rather take a look at the more general statement: The current account balance X′ therefore is given by

$$X' = \left\{Y'^{\eta}q^{*\alpha}Y^{*} + \left[\varphi q^{*}ß^{*}Y^{*} - b'Y^{*}q^{-ß'}\right]\right\} - J = q^{*-\alpha'}\left[Y + q^{*}\varphi\beta^{*}Y^{*}\right] - x'K^{*}q^{*-\alpha'} \tag{1.11}$$

$$dX'/dq^{*} = -aq^{*(-\alpha'-1)} + \left[Y + q^{*}\varphi\beta^{*}Y^{*}\right] + q^{*-\alpha}\left[\varphi\beta^{*}Y^{*}\right] + a'x'K^{*}q^{(a'-1)} \tag{1.12}$$

$$dX'\Big/dq^{*(-q'-1)}\left[Y + q^{*}\right] \tag{1.13}$$

This expression holds under the assumption $q^* = 1$,

$$-\alpha\left[Y + q^{*}\varphi\beta^{*}Y^{*}\right] + \left[\varphi\beta^{*}Y^{*}\right] + a'x'K^{*} \tag{1.14}$$

It is positive if it holds,

$$-\alpha\left[Y + q^{*}\varphi\beta^{*}Y^{*}\right] + a'x'K^{*} \tag{1.15}$$

This expression is much more complex than the standard expressions so that we may emphasize the following: In a realistic setup with foreign direct investment, the reaction of the current account with respect to a change in the real exchange rate is somewhat different than the Marshall-Lerner condition suggests.

1.4 Policy Conclusions

The global innovation race has intensified and therefore more flexibility and greater incentives for innovations, learning and flexible adjustment are needed in Europe. There is an adjustment overhang in those countries which have high unemployment rates, and this particularly concerns some of the large Euro zone countries. Germany and Italy are two countries with serious problems, but Germany has adopted broader reforms than Italy. Italy's rising unit labor costs after the start of the Euro zone point to a failure of the country's collective bargaining system to adjust to the new monetary reality. However, Germany is not much better since the Euro zone's largest economy has had enormous problems for many years beginning with German unification. The R&D-GDP ratio which stood at 2.9% in 1989 has fallen over a decade and only after 1999 has there been a modest increase; the goal of the Lisbon Agenda, namely R&D expenditures of 3% of GDP by 2010 can hardly be achieved. Germany's position in ICT is favorable in some sub-sectors, but the overall ICT position is not top in the EU, and recent analysis (Welfens, Jungmittag, & Vogelsang, 2005)Welfens points to problems with technological progress in core regions of both western Germany and eastern Germany. In 2006/2007, a cyclical expansion in Germany facilitated structural adjustment in the EU's largest economy; at the same time, however, it has become apparent that the country suffers from ineffective prudential supervision (no less than three major banks almost went bankrupt in the period 2000–2007, and the federal supervisory agency BaFin which shares competences with Deutsche Bundesbank seems to be rather weak and also largely intransparent so that there are doubtful incentives for managers of banks).

France has emphasized high-technology growth in Western Europe more than other EU15 countries. However, it has recorded both success stories – including Airbus (actually an EU joint venture) and the nuclear industry – and failures (e.g. Bull in the computer sector which was largely a failure) with its industrial policy. Germany's structural adjustment has been relatively slow in industry, the adjustment dynamics in the UK and France were more pronounced. Italy is rather dependent on the international economy and it risks – similar to Germany – facing reduced growth if global economic growth should slow down.

Some of the EU accession countries have shown remarkable structural adjustment; technological upgrading has been an important element in several eastern European countries which have been able to attract high FDI inflows. Given high sustained unemployment rates in most accession countries one must, however, be worried about the unemployment problem. Jobless growth could be one of the new problems in transition countries. To the extent that the mass unemployment problem contributes to social and political conflicts as well as political radicalization, high long-term unemployment could contribute to political destabilization which in turn will raise the political risk premium and weaken growth in the long run. EU25 in which economic divergence across countries should be observed is likely to be highly unstable. Already in 2005, merely a year after EU enlargement, the EU show signs of increasing weakness; failed referenda in France and the Netherlands have shown that the EU has lost broad popular support in EU founding countries. The apparent inability – thus far – to complete the Euro zone as originally planned, namely by including the UK also shows lack of consistency and political consensus. Overblown projects such as Turkish EU enlargement have clearly undermined the political support for EU integration in Germany, the Netherlands, France, Austria and some other EU countries. If EU25 should turn out to be a rather heterogeneous political club with weak economic growth and an inability to organize innovation, growth and sound public finance in combination with low inflation rates and full employment, the EU is unlikely to survive for many decades.

The basic policy conclusion is that the EU25 should be able to benefit from both EU enlargement and from economic globalization. The EU is a major host country of FDI and also a major source country so that it can benefit from outsourcing. In high wage economies of Western Europe it is fairly clear that the accelerated structural change in Europe requires more wage differentiation which should be more in line with differential productivity growth rates in EU countries. The existence of a high minimum wage is a doubtful exercise, in particular if it is a nation-wide minimum wage as in France and Italy which both have high unemployment rate. Germany's social security system is still relatively generous and implicitly defines a minimum wage which is relatively high, namely in the sense that overall labor costs of unskilled labor is high. The specific unemployment rate of unskilled workers should gradually be brought down to the average unemployment rate; greater wage dispersion as well as stronger retraining could be useful in Germany, France and Italy. More wage dispersion could go along with a higher average growth rate of wages provided that greater wage flexibility brings about higher average productivity – partly related to higher regional labor mobility. As regards incentives of firms to invest more in training and retraining, one should clearly point out that the globalization process is weakening such incentives; the average tenure of workers is declining so that the incentive for firms to invest in training and retraining is declining. Here government's tax policy might want to envisage new adequate incentives which stimulate training and retraining.

In an EU in which national R&D programs are increasingly likely to generate cross-border benefits through international technology spillovers, there is some risk that national policymakers will cut innovation promotion expenditures; R&D expenditures could fall below an optimum level – positive external effects of innovation would not be fully internalized in the EU. Shifting more funds in R&D to the supranational policy level might not be a reasonable way to cope with the problem since the poor political control of the European Commission and the established budgetary priorities for agriculture and structural funds does not allow us to expect an efficient EU innovation policy. However, the EU could be quite useful in innovation policy, in particular by regularly analyzing innovation dynamics in EU countries and in various regions of the EU. More transparency could generate stronger incentives towards adequate national policy reforms. In Europe's ageing society, there is also the problem that a majority of increasingly elderly voters might be reluctant to support rising outlays for education and for higher R&D-GDP expenditure ratios. Rather a political bias in favor of spending more on social security could undermine necessary increases for R&D or the education system.

In eastern European accession countries, economic development has been relatively favorable in the decade from 1995 to 2005. However, even with growth rates of about 5% p.a. in several years in the run-up to accession, Poland, Hungary, the Czech Republic and other transition countries witnessed very high stubborn unemployment rates. If joining the EU should bring about more rigid labor markets, there could be a tendency towards sustained mass unemployment in Eastern Europe. As firms in accession countries themselves will face considerable pressure for outsourcing internationally some of the problems of the large Euro zone core countries could soon be a plague in Eastern Europe as well. Eastern European countries face the problems of ageing not less than western Europe; thus policymakers should stimulate innovations and human capital formation, on the other hand it will be important to encourage the creation of new firms which often not only create new jobs but contribute to overall flexibility and innovativeness.

The European Council adopted the Lisbon strategy in 2000 as a political agenda to encourage growth and employment in the EU. While many small EU15 countries and the UK have been rather successful in this respect, Germany and Italy – and to a lesser extent France – have faced slow growth, insufficient innovation dynamics and only modest ICT dynamics outside mobile telecommunications. Here governments should reconsider ICT policies as well as R&D and education policies. New tax incentives for encouraging training and retraining might be useful. At the same time expenditures in the field of social security policies and most subsidies (not for R&D!) should be cut. Inflexibility of labor markets and wage rigidities seem to be problems in those countries so that policymakers, trade unions and employer federations should look for remedies. If the EU15 should be unable to regain sustained growth and full employment, this will undermine the overall integration of EU25, undercut political support for integration and erode Europe's position in the global competition of market systems. As regards the role of the EU itself, it is not clear that Brussels really can stimulate innovation, growth and employment in the Community in the early 21st century.

All in all, policy makers should not understand our argument as broad support for naïve industrial policy. In accession countries it would be wise if policy makers would emphasize education and R&D support in the course of catching-up, which implicitly means strictly controlling social policy. For EU15 countries it is important to set the right incentives for the actors in labor markets to avoid overpricing, especially in market segments for unskilled labor.

Annex 1: A Demand-Side and Supply-Side Growth Perspective

We basically will argue that in a medium term perspective actual output Y is determined according to

$$Y = (1-\alpha)\,Y^{d} + \alpha Y^{pot}; \tag{1.16}$$

Note that α is a weighting parameter in the interval 0,1 and basically is determined by the dominant type of expectations which assign long run output potential

Y^{pot} a certain weight α – under long run full employment equal to unity – and thus present demand conditions a weight (1–α).

Such a joint impact of Y^d and Y^{pot} indeed is obtained if we assume a special variant of the permanent income hypothesis, namely that consumption is determined by the weighted impact of current real income and expected long run income – this is dubbed a hybrid consumption function – which is assumed to coincide with the production potential (for simplicity we have no discounting here):

$$C = c\left(1 - \alpha'\right) Y + c\alpha' Y^{pot} = cY + c\alpha' \left[Y^{pot} - Y\right] \tag{1.17}$$

Thus consumption is proportionate to current real income; if consumers expect long run income to exceed current income – and hence anticipate real income to rise – current consumption is higher than cY. If α′ is flexible variable (not a constant parameter) one may assume that a lasting gap between Y and Y^{pot} will lead to a decline of α′; and the combined impact of a rising gap and a falling α′ could indeed imply a fall of consumption.

Assume that we have aggregate demand in an open economy given by the following simple equation which assumes that consumption C is determined according to a the hybrid consumption function and that investment I and imports J are proportionate to actual income while export X is proportionate to foreign output Y^* (γ is the exogenous ratio of government expenditures to output Y):

$$\begin{aligned} Y^d &= c\left(1 - \alpha'\right) Y + c\alpha' Y^{pot} + b'Y + \gamma Y - jY + xY^* \\ &= \left[c\left(1 - \alpha'\right) + b' + \alpha - j\right] Y + c\alpha' Y^{pot} + xY^* \end{aligned} \tag{1.18}$$

Inserting (1.18) in (1.16), namely dY/dt = (1−α)dY^d/dt+ αdY^{pot}/dt we get:

$$\begin{aligned} dY/dt = {} & (1 - \alpha)\left[c\left(1 - \alpha'\right) + b' + \gamma - j\right] dY/dt + \left[(1 - \alpha)\, c\alpha' + \alpha\right] dY^{pot}/dt \\ & + (1 - \alpha)\, x dY{*}/dt \end{aligned} \tag{1.19}$$

Let g denote growth rates; then we have (note that $Y^{pot}/Y =: u'$ which is the inverse of the degree of capacity utilization U′) in a medium term perspective with a production function $Y^{pot} = K^{ß}(AL)^{1-ß}$ and defining $s' = 1-(1-\alpha)\,[c(1-\alpha') + b' + \gamma - j]$:

$$g_Y = \left[(1-\alpha)\, c\alpha' + \alpha\right] \left[u'/s'\right] \left[ßg_K + (1-ß)\,(g_A + g_l)\right] + (1-\alpha) \left[x/s'\right] \left[Y^*/Y\right] g_{Y^*} \tag{1.20}$$

Annex 2: NACE Rev. 1.1. Classification (in Parts)

D Manufacturing
15 Manufacture of food products and beverages
16 Manufacture of tobacco products
17 Manufacture of textiles

18 Manufacture of wearing apparel; dressing and dyeing of fur
19 Tanning and dressing of leather, manufacture of luggage, handbags, saddlery, harness and footwear
20 Manufacture of wood and of products of wood and cork, except furniture; manufacture of articles of straw and plaiting materials
21 Manufacture of pulp, paper and paper products
22 Publishing, printing and reproduction of recorded media
23 Manufacture of coke, refined petroleum products and nuclear fuel
24 Manufacture of chemicals and chemical products
25 Manufacture of rubber and plastic products
26 Manufacture of other non-metallic mineral products
27 Manufacture of basic metals
28 Manufacture of fabricated metal products, except machinery and equipment
29 Manufacture of machinery and equipment n.e.c.
30 Manufacture of office machinery and computers
31 Manufacture of electrical machinery and apparatus n.e.c.
32 Manufacture of radio, television and communication equipment and apparatus
33 Manufacture of medical, precision and optical instruments, watches and clocks
34 Manufacture of motor vehicles, trailers and semi-trailers
35 Manufacture of other transport equipment
36 Manufacture of furniture, manufacturing n.e.c.
37 Recycling

References

Audretsch, D., & Welfens, P. J. J. (2003). *The new economy and economic growth in Europe and the US*. Heidelberg: Springer.

Barfield, C. E., Heiduk, G., & Welfens, P. J. J. (2004). *Internet, economic growth and globalization – Perspectives on the new economy in Europe, Japan and the USA*. Heidelberg: Springer.

Bretschger, L. (1998). *Wachstumstheorie*. München: Oldenbourg.

Borbély, D. (2004). *Competition among cohesion and accession countries: Comparative analysis of specialization within the EU market*. European Institute for International Economic Relations (EIIW Discussion Paper No. 122). Wuppertal.

Borbély, D. (2005). EU export specialization patterns in selected accession and cohesion countries: Tough competition on the EU15 market? *Papeles del Este, 9*, http://www.ucm.es/BUCM/cee/papeles/

Chinn, M. (2006). Doomed to Deficits? Aggregate US Trade Flows Re-Examined, Weltwirtschaftliches Archiv 2005, *141*(3), 460–485.

Colecchia, A., & Schreyer, P. (2002). ICT investment and economic growth in the 90s: Is the United States a unique case? *Review of Economic Dynamics, 5*, 408–442.

Daly, M. (2002). *Riding the IT wave: Surging productivity growth in the west*. (FRBSF Economic Letter, No. 2002-34, pp. 1–3). San Francisco: Federal Reserve Bank.

Denis, C., McMorrow, K., Röger, W., & Veugelers, R. (2005). *The Lisbon strategy and the EU's structural productivity problem* (Economic Papers No. 221). Brussels: European Economy.

Dunning, J.H. (1977). Traxde, location of economic activity and MNF: A search for an eclectic approach. In B. Ohlin, P. O. Hesselborn, and P. M. Wijkman (Eds.), *The international allocation of economic activity* (pp. 395–418). London: Macmillan.

European Commission. (2003). *Staff working papers*. European Innovation Scoreboard 2003.

Froot, K. A., & Stein, J. C. (1991, November). Exchange rates and foreign direct investment: An imperfect capital markets approach. *Quarterly Journal of Economics, 106*, 1191–1217.

Grossman, G. M., & Helpman, E. (1991). Innovation and growth in the global economy. Cambridge, MA: MIT Press.

IMF (2007). *World economic outlook 2007*. Washington, DC: IMF.

Inklaar, R., O'Mahony, M., Timmer M. (2003). *ICT and Europe's productivity performance: Industry level growth accounting comparisons with the United States*. Groningen: Groningen Growth and Development Centre, Research Memorandum GD-68.

Jorgensen, D. W., & Stiroh, K. (2000). Raising the speed limit: US economic growth in the information age, Brookings Papers on Economics Activity (2000). US economic growth and development at the industry level. *American Economic Review, 90*(2).

Jungmittag, A. (2004). Innovations, technological specialisation and economic growth in the EU. *International Economics and Economic Policy: JIEEP, 2/3*, 247-273. Berlin: Springer, Bd. 1.

Klodt, H. (1993). Technology-based trade and multinationals' investment in Europe: Structural change and competition in Schumpeterian goods. In M. W. Klein & P. J. J. Welfens (Hrsg.) (Eds.), *Multinationals in the new Europe and global trade*. Berlin: Springer.

Lucas, R. E. (1988). On the mechanics of economic development. *Journal of Monetary Economics, 22*(I), 3–42.

McMorrow, K., & Roeger, W. (2004). *The economic and financial market consequences of global ageing*. Berlin: Springer.

Oliner, S., & Sichel, D. (2002). *Information technology and productivity: Where are we now and where are we going?* (Federal Reserve Board FEDS Paper 2002-29). http://www.federalreserve.gov/pubs/feds/2002/200229/200229abs.ttm.

Romer, P. (1990). Human capital and growth: Theory and evidence. *Carnegie Rochester conference series on public policy: A bi-annual conference proceedings* (pp. 251-286). Amsterdam [u.a.]: Elsevier, S.

Sinn, H.-W. (2005). *Basar-Ökonomie Deutschland*. Ifo Schnelldienst 6.

Stiroh, K. (2003). Information technology and the US productivity revival. *American Economic Review, 92*(5), 1559–1576.

Van Ark, B., & Piatkowski, M. (2004). Productivity, innovation and ICT in old and new Europe. *International Economics and Economic Policy, 1*(2+3), 215–246.

Welfens, P. J. J. (1997). Privatization, structural change and productivity: Towards convergence in Europe? In S. Black (Ed.), *Europe's economy looks east*. Cambridge, NY: Cambridge University Press.

Welfens, P. J. J. (2007). *Innovations in macroeconomics*. New York: Springer-Verlag.

Welfens, P. J. J., Jungmittag, A., & Vogelsang, M. (2005). Innovation, Regulierung und Wirtschaftswachstum in Digitalen Marktwirtschaften: Nationale und regionale Perspektiven. Studie für das Ministerium für Wissenschaft und Forschung des Landes Nordrhein-Westfalen.

Welfens, P. J. J., Zoche, P., Jungmittag, A., et al. (2004). *Internetwirtschaft 2010*. Heidelberg: Springer.

Chapter 2
Wage Dispersion in Germany and the US: Is There Compression from Below?

Joachim Möller

2.1 Introduction

The reasons behind the opening up of striking differences in labor market performance between the US and (continental) European countries since the seventies are still open to debate.[1] Many observers blame European labor market rigidities for the employment gap. According to the standard view the USA is the shining example of flexibility and freedom of the market, while Germany and France are bulwarks of institutional rigidities, high union coverage and comfortable welfare states floors. With respect to the wage structure it is presumed that continental European countries are characterized by much less differentiation compared to the US. The deformation of the wage distribution is considered to be one of the main reasons for the labor market crisis especially in Germany. A lack of differentiation is mainly diagnosed in the low-wage segment of the wage distribution. Following this line of reasoning, many employment possibilities in low-pay service industries are given away. As a consequence, a high share of low-ability/low-productivity workers stays unemployed and increasingly becomes a heavy burden on the social security system. The ongoing trends in skill-biased technical progress (Krugman, 1994) and/or globalization (Wood, 1994, 1998) tend to deteriorate the position of the low-skilled in the developed countries even further. As an obvious remedy it is recommended to abandon institutional and other regulations meaning that wage differentiation at the low end of the wage distribution is substantially extended. (e.g. Siebert, 1997, 2003; Sinn, 2005).

Looking more closely at the data, however, reveals that it is by no means clear to what extent the observed employment and pay patterns are compatible with explanations building solely on labor market rigidities. Already the discussion of micro-data evidence in Krueger and Pischke (1998) and Card, Kramarz, and Lemieux. (1999) cast some doubts on explanations of the employment gap which stress the lack of wage flexibility on the labor market.

[1] See, for instance, Acemoglu (2003) and Pischke (2005).

J. Möller (✉)
Institute of Economics, University of Regensburg, Germany

P.J.J. Welfens, J.T. Addison (eds.), *Innovation, Employment and Growth Policy Issues in the EU and the US*, DOI 10.1007/978-3-642-00631-9_2,

Despite the importance of the topic only a few studies based on large micro-data sets scrutinize the differences in the wage structure from an international perspective, although the availability of adequate data is steadily improving.[2] Of course, international comparisons of the wage structure are difficult for a number of reasons. Typically the collection of data in different countries does not obey the same principles. For example, main wage data sources in the US are survey data, while in Germany social security data plays a more prominent role. In some instances information on hours worked is available, in others not. Definitions of skill types also differ considerably between countries. Top-coding in the data in one case is more severe than in the other, and components included in the remunerations vary considerably (tips, commissions, social security payments of workers and firms). There are even deep conceptual issues rendering direct comparisons of the wage distribution in different countries rather problematic. If the distributional patterns reflect the dispersion of productive abilities of the employed and ability is mainly determined by skills, then the wage distribution should more or less reflect the distribution of skills in the population. The argument of Nickell and Bell (1996) and Freeman and Schettkat (2001) is that countries differ markedly in the dispersion of productive abilities in the population of workers. Using abilities scores Freeman and Schettkat show that the distribution of productive capabilities is more compressed in the German workforce than is in other countries (Britain and the US, for instance). As a consequence of "skill compression", the wage dispersion should be more compressed as well.[3]

In the following paper we argue that, although there are good reasons for being skeptical about direct comparisons of the wage distribution, there are several possibilities for investigating hypotheses on distributional patterns using difference-in-differences techniques. Our main interest is to investigate the case for or against the assertion of wage compression from below. In the next section we first analyze the issue theoretically and derive several hypotheses. In Sect. 2.3 we describe our data sets and harmonization methods. The results of our empirical investigation are contained in Sect. 2.4 and 2.5 concludes.

2.2 Theoretical Considerations

2.2.1 Wage Compression Form Below: The Accordion Effect

In their influential work, Blau and Kahn (1996, 2002) have argued that the differences between (continental) European countries and the US and the UK is mainly

---

[2]Examples of micro-data based international comparisons include Davis (1992), Blau and Kahn (1996, 2003), Freeman and Schettkat (2001), Krueger and Pischke (1998), Card et al. (1999). For a comparison Germany vs. US see Kohnz and Erber (2000).

[3]Newly available measures of school achievements (PISA) in different countries cast some doubts on the hypothesis that dispersion of productive capabilities among younger workers is lower in Germany compared to other countries. According to the PISA results, Germany is one of the countries with the highest variance in the measured ability scores.

in the low tail of the wage distribution. Above the median, wage dispersion among the countries in their study is more or less comparable. Hence their diagnosis for the labor market in continental European countries is "wage compression from below".

In the German context, Sinn (2005) has coined the notion "accordion effect" for this phenomenon. German wage – setting institutions and regulations, so the argument goes, do not allow for adequate wage differentiation at the low end of the wage distribution. This one-sided pressure leads to an asymmetric deformation of the wage distribution. The accordion effect implies that the percentiles of the wage distribution below the median are moving closer together while the upper tail of the distribution is not (or is much less) affected.

The reasons behind this specific form of wage compression lie in the structure of wage-setting institutions in Germany. Although no general statutory minimum wage exists, several mechanisms can be identified that are presumably working in direction of the accordion effect. Negotiated standard wages establish de-facto minimum wage floors. Therefore, the institutional framework is likely to generate a fundamental asymmetry. Firms have always the freedom to pay a wage premium on top of the negotiated standard wage, but they are facing some constraints if they want to pay less.

The second main mechanism behind the accordion effect can be seen in the social security system. Social assistance schemes create a lower bound for wages. The more generous the social benefits are, the higher is the average reservation wage. A successful match between a worker and a workplace, however, requires that the wage offer exceeds the reservation wage which is less likely for low wage offers.

2.2.2 *Actual and Counterfactual Distribution*

Several hypotheses can be derived from the arguments for asymmetric deformation of the wage distribution given above. Consider the extreme case of an effective unique minimum wage. Then no wage should be observed below the minimum wage. Empirically this would correspond to a sharp truncation of the wage distribution at the minimum wage floor. However, such a situation is hardly realistic for the aggregate wage distribution. At least three arguments can be given for a smoothening of the distribution at the truncation point. First, the individual wage is observed with measurement error. Second, minimum wages might vary across federal states and sectors of the economy. Third, the enforcement of a legal minimum wage may not be always possible.

A truncation of the wage offer distribution can also be generated through the reservation wage. In the case of individuals with identical preferences, endowments and information, this yields a unique reservation wage and hence a sharp truncation of the wage distribution at this point. Smoothing down can be obtained by adding observational errors or heterogeneity of individuals. In the latter case one would expect not a sharp truncation but rather a thinning out of the natural wage distribution which becomes more effective the more it comes to the low end of the distribution.

A simple simulation experiment shows that truncation contaminated by an error process and a thinning out of the wage distribution both lead to observationally equivalent outcomes.

In the first experiment we impose a truncation of a lognormal distribution alternatively at the first to the fourth decile and superimpose a normally distributed error term with mean zero and a standard deviation of 0.3.[4] The kernel density estimates of the resulting distributions are shown in Fig. 2.1 along with the non-truncated "natural" wage offer distribution – the so-called counterfactual.

The second experiment traces a thinning out in the left tail of the wage distribution. It is assumed that a wage offer x drawn from a lognormal distribution is above the reservation wage of a randomly selected person with probability

$$p(x) = \begin{cases} 1 - F(|d - x|) & \text{if } x \leq d \\ 1 & \text{if } x > d \end{cases} \tag{2.1}$$

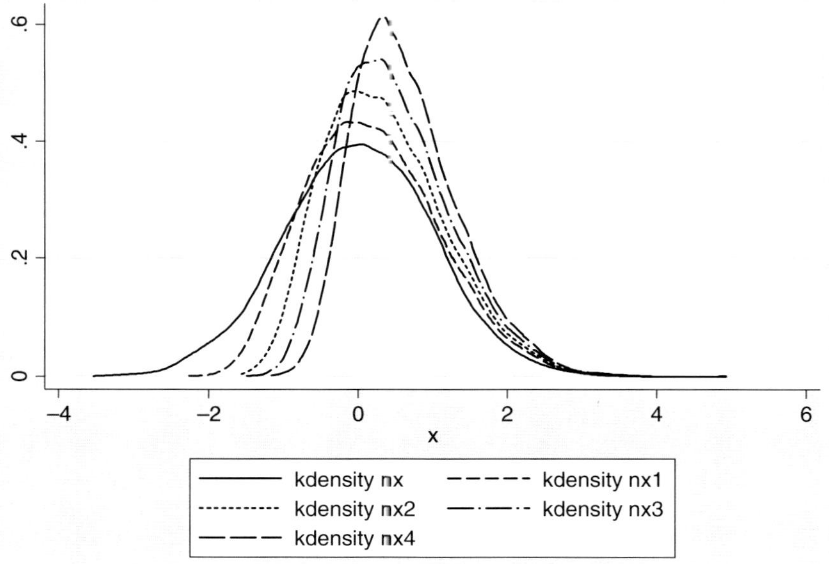

Fig. 2.1 Effect of superimposing an error process on a normal distribution truncated at the first to the fourth decile

[4]The standard error roughly corresponds to the standard error typically obtained for a Mincer type wage equation.

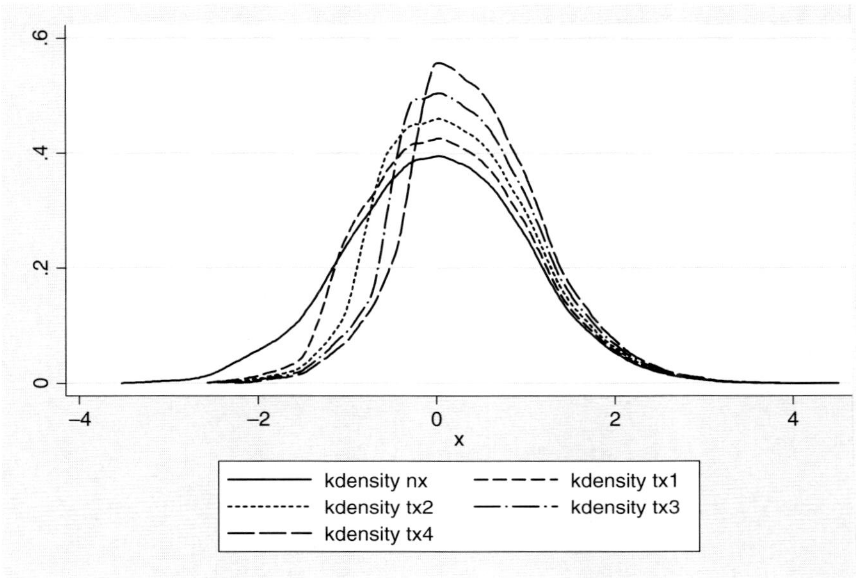

Fig. 2.2 Effect of thinning out of a normal distribution below the first to the fourth decile

where d is a critical quantile of the wage distribution and $F(\bullet)$ is the distribution function of a standard normal. Figure 2.2 depicts the kernel density estimates of the resulting distributions with the first to the fourth decile as alternative choices for d.

It turns out that truncation contaminated by an error and a random thinning-out process lead to roughly the same results. Both simulations generate a steeper increase in the low tail of the distribution compared to the counterfactual. As verified by the two figures, the actual distribution is affected over the whole range of the distribution. Moreover, the mean and median will shift to the right. However, the deviation between the actual and the counterfactual distribution fades away at the high end of the distribution. Hence wage compression from below gives rise to the accordion effect.

2.2.3 The Accordion Effect and Inter-Quantile Distances

The different shape of the actual distribution compared to the counterfactual has consequences for the inter-quantile distances. To express this in a more formal way, let us consider deciles as specific quantiles of the observed wage distribution, denote them as D_i $(i{:=}1,2,\ldots,9)$ and define $d_i{:=}\mathrm{E}(\ln D_i)$. Then the expected relative distance between the deciles and the median is given by $\tilde{d}_i := |d_i - d_5|$ for $(i{:=}1,2,\ldots,9)$. If the log transformation of the distribution generating the sample was symmetric, then the corresponding relative distances below and above the median would be equal, or, $\tilde{d}_{5-i} = \tilde{d}_{5+i}$ for $i{=}1,\ldots,4$.

For the moment assume symmetry of the "natural" or counterfactual wage distribution. Due to the accordion effect described in Sect. 2.1, compression in the left tail of the distribution then simply means that the log decile distances below the median are smaller than the corresponding ones above the median. This can be stated as

Hypothesis 1 (Accordion Effect)

$$\tilde{d}_{5+i} > \tilde{d}_{5-i} \qquad \text{in ascending magnitude for } i = 1,.,4.$$

Of course, the assumption of symmetry in the counterfactual log wage distribution might be seen critically. Therefore we develop alternatives in the following subsection.

2.2.4 The US as a Reference Case

A second type of hypotheses can be derived if one uses the US distribution as a reference case for an unconstrained (flexible) distribution. The underlying assertion is that the labor market in the US can be considered as a prototype of a market with a low level of regulations. Let D_i^G and D_i^{US} be the i-th decile in Germany and US. Assume that the natural or counterfactual wage distribution for a homogenous group of workers has the comparable inter-decile distances in both countries. Assume further that wage compression from below is more relevant in the German case. Then, using the same notation as before, one can formulate

Hypothesis 2a

$$\tilde{d}_i^D < \tilde{d}_i^{\mathrm{US}} \text{ for } i = 1, \ldots ,4, 6, \ldots ,9.$$

Note that the asserted constraints on the wage distribution in Germany would reduce all quantile distances from the median over the full range of the distribution. As outlined above, however, the effect of wage compression from below on the inter-quantile distances fades out in the right tail of distribution. Hence one would expect the log distance from D9 to D5 in the two countries to be more similar than the distance from D5 to D1. This leads to a hypothesis which is closely related to what Blau and Kahn (1996, 2002) have postulated in their comparison between the US on the one hand, and Germany and other continental European countries like France on the other. According to their findings, the log decile distances below the median appear to be substantially lower in Germany compared to the US, while the inequality measures are more or less identical in the right tail of the distribution. Correspondingly, one can formulate

Hypothesis 2b (Blau and Kahn)

$$\begin{aligned} \tilde{d}_i^D &< \tilde{d}_i^{\mathrm{US}} \text{ for } i = 1, \ldots ,4, \\ \tilde{d}_i^D &\simeq \tilde{s}_i^{\mathrm{US}} \text{ for } i = 6, \ldots ,9. \end{aligned}$$

The disadvantage of Hypotheses 2a and 2b is that it relies on the rather strong assumption of identical inter-decile of the counterfactuals in the two countries. Corresponding to the skill-compression argument, for instance, the spread in the ability distribution might differ considerably between the two countries. Therefore, it seems to be preferable to use a difference-in-difference approach.

Define the difference in the corresponding log distances of the deciles from the median in the upper and lower tail of the distribution as

$$\Delta_{5+i,5-i} := \tilde{d}_{5+i} - \tilde{d}_{5-i} \text{ for } i = 1, \ldots 4$$

Note that a symmetric spread component of the counterfactual cancels out. The hypothesis stating that the US wage distribution is more close to the counterfactual can then be formulated as

Hypothesis 3 (Blau and Kahn, Differences-in-Differences)

$$\Delta^{D}_{5+i,5-i} > \Delta^{\mathrm{US}}_{5+i,5-i} \text{ for } i = 1,2,\ldots,4.$$

This means that in Germany there should be more "excess inequality" in the right tail over the left tail of the distribution than in the US It should be stressed that Hypothesis 3 does not necessarily assume the symmetry of the counterfactual log wage distribution. It only requires that the deviation from symmetry is not too dissimilar in the two countries.

2.2.5 Comparing Different Groups of Workers

Another consideration concerns the effect of wage compression for different groups of workers. If the accordion effect were caused by minimum wages and/or reservation wages that are determined by social security standards, the deformation of the actual wage distribution compared to the counterfactual should be more relevant for low-income groups. Hence low-skilled workers should be more affected by wage compression than skilled or high-skilled workers. By the same argument, one would expect a higher effect in an industry with relative low average wages and vice versa.

Again using a difference-in-difference approach to get rid of differences in the general spread of the distribution and denoting low-skilled (high-skilled) workers by superscript u (s, respectively) one can formulate:

Hypothesis 4 (Skill-Specific Deformation)

$$\Delta^{s}_{5+i,5-i} < \Delta^{u}_{5+i,5-i} \text{ for } i = 1,2,\ldots,4.$$

Finally, combining the intra- and international comparisons, one would obtain.

Hypothesis 5 (Skill-Specific Deformation in International Comparison)

$$\Delta^{D,s}_{5+i,5-i} - \Delta^{\mathrm{US},s}_{5+i,5-i} < \Delta^{D,u}_{5+i,5-i} - \Delta^{\mathrm{US},u}_{5+i,5-i} \quad \text{for } i = 1,2,\ldots,4.$$

This hypothesis states that the international differences in the left-tail deformation of the actual wage distribution should be less severe for the distribution of skilled or high-skilled workers compared to the low-skilled.

2.3 Data

2.3.1 Data Sources

For the empirical analysis we use the *IAB-Beschäftigtenstichprobe* (IAB-BST) for Germany and the *Current Population Survey/Outgoing Rotation Group* (CPS-ORG) for the US IAB-BST is a 2% random sample from the Employment Statistics of the Institute of Employment Research, Nuremberg. It includes all workers, employees and trainees being obliged to pay social insurance contributions. Not included in the data are self-employed persons, civil servants and students enrolled in higher education. Marginal employed persons are in the data set only since the year 1999.

We consider all workers who were employed on June 30th of each year. Because there are still large structural differences in labor market and migration patterns between the eastern and the western part of Germany we constrict the analysis to workers in West Germany.[5] We exclude part-time workers, those in an apprenticeship or with more than one employment contract. Moreover, we drop all observations with no valid information on earnings, age, skills or the region of the workplace.

The US data is from the Current Population Survey (CPS), which is a monthly survey of 50,000–60,000 households, conducted by the Bureau of the Census for the Bureau of Labor Statistics.[6] The particular version of the CPS data analyzed here is the Economic Policy Institute's (EPI) extract of the "Outgoing Rotation Group" (ORG) of the CPS. The ORG is a one-quarter subset of the CPS that, in addition to answering detailed questions about their labor market circumstances has, since 1979, also provided information on earnings from work.[7]

The extract of the ORG sample used here attempts to compensate for several problems with the raw CPS data. First, the hourly wage concept in the CPS is somewhat inconsistent within each annual survey. Hourly wages for "hourly workers" (those paid by the hour or who report their earnings on an hourly basis) exclude overtime, tips, and commissions. Hourly wages for "non-hourly workers," however,

[5]For changes of the wage structure in East Germany see Franz and Steiner (2000).

[6]I am grateful to John Schmitt (Washington) for the detailed description of the US data.

[7]See Webster (2000) and Gao (2003) for more thorough descriptions of CPS-ORG.

Table 2.1 Number of observations and data selection for Germany in 1984, 1990, 1992 and 2001

	2001	
Total (excl. those in apprenticeship)	568,233	
East		91,097
Multiple jobs		16,988
Part-time < 50%		61,563
Part-time > 50%		55,645
Age < 25 or age >55		52,227
Minor employment		3,910
Minor employment		64,839
N	286,803	
Thereof		
Male		192,427
Female		94,376
Low-skilled		37,231
Skilled		184,815
High-skilled		34,034
Skill missing		30,723

Source: Own calculations with IAB-BST data.

are calculated as usual weekly earnings (a variable constructed by the BLS from more detailed responses) by usual weekly hours, and includes overtime, tips, and commissions. Within a given cross-section, therefore, the hourly earnings concept is not consistent across hourly and non-hourly workers. From 1994, when the Census and BLS thoroughly redesigned the CPS, respondents were allowed to answer that their "usual weekly hours" at work varied, a response not previously permitted. Each year since 1994, about 6% of workers chose to report that their hours varied, making it impossible to calculate an hourly wage for these workers. The EPI extracts uses information on these workers' employment characteristics to impute their "usual weekly hours."

2.3.2 Harmonization of Data

A sensible comparison of micro-data evidence for different countries requires a careful harmonization of variables. Several adjustments were necessary to render the US data as similar as possible to the German data. First, minor employment and self-employed were excluded from the US data since information on latter group was not available in the German data and the same is true for the former group before 1999. Second, because IAB-BST contains qualitative information on working time only, all comparisons were based on *earnings* rather than *on hourly wages*.[8] The information in the two variables PT_1 (part-time with more than 50% of normal full time hours) and PT_2 (part-time with less than 50% of normal full time hours)

[8]In the following we use the notions "wages" and "earnings" interchangeable.

that were available in the German data set were used to exclude part-time workers. Accordingly, all workers with less than 35 usual hours per week were excluded from the US data. Third, in both countries three skill levels were defined, low-skilled, skilled and high-skilled workers applying ISEC codes as far as possible.

The category *low-skilled* was taken as equivalent to *less than high-school* in the US, the category *high-skilled* with *college plus*. All other groups were attributed to

Table 2.2 Deciles of the wage distribution for full-time workers in the US and Germany (in PPP adjusted US Dollars, 2001)

	US			*D*		
	Value	95% confidence limits		Value	95% confidence limits	
Decile		Lower	Upper		Lower	Upper
	All workers					
D1	46.8	46.4	47.1	45.6	45.6	46.6
D2	58.9	58.9	58.9	58.9	58.9	59.9
D3	69.9	69.2	70.5	67.5	67.5	67.5
D4	80.9	80.9	81.2	75.1	74.1	75.1
D5	93.3	92.8	94.2	80.8	80.8	81.7
D6	108.6	107.4	108.9	89.3	89.3	89.3
D7	127.3	126.5	127.3	99.8	98.8	99.8
D8	152.7	150.1	152.7	115.0	114.0	115.0
D9	200.3	198.1	203.6	#	#	#
N	110,954			286,803		
	Male workers					
D1	51.5	51.5	52.5	56.1	56.1	57.0
D2	66.2	65.0	66.2	66.5	66.5	66.5
D3	79.2	78.0	79.5	74.1	73.2	74.1
D4	91.7	90.5	92.1	79.8	79.8	79.8
D5	105.9	105.9	107.4	86.5	86.5	87.4
D6	121.5	121.2	123.2	95.0	95.0	96.0
D7	141.4	141.4	144.2	107.4	106.4	107.4
D8	170.7	169.7	173.8	124.5	124.5	125.4
D9	226.3	226.3	226.3	#	#	#
N	61,442			192,427		
	Female workers					
D1	41.5	41.2	42.4	34.2	33.3	34.2
D2	51.5	51.5	51.9	45.6	44.7	45.6
D3	59.6	59.2	60.3	54.2	54.2	55.1
D4	70.6	69.9	70.6	62.7	62.7	62.7
D5	79.5	79.2	80.6	69.4	69.4	69.4
D6	90.5	90.5	91.7	76.0	76.0	77.0
D7	106.1	105.9	107.4	84.6	84.6	84.6
D8	127.3	126.5	127.3	95.0	95.0	96.0
D9	164.1	161.9	165.5	113.1	113.1	114.0
N	49,512			94,376		

Source: Calculations by the author on the basis of IAB-BST and CPS data (see text);
Note: #, Decile not available because of censored data.

the intermediate skill category.[9] Fourth, in order to avoid specific selectivity problems we also excluded very young workers (<25) and old workers (>55).[10]

2.4 Results

2.4.1 Comparing the Aggregate Wage Distribution Between Germany and the US

Table 2.3 contains the deciles of the 2001 wage distribution for full-time workers in the US and Germany in PPP adjusted US dollars.[11] The table also gives the 95% lower and upper bounds for the deciles.[12] As can be expected by the high number

Table 2.3 Distances of deciles from the median and differences between corresponding measures in the lower and upper tail of the distribution for full-time workers in the US and Germany (log differences in absolute value, 2001)

	Distance to D5 (Left tail)			Distance to D5 (Right tail)			Difference (Right tail–left tail)	
	US	*D*		US	*D*		US	*D*
	All workers							
$\tilde{d}_4$	0.142	0.073	$\tilde{d}_6$	0.152	0.101	Δ_{64}	0.010	0.027
$\tilde{d}_3$	0.289	0.180	$\tilde{d}_7$	0.311	0.211	Δ_{73}	0.022	0.031
$\tilde{d}_2$	0.461	0.316	$\tilde{d}_8$	0.493	0.353	Δ_{82}	0.032	0.038
$\tilde{d}_1$	0.690	0.571	$\tilde{d}_9$	0.764	#	Δ_{91}	0.074	#
	Male workers							
$\tilde{d}_4$	0.145	0.080	$\tilde{d}_6$	0.137	0.094	Δ_{64}	−0.007	0.014
$\tilde{d}_3$	0.291	0.154	$\tilde{d}_7$	0.289	0.217	Δ_{73}	−0.003	0.062
$\tilde{d}_2$	0.470	0.262	$\tilde{d}_8$	0.477	0.364	Δ_{82}	0.007	0.102
$\tilde{d}_1$	0.721	0.433	$\tilde{d}_9$	0.759	#	Δ_{91}	0.038	#
	Female workers							
$\tilde{d}_4$	0.118	0.101	$\tilde{d}_6$	0.130	0.092	Δ_{64}	0.012	−0.009
$\tilde{d}_3$	0.288	0.247	$\tilde{d}_7$	0.289	0.198	Δ_{73}	0.001	−0.049
$\tilde{d}_2$	0.434	0.419	$\tilde{d}_8$	0.471	0.315	Δ_{82}	0.038	−0.105
$\tilde{d}_1$	0.650	0.707	$\tilde{d}_9$	0.725	0.489	Δ_{91}	0.075	−0.218

Source: Calculations by the author on the basis of IAB-BST and CPS data (see text);
Note: #, Decile not available because of censored data.

[9] These categories roughly correspond to ISCED classification's levels 0–2, 3–4 and 5–7, respectively. See also the equivalence table given in Freeman and Schettkat (2001).

[10] Note that a large share of German workers below age 20 is in the apprenticeship system which obeys specific remuneration rules.

[11] According to IMF data, the PPP US dollar exchange rate in 2001 was 1.03 for the US and 0.95 for Germany.

[12] The confidence bounds are calculated on the basis of binomial interpolations.

of observations in the two samples, the confidence bounds indicate that the deciles are estimated with high precision.

Comparing all workers in the two countries, one observes that the lowest deciles (D1 and D2) in the US almost exactly correspond to those in Germany. By contrast, the US median exceeds the German one by roughly 15%, and the eighth decile (D8) by about one third. There are remarkable differences in the wage distribution by gender. Male full-time workers at the low end of the distribution are apparently better off than their US colleagues. PPP adjusted D1 earnings in Germany exceed those in the US by about 8%. This relation is reversed for the median and for higher deciles. For example, in the US the median is 23% and D8 is even 37% higher than the corresponding value for Germany. By contrast, full-time female workers in Germany are worse off than their American colleagues over the whole range of the distribution. For this group D1, D5 and D9 for the US exceed the corresponding values for Germany by 21, 15 and 45%, respectively.

In order to obtain a comprehensive picture of the shape of the wage dispersion, Fig. 2.3 draws the distances of the deciles from the median (in logs) in the US and Germany for all workers and by gender. For all workers the figure corroborates the perception of a substantially higher spread in the overall US wage distribution compared to Germany. Wage inequality as measured by log decile ratios in the US markedly exceeds the wage inequality in Germany *over the entire range of the distribution*.

In both countries the log decile distances from the median in the upper tail of the distribution surpass those in the lower tail. Hence there is some evidence for the accordion effect (Hypothesis 1) not only in Germany but also in the US. The measure of excess inequality in the upper tail of the distribution appears to be rather similar in the two countries. For example, $\Delta_{82} := \tilde{d}_8 - \tilde{d}_2$ is 3.2 log percentage points in the US and 3.8 in Germany.

Moreover, we observe that the relative distances from the median are always lower in Germany than in the US as suggested by Hypothesis 2a. However, comparing the differences of this measure between the two countries in the lower and upper tail of the distribution, we find only minor differences. For example, $\tilde{d}_2^{\mathrm{US}} - \tilde{d}_2^{D} = 0.145$ and $\tilde{d}_8^{\mathrm{US}} - \tilde{d}_8^{D} = 0.140$ (see Table 2.3). If anything, the evidence for Hypothesis 2b (Blau and Kahn) finds only weak support for the overall distribution of full-time workers.

Differentiating by gender gives further insights. For male workers in the US, Fig. 2.3 shows a remarkable symmetry in the log decile distances below and above the median. Only the distances of the lowest and highest decile from the median is slightly higher for the latter $\Delta_{91}^{\mathrm{US}} = 0.038$. Hence one can conclude that the accordion effect is only visible for full-time male workers in the US at the very low end of the distribution. In contrast to this, the accordion effect for this group is strongly present in the German case. For instance, the relative distance of D8 from the median exceeds that of D2 by about 10 log percentage points.

When it comes to Hypothesis 2, we do not find evidence for the Blau and Kahn hypothesis in its strong form (implying that the wage distributions above the median have more or less the same shape in both countries). However, as can be

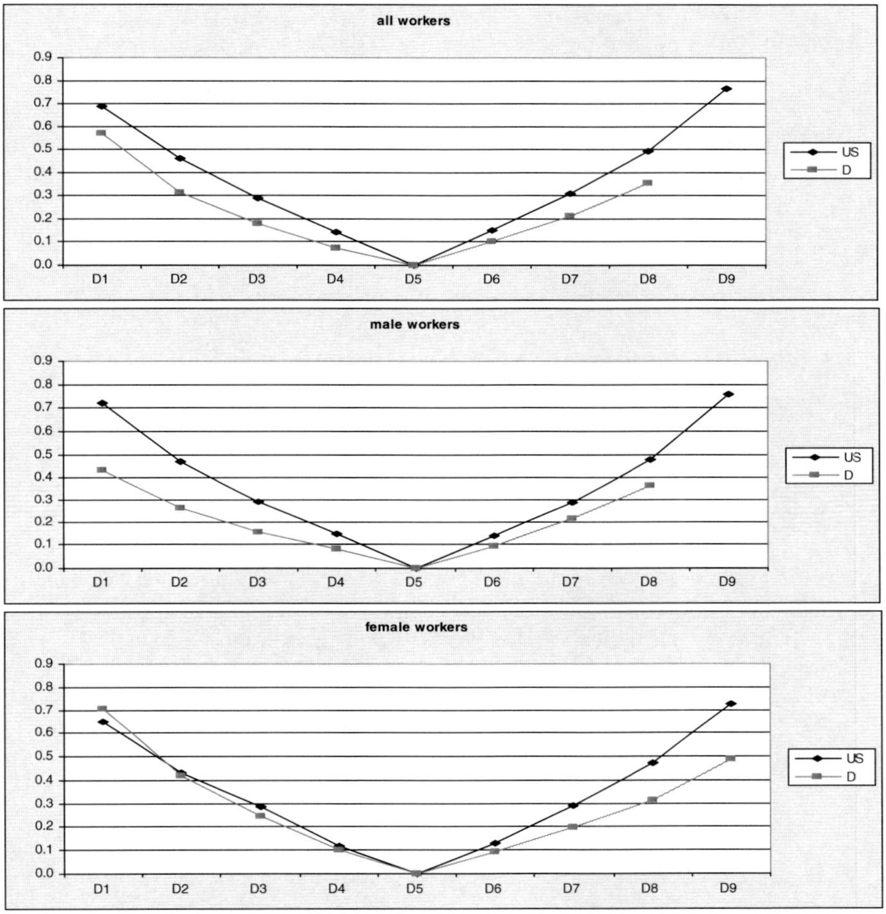

Fig. 2.3 Distances of deciles from the median of the earnings distribution in the United States and Germany (full-time workers 2001, log differences in absolute value)

seen from Fig. 2.3, the German distribution for full-time male workers deviates from the US distribution especially in the left tail. As we find strong confirmation of $\Delta^{D}_{5+i,5-i} > \Delta^{\mathrm{US}}_{5+i,5-i}$ for i=1,2,...,4., Hypothesis 3 (i.e. Blau and Kahn in difference-in-difference form) is corroborated. All in all, the evidence supports the hypothesis that wage compression from below affects the earnings distribution for German male full-time workers significantly in the entire left tail of the distribution, while for the US this is the case at the very low end only.

For full-time female workers the results for the US are quite similar to those of male workers, although wage compression from below here also affects the second decile. In the German case, however, the picture is remarkably different. Wage inequality for female workers with earnings below the median comes very close to the amount of inequality found for low-pay female workers in the US. In the upper tail of the distribution the log decile distances from the median in the two countries are even more pronounced than for male workers. For example, we find $\tilde{d}_8^{US} - \tilde{d}_8^{G} = 0.113$ for males, while the corresponding difference for female workers is 0.156. Therefore, one has to conclude that for female workers there is an accordion effect at the low end of the wage distribution in the US, while there is no evidence for a corresponding phenomenon in the German case. Perhaps somewhat astonishingly, the German results would be in line with wage compression from *above* rather than with wage compression from *below*. This is clearly at odds with all variants of the Blau and Kahn hypothesis.

2.4.2 Results for Different Skill Groups

We now differentiate between skill groups. As pointed out above, excess inequality in the right tail of the distribution should be more pronounced for low-skilled rather than for skilled or high-skilled workers. Figure 2.4 gives an overview of the results. We start with the findings for the US and Germany separately before comparing the two countries. There are several remarkable features. First, for skilled workers in the US we find almost perfect symmetry for both genders, only at the very low end of the wage distribution for females there seems to be a minor deformation. In contrast to this, there is a clear indication that the US wage distribution for low-skilled male and female workers exhibits the accordion effect. For the former, excess inequality

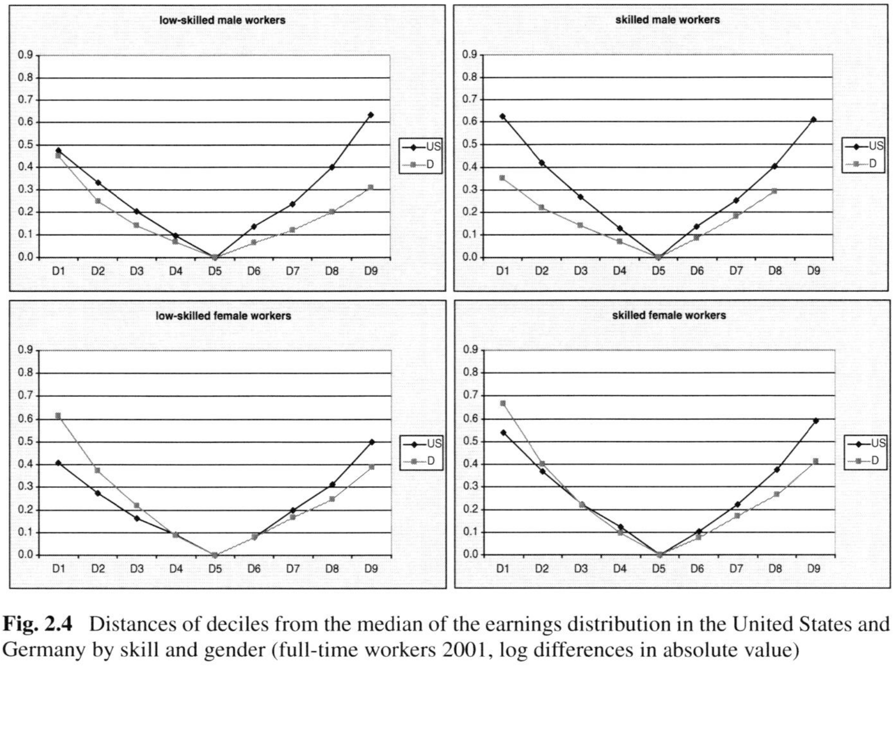

Fig. 2.4 Distances of deciles from the median of the earnings distribution in the United States and Germany by skill and gender (full-time workers 2001, log differences in absolute value)

Table 2.4 Distances of deciles from the median and differences between corresponding measures in the lower and upper tail of the distribution for full-time workers in the US and Germany by skill and gender (log differences in absolute value, 2001)

	Distance to D5 (left tail)			Distance to D5 (right tail)			Difference (right tail–left tail)	
	US	*D*		US	*D*		US	*D*
	Low-skilled male workers							
$\tilde{d}_4$	0.095	0.067	$\tilde{d}_6$	0.138	0.063	Δ_{64}	0.043	−0.004
$\tilde{d}_3$	0.207	0.139	$\tilde{d}_7$	0.237	0.122	Δ_{73}	0.031	−0.017
$\tilde{d}_2$	0.332	0.249	$\tilde{d}_8$	0.401	0.199	Δ_{82}	0.069	−0.050
$\tilde{d}_1$	0.475	0.452	$\tilde{d}_9$	0.629	0.310	Δ_{91}	0.154	−0.142
	Skilled male workers							
$\tilde{d}_4$	0.127	0.068	$\tilde{d}_6$	0.137	0.084	Δ_{64}	0.010	0.016
$\tilde{d}_3$	0.268	0.141	$\tilde{d}_7$	0.250	0.180	Δ_{73}	−0.018	0.039
$\tilde{d}_2$	0.419	0.220	$\tilde{d}_8$	0.404	0.293	Δ_{82}	−0.015	0.073
$\tilde{d}_1$	0.625	0.352	$\tilde{d}_9$	0.607	#	Δ_{91}	−0.018	#
	Low-skilled female workers							
$\tilde{d}_4$	0.093	0.086	$\tilde{d}_6$	0.079	0.079	Δ_{64}	−0.014	−0.007
$\tilde{d}_3$	0.162	0.219	$\tilde{d}_7$	0.200	0.166	Δ_{73}	0.038	−0.053
$\tilde{d}_2$	0.274	0.373	$\tilde{d}_8$	0.313	0.246	Δ_{82}	0.038	−0.127
$\tilde{d}_1$	0.409	0.614	$\tilde{d}_9$	0.498	0.389	Δ_{91}	0.089	−0.225
	Skilled female workers							
$\tilde{d}_4$	0.122	0.097	$\tilde{d}_6$	0.103	0.076	Δ_{64}	−0.019	−0.021
$\tilde{d}_3$	0.223	0.220	$\tilde{d}_7$	0.223	0.169	Δ_{73}	0.000	−0.051
$\tilde{d}_2$	0.369	0.399	$\tilde{d}_8$	0.377	0.264	Δ_{82}	0.009	−0.135
$\tilde{d}_1$	0.539	0.667	$\tilde{d}_9$	0.589	0.414	Δ_{91}	0.050	−0.253

above the median amounts to roughly 15 log percentage points and for the latter roughly 9 percentage points if $\Delta_{9,1}$ is considered (see Table 2.4). What the data for low-skilled workers in the US suggests is exactly in accordance with Hypothesis 4 (i.e. a marked left-tail deformation of the wage distribution for low-skilled workers due to wage compression from below).

Now consider the German earnings distributions. Quite surprising, the only case that corresponds to a priori expectations of excess inequality in the right tail of the distribution is that of *skilled* male workers. In all other cases, the distributions exhibit excess inequality not above, but *below* the median, i.e. $\Delta_{5+i,5-i} < 0$ for i=1,...,4. Hence for male workers in Germany, Hypothesis 4 is clearly rejected.

The case of female workers is even more at odds with common perceptions. For both skill groups the relative distances of the lower deciles from the median in Germany exceed the corresponding measures for the US According to these results, Hypotheses 1–4 are all rejected and it comes as no surprise that also Hypothesis 5 is not supported by the empirical evidence. More precisely, one obtains

$\Delta_{9,1}^{D,s} - \Delta_{9,1}^{\text{US},s} = -0.303$ and $\Delta_{9,1}^{D,u} - \Delta_{9,1}^{\text{US},u} = -0.315$ for female workers which clearly contradicts the hypothesis.

In Germany the highest decile for skilled male workers is censored. Calculating the measures for D8 instead gives: $\Delta_{8,2}^{D,s} - \Delta_{8,2}^{\text{US},s} = 0.088$ and $\Delta_{8,2}^{D,u} - \Delta_{8,2}^{\text{US},u} = -0.119$, which again rejects Hypothesis 5.

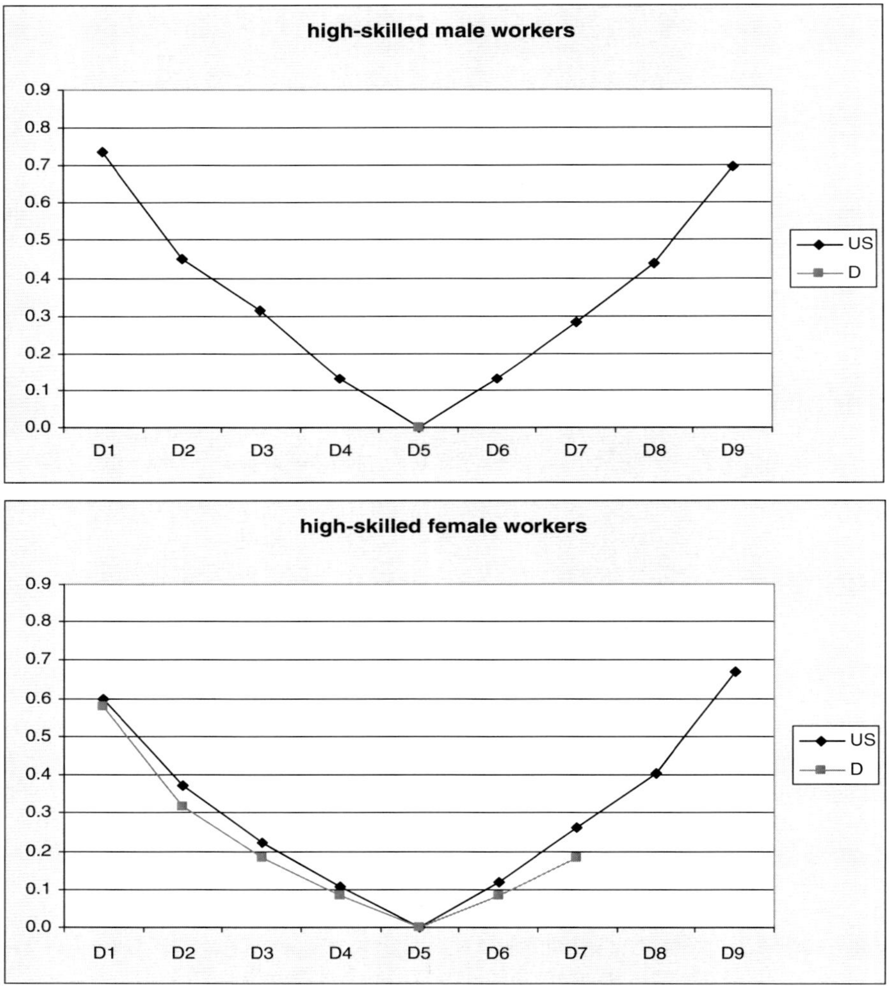

Fig. 2.5 Distances of deciles from the median of the earnings distribution in the United States and Germany for high-skilled workers by gender (full-time workers 2001, log differences in absolute value)

Note: The measures for male workers cannot be calculated in the German case because of censoring. The same applies to D8 and D9 for female workers of this skill category.

When it comes to the group of high-skilled workers in Germany, even the median is censored. Therefore we are not able to calculate the log decile distances from the median in this case. For high-skilled female workers, the highest decile available is D7. Figure 2.5 shows for US male workers of this skill category almost perfect symmetry of the log distances as should be expected in the absence of wage compression. By contrast, for female high-skilled in the US there is a certain indication of compression at D1 and D2. As far as these measures can be calculated in the German case, the log distance of the deciles from the median below and above the median are roughly identical. If comparison is possible, our findings show no marked differences between the two countries for high-skilled females below the median. In the right tail of the distribution, however, the spread in the US distribution seems to surpass the German one.

2.5 Conclusions

Wage compression from below is a common diagnosis for the German wage distribution. Although no general statutory minimum wage regulation exists in Germany, tariff wages or high reservation wages could lead to a deformation of the wage distribution. In the theoretical part, we show through simulations that a "truncation-plus-error" model and a "random thinning-out" approach qualitatively lead to the same results. Although the entire distribution is affected, the impact of wage compression from below is mainly found in the left tail. Compared to an unconstrained distribution, the relative inter-quantile distances should shrink, especially below the median. We develop alternative hypotheses in order to identify this so-called accordion effect. It is argued that the phenomenon should be more visible for low-skilled workers than for skilled and high-skilled workers. Moreover, the wage distribution in countries with a more flexible labor market, such as the US, should be closer to the counterfactual than in countries with high standards of social security systems, higher union coverage and more institutional regulations such as Germany.

In the empirical investigation large micro data sources for the US and Germany are used. We confine the analysis to full-time workers and harmonize the data as far as possible.

For skilled and high-skilled full-time male workers in the US we find almost perfect symmetry in the inter-decile distances below and above the median. This can be taken as evidence for the presumption that measures of inequality are close to symmetry in the left and right tail of the counterfactual distribution. This is so because high pay groups are not affected by institutional restrictions.

For other groups in the US we find evidence of an accordion effect, at least at the very low end of the distribution. The effect tends to be higher for females than for males and is more significant for low-skilled rather than for skilled workers. Hence the findings for the US support our expectations because there is a (moderate) deformation of the wage distribution due to the accordion effect exactly where it was supposed to work (i.e. at the low deciles of groups which tend to be ill-paid).

The results for Germany, however, are contrary to what one would have expected in case of marked wage compression from below. Skilled male workers are the only group for which the differences of the inter-decile distances fit the pattern of the accordion effect. For low-skilled workers of both genders and for female workers in general we find higher inter-quantile distances *below* rather than *above* the median. This is clearly in contrast to the interpretation given in the famous work by Blau and Kahn (1996 and 2002). Put differently, in these cases we simply find the reverse of what one would expect if the "wage-compression-from-below hypothesis" did hold.

A further puzzling phenomenon is the strong evidence for the accordion effect for skilled male workers in Germany rather than for low-skilled. A tentative explanation is that the bargained standard wages are especially important for this group. *Facharbeiter* – the German notion for skilled workers – are the main target group of the unions. If unions typically opt for wage compression, this could produce the presented results.

All in all our analysis casts some doubts on whether within-group wage dispersion plays some role for the clearly unsatisfactory employment situation of low-skilled workers in the German labor market. A possible alternative hypothesis is that between-group wage differentials (between groups of workers defined by skill, experience or industry) are inadequate. In this context, a recent study by Beaudry and Green (2003) seems relevant. These authors have investigated the wage-education relationships in the United States and Germany. They state that the skill premium "...evolved very differently in the two countries while the education composition of employment differences evolved in a parallel fashion." (Beaudry and Green 2003:773) develop an endogenous organizational choice model where they assume that new technology is skill-intensive and is complementary to physical capital. Then, high accumulation of physical capital relative to human capital is associated with a flattening and upward shift of the wage education profile.

References

Acemoglu, D. (2003). Cross-country inequality trends. *The Economic Journal, 113*(485), F121–F149.

Beaudry, P., & Green, D. A. (2003). Wages and employment in the United States and Germany: What explains the differences? *American Economic Review, 93*(3), 573–602.

Blau, F. D., & Kahn, L. M. (1996). International differences in male wage inequality: Institutions versus market forces. *Journal of Political Economy, 104*(4), 791–837.

Blau, F. D., & Kahn, L. M. (2002). *At home and abroad: US labor market performance in international perspective*. New York: Russell Sage Foundation.

Card, D., Kramarz, F., & Lemieux, T. (1999). Changes in the relative structure of wages and employment: A comparison of the United States, Canada, and France. *Canadian Journal of Economics, 32*(4), 843–877.

Davis, S. J. (1992). Cross-country patterns of change in relative wage. *NBER macroeconomics annual* (pp. 239–292). Cambridge, MA: MIT-Press.

Franz, W., & Steiner V. (2000). Wages in the East German transition process: Facts and explanations. *German Economic Review, 1*(3), 241–269.

Freeman, R., & Schettkat, R. (2001). Skill compression, wage differentials, and employment: Germany vs. the US. *Oxford Economic Papers, 53*(3), 582–603.

Gao, D. (2003). *Wage Analysis Computations, in Mishel, Lawrence, Jared Bernstein, and Heather Boushey, The State of Working America 2002–2003*, Ithaca, New York: Cornell University Press.

Kohnz, S., & Erber, G. (2000). *Lohnspreizung und Arbeitslosigkeit*. DIW Diskussionspapier 194/2000.

Krugman, P. (1994). Past and prospective causes of high unemployment. *Reducing unemployment.* Current issues and policy options, proceedings of a symposium in Jackson Hole (pp. 68–81). Wyoming, Kansas City: Federal Reserve Bank of Kansas.

Krueger, A., & Pischke J.-S. (1998). *Observations and conjectures on the US employment miracle*. Third public GAAC symposium: Labor markets in the USA and Germany (pp. 99–126). Bonn: German-American Academic Council.

Nickell, S., & Bell, B. (1996). Changes in the distribution of wages and unemployment in OECD countries. *American Economic Review, Papers and Proceedings, 86*(2), 302–308.

Pischke, J.-S. (2005). Labor market institutions, wages, and investment: Review and implications. *CESifo Economic Studies, 51*(1), 47–75.

Siebert, H. (1997). Labor market rigidities: At the root of unemployment in Europe. *Journal of Economic Perspectives, 11*(3), 37–54.

Siebert, H. (2003). Why Germany has such a weak growth performance (Kieler Working Paper No. 1182). Kiel: Institut für Weltwirtschaft.

Sinn, H.-W. (2005). *Ist Deutschland noch zu retten?* München/Berlin: Econ/Ullstein.

Webster, D. (2000). *Economic Policy Institute CPS ORG Labor Extracts, 1973–1999*, Version 1.31, Washington, DC: Economic Policy Institute.

Wood, A. (1994). *North-south trade, employment and inequality: Changing fortunes in a skill driven world*. Oxford: Clarendon.

Wood, A. (1998). Globalisation and the Rise in Labor Market Inequalities. *Economic Journal 108*(450): 1463–82.

Chapter 3
The Performance Effects of Unions, Codetermination, and Employee Involvement: Comparing the United States and Germany (with an Addendum on the United Kingdom)*

John T. Addison

3.1 Introduction

In this paper we examine the associations between worker representation (i.e. in unions and works councils), employee involvement/innovative work practices, and various indicators of firm performance.[1] The two themes of workplace representation and innovative workplace practices have historically been analyzed separately, although in recent years they have shown more overlap. There are a number of reasons for this historical *apartheid*. First, the (linking) theme of collective voice was first applied to unions, where it shared equal billing with unionism's monopoly face.[2] Second, employee involvement has tended to be seen as management-led and thus as a human resource management technique emphasizing high-commitment employment practices. (At the price of some imprecision, we will equate innovative work practices with high performance work practices and use both interchangeably with employee involvement.) Third and coinciding with the decline in unionism and growth in human relations practices, there has been the suggestion that the unions and employee involvement are alternatives.[3]

On the other hand, several other forces have encouraged a more integrationist approach. First, developments in the collective voice model have emphasized

*The main body of this chapter draws in large part on Addison (2005).

[1]The composite employee involvement/innovative work practices has no settled meaning. I will use it to include profit-sharing/share ownership arrangements, consultative bodies, team working, quality circles and problem-solving groups, briefing groups, systematic use of the management chain to communicate, and regular meetings with senior management plus benchmarking, total quality management (TQM), training, job security, job rotation, and payment for skill acquisition, *inter al.*

[2]There is inevitably some artificiality in this separation. For example, some models in the spirit of collective voice stress that successful employee involvement requires the presence of a union before workers will practice consummate as opposed to perfunctory cooperation (Levine & Tyson, 1990).

[3]For an exhaustive set of tests of the hypothesis for Britain, see Machin and Wood (2005).

J.T. Addison (✉)
Moore School of Business, University of South Carolina, Columbia, SC, USA
e-mail: ecceaddi@moore.sc.edu

P.J.J. Welfens, J.T. Addison (eds.), *Innovation, Employment and Growth Policy Issues in the EU and the US*, DOI 10.1007/978-3-642-00631-9_3,

cooperation and the possibility that institutional innovation might allow a decoupling of distribution from production and workplace issues. Second, the reduced bargaining power of unions might have produced the same result and elevated the pro-productive aspects of unionism subsumed under collective voice. The bottom line is that although most analyses of worker representation and employee involvement/high performance work practices have been conducted in isolation – while sometimes including the other as a control – research is beginning to consider their interaction.

In the present treatment, we will consider developments in the United States and Germany, only noting the British literature in qualification and *en passant*. The US experience is of interest because it provides one of the least positive assessments of union impact while yet offering some early evidence of there being a mutually supportive relationship between union presence and employee involvement. The German experience is of interest because workplace representation occurs through the mechanism of the works council rather than the union. Historically, the works council has been viewed as the exemplar of collective voice because of its array of information, consultation and participation/codetermination rights. Latterly, with the recognition of the bargaining problem, the German institution has become ever more closely identified with the expression of pro-productive voice by reason of its "peace obligation" and the dual system of industrial relations within which it is embedded. That said, there has been comparatively little analysis of the interaction of workplace representation with more direct employee involvement mechanisms, although analysis of the training function might provide a promising bridge.

The plan of the paper is as follows. To set the scene for our discussion of the empirical evidence, we first offer some theoretical conjectures on the efficiency properties of worker representation and employee involvement mechanisms. We next discuss for the United States the impact on firm performance of unions, employee involvement/high performance work practices and their interaction. We follow this with a broadly parallel analysis for Germany, now with works councils substituting for the union institution. An interpretative section concludes.

3.2 Theoretical Conjectures

The arguments favoring unionism, works councils and employee involvement/high performance work practices (EI/HPWP) are closely linked. Thus, the principal works council model is unambiguously an extension of the collective voice model, first advanced to make the case for unionism; the scope for EI/HPWP to improve performance rests on many of the same arguments used in the voice model; and the innovative work practices identified in the EI/HPWP literature may pierce the veil of the collective voice model which is opaque on mechanisms.

The collective voice model advanced by Freeman and Medoff (1984) is properly represented as a union voice-institutional response model. While arguing that voice is synonymous with (autonomous) unions, the authors are also concerned to make the point that voice cannot succeed without an appropriate response from management (and then from unions in response to any changes proposed by management). Freeman and Medoff (1984, p. 165) write: "Some managements will adjust to the union and turn unionism into a positive force at the workplace; others will not. Over the long run, those that respond positively will prosper while those that do not will suffer in the market place."

The centrepiece of collective voice is the union role in providing information. The labor market context is important here: it is characterized by continuity rather than spot market contracting because of on-the-job skills specific to the firm and the costs attaching to worker mobility and labor turnover. Given the information problem in such complex and multidimensional continuity markets, there are two basic mechanisms to elicit information on worker preferences or discontent. Quit behavior can provide such information either inferentially or directly (via exit interviews). However, the collective voice model contends that such information is likely to suffer from selection biases, from problems of motivating the worker to disclose information when there is no benefit from doing so (and the certainty of some positive cost) and finally from the sheer cost of the process of trial and error in determining the efficacy of contract innovations.

The other mechanism is voice. Collective voice through the agency of a union may outperform individual voice for several reasons. One reason is that non-rival consumption of shared working conditions (e.g., safety conditions, line speeds and grievance procedures) and common workplace rules create a public goods problem of preference revelation. Without some collective form of organization, there will be too little incentive for the individual to reveal his or her preferences since the actions of others may produce the public good at no cost to that individual. Unions collect information about the preferences of all workers and "aggregate" them to determine the social demand for such public goods. Substituting average preferences for marginal preferences and arbitraging of worker preferences may be efficient in such circumstances. Unions can determine the social demand for such goods, thereby enabling firms to choose a more efficient mix of personnel policies.

A second public goods dimension of the workplace stems from the nature of the input of effort. Without some form of collective organization, so this argument runs, the individual's incentive to take into account the effects of his actions on others may be too small, just as with preference revelation. This problem will only arise where there are significant complementarities in worker effort inputs, so that output may depend on the lowest level of input by any one worker. In short, collective organization may potentially increase output through a joint determination of effort inputs and perhaps more so through increased cooperation between workers in continuity labor markets.[4]

---

[4]For the public goods argument to have force, two further conditions must be met – both of which are contained in Freeman (1976, p. 362). First, costs must be incurred in using external markets: if quitting were costless, the individual worker could simply choose the employer whose working conditions most closely approximated his/her own preferences. Second, the workplace must continue to be buffeted by unforeseen shocks that change the nature of the workplace in an infor-

The expression of collective voice is expected to reduce exit behavior: quits, absenteeism, malingering and even "quiet sabotage." The reduction in quits in particular is expected to lower hiring and training costs and increase firm-specific investments in human capital. Apart from training effects, lower quits (*inter al.*) should also occasion less disruption in the functioning of work groups. Reduced exit behavior is the most tangible source of potential efficiency gain (to the parties) in the model.

The other main aspect of collective voice is governance. In the context of the continuity relation, governance refers to the policing or monitoring of incomplete employment contracts. Governance will involve the use of grievance and arbitration procedures and other mechanisms to mitigate problems stemming from the authority relation. Such arrangements should help improve the flow of information between the two sides. The difficulty is that these specialized procedural arrangements are not unique to union settings, so that the argument presumably must be that unions make it easier (i.e. less costly) to negotiate and administer these arrangements. Unions may then facilitate long-term efficient contracting. A union specializing in information about the contract and in the representation of workers can prevent employers from engaging in opportunistic behavior. Workers may withhold effort and cooperation when the employer cannot credibly commit to take their interests into account. Fearing dismissal, workers may be unwilling to invest in firm-specific skills or disclose information facilitating pro-productive innovations at the workplace.[5] The formation of a union and the introduction of a system of industrial jurisprudence is one way of protecting employees' interests. In this way, unions may generate worker cooperation, including the introduction of efficiency-enhancing work practices. In other words, if the reputation effects mechanism is weak, there is scope for unionism to be pro-productive.

If we assume that there is a commitment problem in regular markets, an interesting side issue is whether the divorce of ownership and control in the modern corporation could make self-enforcing contracts more feasible. More feasible in the sense that management might be less interested in reneging on an implicit contract in the interest of short-term profit maximization than the owner principal; and conversely where the interests of managers and shareholders are more closely aligned by, say, profit- sharing schemes for managers. In this case much might hinge on whether unions and self-enforcing contracts are substitutes or complements in establishing workplace cooperation. If they are substitutes, any positive effect of unions

---

mational context; otherwise, there would be no need for the union's demand-revealing function after the formative match between employer and worker.

[5]The notion that unions might help increase training has formally been allied to contract enforcement by Dustmann and Schönberg (2004), who argue that the unavailability/infeasibility of long-term wage agreements means that training will be underprovided in regular markets and that unions move (apprenticeship) training closer to the social optimum by guaranteeing trained workers at least the union wage in the future. Since wage compression is also involved, unions facilitate firm-financed general training in this model (see also Acemoglu & Pischke, 1999). The most recent British evidence is provided in Addison and Eelfield (2007)

on performance will be stronger in firms with less severe agency problems. If they are complements, unions will be more effective in firms where agency stimulates self-enforcing contracts. This argument is of course based on a very narrow view of the agency problem in corporations and must be widened to incorporate rent-seeking behavior by managers which may detract from trustful and cooperative industrial relations and may decrease the range of feasible self-enforcing contracts. Jirjahn (2003) has recently examined the relationships between unions (actually works councils) and self-enforcing contracts and also those between agency and trustful, cooperative industrial relations using information on management profit-sharing schemes. We shall report some of his findings when looking at the German evidence.

While governance receives emphasis, there is virtually no discussion of bargaining power in the original collective voice model. But in contract theory models in which the union can make credible the employer's *ex ante* promises there must be some threat of credible punishment by the union (e.g., Malcomson, 1983). So it seems that the governance argument also depends on power: union monopoly power. Unfortunately, such power generally involves a hold-up problem, with unions taxing the returns on tangible and intangible capital.

The standard collective voice model treats the exertion of bargaining power and the expression of voice as distinct and offsetting facets of unionism (hence the reference to the "two faces" of unionism in Freeman and Medoff, 1984, p. 5). Recognition of bargaining power is integral to the main theoretical justification for *works councils* in a model offered by one of the architects of collective voice. Thus, Freeman and Lazear (1995) argue that participation/codetermination will be under-provided by the market because institutions that give power to workers will affect the distribution as well as the size of the surplus. The ideas behind collective voice are fleshed out and set in a continuum bounded by information provision at one end and by participation or codetermination at the other.

Freeman and Lazear argue that the joint surplus of the enterprise will increase as one moves cumulatively from information exchange through consultation to participation. Among other things, information rights can help verify management claims as to the state of nature, rendering them credible to the workforce and avoiding costly disputes that might even threaten the survival of the enterprise. Consultation for its part allows new solutions to production and other problems by reason of the non-overlapping information sets of the two sides and the creativity of discussion. Finally, participation or codetermination rights increase the joint surplus by providing workers with more job security and encouraging them to take a longer-run view of the firm and its prospects.

However, Freeman and Lazear recognize that, unless constrained, these rights will give rise to a bargaining problem. Specifically, they argue that the workers' share in the joint surplus grows with the surplus while that of capital declines both relatively and absolutely. The workers' share rises because knowledge and involvement are power, so that the very factors that cause the surplus to rise also cause profitability to fall, with the result that workers will demand too much power/involvement because their share will continue to rise after the joint surplus has peaked. Symmetrically, employers will either oppose works councils or vest them with too little power because profits decline even as the surplus is increasing.

Some means of third-party regulation limiting bargaining power has to be found if the societal benefits of worker voice are to be realized. Here, Freeman and Lazear see the German institution as attractive in two respects. First, the German works council cannot strike. Second, neither can it formally engage in bargaining over wages and other conditions of employment unless expressly authorized to do so under the relevant industry-level or regional collective bargaining agreement. In this respect, the authors speak of a potential "decoupling" of the factors that determine the size of the surplus from those that determine its distribution. Left open is whether or not there is a *sufficient* decoupling in practice. Thus, even if the works council is an exemplary collective voice institution, theory does not provide an unambiguous answer as to its consequences for efficiency.

Finally, the starting point of the EI/HPWP model is the notion that "workplace innovations change the production function in such a way as to increase the productivity of a firm's inputs, in particular labor" (Cappelli & Neumark, 2001, p. 739). The basic premise is by now familiar: workers have important private information and valuable insights into how to improve their jobs. There is therefore scope for beneficial trades once workers are trained and presented with better opportunities to exercise their skills through job redesign, decreased supervision and involvement in decision making and motivated to contribute through productivity bonuses (Handel & Levine, 2004, p. 2). Recognition of these potentialities marks a shift in management philosophy from the *status quo ante* of traditional work systems and labor-management relations, even if there is considerable variation in how the new practices have been adopted in practice (for one taxonomy, see Godard, 2004).

EI/HPWPs "work" by encouraging workers to work harder and smarter and by inducing structural changes that improve performance. The idea that workers will work harder is based on their enjoying work more when the job is interesting and where it provides feedback and rewards (i.e. job enrichment). The second element more clearly reflects the distinct input of the worker in efficient job (re)design, not least when there are complementarities in production. As noted by Ichniowski, Kochan, Levine, Olson, and Strauss (1996, pp. 301–302), the final element reflects such factors as cross training and flexible job assignment, which can reduce the cost of absenteeism; decentralized decision making to self-managed teams, permitting a reduction in line management while benefiting communication; and training in problem solving and computer skills, which can increase the benefits of new information technologies. In other words, innovative work practices beget other changes that improve productivity independent of their effects on motivation.

The suggestion is that the various strands of employee involvement are interdependent and mutually reinforcing, Edgeworth complementarities. This is the notion of "bundling," namely, the idea that innovative high performance work practices are more effective when combined with supporting management practices (e.g., Milgrom & Roberts, 1995). As suggested earlier, compensation schemes such as profit sharing can incentives employee involvement and information disclosure and training can improve worker decisions at the same time as job security encourages them to take a long-run view of the firm and make suggestions. Another example is the potential synergies between job rotation, self managed teams and pay-for-skill plans. However, this should not be taken to imply that the theory provides unambiguous guidance on either the identification or measurement of the bundles.

One other point needs to be made. Although many of these arguments are consonant with the voice model, the conjunction of the growth in EI/HPWP and the decline in unionism at least raises the possibility that the two may be substitutes, either by design or by effect (i.e., by enhancing worker satisfaction they may reduce the demand for unionism). This is not without consequence for the empirical work reviewed below in which the maintained hypothesis is often that the two institutions are complementary.

In sum, there are theoretical grounds for supposing that both collective voice and EI/HPWP can improve the productivity of the workplace. Some observers profess to see little difference in the arguments for workplace representation and EI/HPWP when the industrial relations system offers some means of some means of decoupling production from distribution issues. Others might still argue that traditional workplace representation through unions is important to the success of innovative work practices – by, say, providing greater assurance that a serious hearing will be given to employees' suggestions, or by virtue of union access to higher levels of management – or, backhandedly, that such practices offer a bigger payoff in union regimes with the elimination of restrictive practices, reminding us that in all such cases the monopoly face of unions has to be taken into account. Study of the effect of workplace representation on firm performance would seem to be quite closely bound up with the EI/HRWP issue. Moreover, comparisons between the United States and Germany might prove especially instructive.

3.3 US Evidence

We next review US findings on the determinants of firm performance, first distinguishing between union and employee involvement effects and then addressing the comparatively few treatments that have attempted to integrate the two mechanisms.

3.3.1 Union Effects on Firm Performance

The starting point for all analyses of union effects on firm performance must be Brown and Medoff's (1978) pioneering production function study of US manufacturing, using aggregate two-digit industry data cross classified by state groups for 1972. Using an augmented Cobb-Douglas production function, Brown and Medoff estimate

$$\ln\left(Y/L\right) = \ln A + \lambda \ln\left(K/L\right) + (1-\lambda)\,(c-1)\,P$$

where Y is a value-added measure of output, L is total labor (comprising union labor L_u and nonunion labor L_n), A is a constant of proportionality, K is capital, λ and

$(1 - \lambda)$ are the output elasticities of K and L, c indexes productivity differences between union and nonunion labor ($c > 1$ indicating union labor is the more productive and conversely for $c < 1$) and P is union density, L_u/L.[6]

In this framework, the coefficient estimate for P yields the logarithmic productivity differential of unionized establishments. Assuming this derives solely from labor inputs, dividing this coefficient by $(1-\lambda)$ gives the union labor productivity effect. Brown and Medoff estimate that the productivity of unionized establishments is between 25 and 27% higher than that of comparable nonunion plants, implying a union productivity effect of 35–36%. Interesting, as we shall see, some of the first production function studies for German works councils obtained similarly high estimates.

In the years since Brown and Medoff's study, evidence has steadily accumulated to the effect that average union effects in the United States are nowhere near as large as this. To start with and abstracting from issues of restrictive functional form and limitations of value added as a measure of output, other similarly broadly-based studies have failed to confirm large effects. Perhaps the "closest" study is by Hirsch (1991) who estimates a variant of the Brown-Medoff model using data for a sample of around 570 publicly quoted companies over the sample period 1968–1980, matched to union data for 1977. Hirsch's OLS value-added production function includes in addition to capital, labor, the stock of R&D per employee, union density, variables measuring firm and industry growth, industry concentration and import competition. He reports negative and statistically significant coefficients for union density (of around −17%). These estimates fall in absolute magnitude, the more detailed the industry dummies included: from −12.3% (2-digit level) through −8% (3-digit) to 3% (4-digit). Re-estimating the production function for each of 19 industries yields some positive coefficient estimates for union density – examples include textiles and apparel and fabricated metal products – but the majority of union effects remain negative. While conceding that union effects will likely vary considerably by industry and cannot be estimated precisely with existing techniques and data bases, Hirsch (1991, pp. 104–105) nevertheless concludes that there is nothing to support the contention of large and statistically significant positive union productivity effects. Rather, union productivity effects are small on average and insufficient to offset the union wage differential.

We will return to the question of estimation technique (and the issues of omitted variables and union endogeneity) later in this discussion, but for completeness we should also note some findings from an earlier manufacturing industry study that are quite consistent with those of Hirsch. Using data from the Profit Impact of Market Strategy (PIMS) on 902 manufacturing businesses for 1970–1980, Clarke (1984) estimates similar value-added equations and finds small but well determined coefficient estimates for the union variable in the range −2 to −3%. On this occasion, however, the point estimates are not sensitive to inclusion of detailed controls.

[6]The equation assumes constant returns to scale, which constraint is relaxed by adding a measure of establishment size, ln$L$, to the regressors.

The bottom line is that, contrary to Brown and Medoff, *average* union productivity effects are small and as likely to be negative as positive. This interpretation is consistent with other pieces of information having to do with union effects on profitability (and employment).

Before turning to these other performance outcomes, it is important to emphasize again that we are speaking of average union effects and not results from individual industry and firm-specific studies that in principle can avoid many of the econometric and data problems inherent in more aggregative studies (e.g. in the measurement of output and in tackling input endogeneity problems). Such advantages are achieved at the price of a loss in generality, however and so the goal must be to use the unique perspective of such studies to help inform more aggregative exercises.

Perhaps the most important indirect piece of evidence against large pro-productive union effects of the magnitude suggested by Brown and Medoff's (1978) study is the US evidence on profitability. Unlike the recent British evidence, *every* US study points to reduced profitability in union settings. This result holds irrespective of the financial indicator used (price-cost margin, rate of return on sales or capital, or market valuation of the firm's assets/Tobin's q), unit of observation (aggregate industry, firm, or line of business), or methodology. On the basis of the 16 studies reviewed in Addison and Hirsch (1989), unions are associated with 10–15% lower profitability. More recent studies provide confirmation of this central tendency and, again inconsistent with British results, contain little indication of any material change in the magnitude of the union effect over time (Hirsch, 1991; Hirsch & Morgan, 1994).

However and again abstracting from issues of statistical inference, the profitability evidence is not without controversy because the union effect might just be a transfer and thus neutral from an efficiency perspective (assuming perfect capital markets). Indeed, this is the take of Freeman and Medoff (1984, p. 186). Whatever the indications to the contrary in the earliest US literature, however, subsequent research has indicated that unions do not capture a significant share of potential monopoly profits. For example, in his evaluation of profitability (two measures: Tobin's q and the rate of return on capital) in 513 firms, 1968–1980, Hirsch (1991) reports that although the four-firm industry concentration ratio is positively associated with profitability, the interaction between union density and concentration is both positive and statistically significant. In other words, the suggestion instead appears to be that union-nonunion differences in profitability are most substantial in highly competitive industries.

There are clearly other sources of union rents such as sales growth and protected markets. However, the source that has most exercised US researchers is distinctly competitive: the current and future quasi rents – or normal returns – on firm investments. US research consistently points to a strong negative association between unionism and investment in physical and innovation capital (Hirsch, 1991; Bronars & Deere, 1993; Bronars, Deere & Tracy, 1994; Cavanaugh, 1998; Fallick & Hassett, 1999). The fullest analysis is by Hirsch (1991), who again presents pooled cross-section/time-series results for both types of investment in his sample of more than 500 firms. For capital investment he reports that the union firm with average coverage has annual capital investment that is 13% lower than its nonunion counterpart. The direct effect of unionism or the *union tax* on the returns to long-lived capital contributes a little under one-half of this effect. The other 7% is an indirect effect operating via the union impact on profitability, profits being an important determinant of capital investment. For R&D expenditures, unionized companies invest some 15% less than do comparable nonunion firms. Well over three-quarters of this effect is now direct, resulting from the union tax. We should also note that Hirsch reports that union coverage is negatively associated with the ratio of advertising expenditures to sales. Furthermore, union coverage is also positively related to the propensity to patent, which should reduce the liability of the firm to hold up, ceteris paribus. That said, union coverage is unrelated to the ratio of debt to equity, higher values of which might be expected to lessen exposure to hold up (see, inter al., Baldwin, 1983).

The US evidence on investments in tangible and intangible capital does not augur well for long-run productivity growth in unionized companies. However, this implication is not corroborated in the productivity growth literature, where the slower growth in unionized firms is due to a disproportionate presence of unionization in industries with slower growth. Hirsch (2004) contends that the implication of slowed growth in union companies is not inconsistent with this result because productivity growth equations typically control for both tangible and intangible capital. *Vulgo*: they address only the direct effects of unions, neglecting the important indirect effect operating through reduced investment. Accordingly, Hirsch directs our attention to the emerging employment growth literature (e.g. Lalonde, Marschke, & Troske, 1996; Dinardo & Lee, 2002). He interprets the evidence on slowed employment growth in unionized firms as the teleological outcome of their reduced investments in physical and intangible capital. That said, there is some lingering ambiguity in the notion of an employment maximand, as well as the potentially awkward result that unionized firms in the United States are apparently no more likely to fail than their nonunion counterparts.[7]

Summarizing, we see that *on average* union effects on productivity are small, that profitability is significantly reduced in union settings and that investments in physical and intangible capital are also materially lower. The studies on which these findings are based attempt to deal with potential biases stemming from omitted variables and union endogeneity, but they not address mechanisms that might allow unionized establishments to do better than this, even outperform (some) nonunion plants.

3.3.2 The Employee Involvement/High Performance Work Practice Literature

As Cappelli and Neumark (2001, p. 738) note, the literature on work organization and attempts to reform it is vast. Here we offer an eclectic review of the impact of

[7] But for evidence of higher closure rates among union establishments in Britain, see the Addendum. The German evidence is noted below.

innovative work practices on several measures of firm performance that always has an eye to unions, reflecting the conjecture that participatory systems might work better in union regimes (Levine & Tyson, 1990). We preface our summary of the empirical evidence with some cautionary remarks on methodology that arise in this literature and are additional to the classic problem of omitted variable bias encountered in the union literature.

One important preliminary issue is what constitutes high performance work practices. Although different researchers have used different measures (and terminology), there is broad agreement that employee involvement is central to the definition (e.g. Applebaum & Batt, 1984). More concretely, TQM programs, quality circles, functional flexibility and teamwork are core elements of HPWP. Aiding such practices are individual and organizational supports (the terminology is that of Forth & Millward, 2004, p. 100). Examples of the former are information disclosure, job rotation and training, while the latter include job security guarantees, internal labor market structuring and financial participation. Management practices that are difficult to fit into this mould, but are no less central, include benchmarking and computer usage. Despite this agreement, as Handel and Levine (2004, pp. 14–15) note, measurement problems arise because there is no theoretical guidance on which combination of practices might be more effective and no unambiguous way of measuring the bundles. As we shall see, researchers have used interaction effects, additive indices, factor analysis and combinations supported by a priori reasoning. A potentially major complicating factor is that survey data may not go beyond identifying the presence of a practice, neglecting its reach, coverage and intensity.

Another important issue that needs to be raised at the outset is the diffusion of high performance work practices. Researchers typically do not have information on when a particular practice was initially introduced and typically do not have information on when it was discontinued. If researchers are observing a situation in cross section that is late in the diffusion of the practice in question, then any pro-productive effect will likely be biased downward. Also, there is a potentially serious loss of information arising from the failure to observe situations in which practices were discontinued. Even if the researcher has information on changes in practice (in either direction) and abstracting from measurement error, the panels are typically too short to accommodate learning effects. The bigger problem is of course that the number of changes in innovative practices in the typical panel is simply too small to take advantage of panel estimation techniques.

The studies we next examine in detail are among the most recent US empirical inquiries of workplace innovations. Since we focus on direct performance measures they do not include work on the relationship between EI/HPWP and earnings (for a review of which see Handel & Levine, 2004, Table 3.2). They constitute part of what Cappelli & Neumark (2001, p. 739) call the "second wave" of research on organizational performance are very much in the spirit of the wider employee involvement/high performance work practice literature. Table 3.1 provides the bare bones.

Unionism is almost incidental in the first three studies in the table and even when a union measure is included in the array of control variables its influence

Table 3.1 Selected US studies examining the effect of employee involvement/high performance work practices on firm performance

Study	Data	Methodology	EI/HWPA measure	Union variable	Findings
1. Huselid and Becker (1996)[a]	1991 sample of 820 publicly-quoted companies with more than 100 employees; and a panel of 218 companies, 1991–1993, also used in cross section for 1991 and 1993	Cross section, fixed effects and random effects models. Dependent variable: financial performance, measured by Tobin's q and the gross rate of return on assets	13 high performance work practices. Factor analysis used to identify 2 factors or groupings and scales constructed for each factor by averaging those questions loading clearly on each respective factor. However, since the effects of a change in the human resource system was found to be the same whether it occurred through a unit change in either factor, the authors subsequently sum the two dimensions	Union coverage included in the control set, but its effect not separately identified	Strong effects of human resource strategy (factors and summed factors) on financial performance reported in cross section. Specifically, a one standard deviation increase in the firm's usage of high performance work systems increased Tobin's q by 14% and the accounting rate of return by between 13 and 28%. In the fixed effects specifications, however, the effects were only one-fourth as large and were statistically insignificant throughout. However, correction for measurement error in the panel is claimed to yield impact effects that are in line with the cross-section estimates

Table 3.1 (continued)

Study	Data	Methodology	EI/HWPA measure	Union variable	Findings
2. Ichniowski, Shaw, and Prennushi (1997)[b]	36 production lines of 17 steel producers. Up to 2190 monthly observations in the panel	OLS and fixed effects engineering production functions. Dependent variable: production uptime, namely, the fraction of scheduled operating time that the line actually runs	Up to 15 individual human resource management variables are identified. Grouped into 4 HRM "systems" from HRM1 ("traditional") where none of (11) practices were encountered to HRM4 ("innovative") where all practices are found. Identification of grouping is on the basis of inspection, but also supported by/robust to statistical classification procedures. Models include up to 25 controls for detailed features of the line (e.g. line width and speed)	Meetings with unions and union status of the production line	OLS results show that uptime increases monotonically in degree of HRM innovation. Fixed effect estimates – based on HRM group changers – also indicates this positive hierachical pattern. Estimated productivity effects very similar across specifications. No real indication that individual HRM practices or unionism have an additional (to HRM group) effect on the production measure in either OLS or fixed effect specifications. In short. innovative employment practices are complements

(continued)

Table 3.1 (continued)

Study	Data	Methodology	EI/HWPA measure	Union variable	Findings
3. Cappelli and Neumark (2001)	National Employer Surveys of 1994 and 1997 matched to the 1977 Longitudinal Research Database (LRD) to provide panels for 1977–1993 and 1977–1996. Number of observations is 443 and 666 respectively. It is assumed that EI/HPWP measures were only introduced some time after 1977 so that their "levels" observed in 1993 and 1996 can be used in first difference estimates	OLS regressions for 1993 and 1996 cross sections and for models in first differences. Dependent variables: log sales per worker; log labor costs per worker; log of the inverse of unit labor costs	11 work practices including benchmarking and computer usage in addition to "employee involvement mechanisms" (team working and TQM) and supportive measures (job rotation, gain-sharing, etc.) These considered singly and severally in 6 bundles suggested by a priori reasoning	No union argument is included in the control variables	The results for productivity indicate a positive association between individual practices and log sales per worker. But statistical significance is weak and the effect is further attenuated in first difference estimates. Some evidence of potential synergies in the bundles but the balance of the main and interactive effects is seldom statistically significant. The results for labor costs suggest that several practices increase costs in roughly similar amounts to the productivity effects. No indications of any efficiency effects (i.e. reductions in unit costs)

Table 3.1 (continued)

Study	Data	Methodology	EI/HWPA measure	Union variable	Findings
4. Cooke (1994)[c]	841manufacturing firms in Michigan, 1989	OLS cross section estimates. Dependent variable: company performance, defined as value added net of labor cost per employee. Three separate regressions for value-added per employee, average hourly wages and labor costs as a share of total cost. Firm performance is thus the estimated value added per employee for the relevant EI/HPWP combination less the estimated wage difference adjusted for the labor cost share differential	Employee participation, as proxied by presence of work teams and group-based incentives, as measured by existence of profit/gain sharing plans	Union status of the firm is interacted with team and profit/gain sharing variables. Reference category: nonunion firms with neither team working nor profit/gain sharing	Union firms are 13% more efficient than nonunion firms. If union firms have teams they are 35% more efficient than the nonunion no-team counterfactual. The corresponding advantage of nonunion firms with team working is actually negative. But the addition of profit/gain sharing has a much bigger relative payoff in nonunion than union firms (18% versus 6.5%). The interaction of teams and profit/gain sharing is modest in nonunion firms and negative in union firms. No significance levels of these effects are provided

(continued)

Table 3.1 (continued)

Study	Data	Methodology	EI/HWPA measure	Union variable	Findings
5. Black and Lynch (2001)	Educational Quality of the Workforce National Employers Survey (EQW-NES) data for 1994 matched to information from the LRD for 1987–1993 ($n = 638$)	Cobb-Douglas production function. Dependent variable: log sales per employee. Cross-section estimates for 1994 plus panel analysis for 1987–1993 period. The panel analysis uses within and GMM estimators to first derive average residuals for each establishment. These residuals are then regressed on the EI/HPWP measures and other time-invariant variables from the 1994 EQW-NES (technology, worker characteristics, etc.)	6 variables proxying HPWP (TQM, benchmarking, number of managerial levels, number of employees per supervisor, proportion of workers in self-managed teams and log number of workers in training); 2 measures of "employee voice" (unionization and proportion of worker meeting regularly in groups); and 2 types of profit sharing (for management and for production workers)	Union presence plus interactions with nonmanagerial profit sharing and TQM	In the cross sections only one HPWP – benchmarking – is statistically significant. No synergies between the HPWPs are detected. Of the other variables, the proportion of workers meeting in groups and profit sharing schemes for production workers often seem to raise productivity. Of the interactions tested, there is a positive association between unionism and profit sharing for production workers. The panel estimators produce basically similar results except that TQM now enters negatively as does the union variable. The latter effect is offset by the union-profit sharing interaction

Table 3.1 (continued)

Study	Data	Methodology	EI/HWPA measure	Union variable	Findings
6. Black and Lynch (2002)	EQW-NES data for 1994 and 1997. Cross-section sample is 1443 firms; panel sample comprises 284 firms	Cobb-Douglas production function estimates in cross section (1996) and first differences (1993–1996). Dependent variable: log sales per worker. Parallel wage functions are also estimated for 1996 and 1993–1996	Somewhat different mix to Black and Lynch (2001) in row 5. In addition to the proportion of workers in self-managed teams, benchmarking and the number of managerial levels, the HPWP set also includes "re-engineering" and profit sharing. Employee voice variables are as before, namely, union presence and the proportion of workers meeting regularly in groups	Union presence plus interactions of the union measure with profit sharing, re-engineering and the percentage of workers meeting regularly in groups	For the 1996 cross section, of the HPWP measures only the positive effect of profit sharing on productivity is well determined across specifications. The union role, but not the other voice argument is positive and statistically significant. But the union profit sharing interaction term is larger and of opposite sign. The union re-engineering interaction term is positive and well determined. For the panel estimates, the effect of re-engineering (proportions of workers in self-managed teams) is positive (negative) and well determined. Neither voice argument is statistically significant. No interaction term is statistically significant other than that between union presence and the proportion of workers meeting regularly in groups, which is strongly positive. For their part, the wage regressions show further variation

[a]See also Huselid (1995).
[b]See also Ichniowski (1990).
[c]See also Cooke (1992).

on the outcome indicator is scarcely documented. The *leitmotiv* of these studies is the attempt to make the best use of longitudinal and cross section data. In the case of the first row entry of Table 3.1, the goal is indeed methodological: Huselid and Becker (1996) provide a follow-up study of an earlier cross-section analysis by one of the authors (Huselid, 1995), now using panel data for 1993–1996 to control for firm fixed effects while examining the sensitivity of the latter to measurement error.[8]

Huselid and Becker identify no less than 13 "human resource management practices" and consider their impact on two measures of financial performance. Factor analysis is used to construct two factors measuring the extent of use each of the practices. The two factors are termed "employee skills and organizational structures" and "employee motivation," suggesting that they may include constituent elements that others might not consider to be high performance practices (Godard, 2004, p. 354). The control variables comprise capital intensity, employment, union coverage, the R&D sales ratio and sales growth. The upshot of the authors' analysis is that factors having a statistically significant effect on financial performance in (three out of four) cross section(s) were no longer significant using a fixed effects specification. The story does not end here as Huselid and Becker report sizeable measurement error in the workplace practices. Based on independent estimates of the measurement error (Cronbach's alpha), they argue that the fixed effects estimates become closer to the cross section estimates and in particular that a one standard deviation increase in a unidimensional measure (being the sum of the two dimensions noted above) of high performance work systems increases the market valuation of the corporation by $15,000 per worker. However, as noted by Cappelli and Neumark (2001, p. 741), the problem of measurement error is exacerbated in a short panel of this nature because most firms would have had the work practices in question so that we would expect there to be little difference in their use over the two-year interval in question (the practices were observed in 1991 and again in 1993).

The advantage of the study in row 2 of the table is that in examining a single process – steel finishing – within an industry, the problem of output heterogeneity is presumably minimized. And in constructing a long time series, the beginning of which is likely marked by an absence of high performance work practices, the impact of those practices might be better discerned – subject to various caveats on the diffusion of the practices in question. In this study, Ichniowski, Shaw, and Prennushi (1997) identify 15 human resource management practices that cover incentive pay, recruitment and selection, teamwork, flexible job assignment, employment security, skills training, communication and labor relations. As noted in the table, these are subsequently grouped into a hierarchy of four distinct human resource management systems. The delineation is on the basis of a priori reasoning, supported by several statistical procedures. The hallmark of this study is, then, a hypothesized complementarity of workplace practices. The workplace practices are accompanied by

[8] Huselid (1995) reported that his additive indices of HPWPs were associated with reduced labor turnover, higher sales per employee, and improvements in (one measure of) financial performance among his cross-section sample of 986 firms.

detailed controls (among the time-varying regressors included in the fixed effects specifications are age of line, start-up periods, the quality of steel input and the introduction of new equipment). The dependent variable is product uptime.

It is reported that the more innovative human resource systems that significantly increase line uptime in cross section have similar effects when controlling for permanent unobserved line heterogeneity, unlike the study by Huselid and Becker in row 1 of the table. Consistent with that study, however, is the finding of complementarity in employment practices. Individual practices have little or no incremental effect on productivity. That is to say, when individual practices are added to regressions containing the HRM systems of practices, they have little additional effect on productivity; and in comparing these (OLS and fixed effects) regressions with ones containing only the individual practices, the typically well determined estimates for the latter (including in one specification a negative effect of unionism) effectively vanish with the addition of the system effects. Ichniowski et al. (1997, p. 311) conclude that "systems of HRM practices determine productivity ... while marginal changes in individual practices have little effect." That said, there are some lingering ambiguities in the study stemming from the collinearity among practices and no discussion of whether all practices in the round contributed independently of the other bundles (Godard, 2004, p. 355).

The study by Cappelli and Neumark (2001) in row 3 of the table uses information on working practices from a nationally representative sample of establishments based on two surveys (the National Employers Surveys for 1994 and 1997) with a high response rate. These establishments are matched to LRD data to obtain information on their performance, productive inputs and other characteristics. Another distinguishing characteristic of the study is its use of multiple outcome indicators, the goal being to determine whether increases in productivity translate into improved performance by lowering unit costs.

To all intents and purposes, the high performance work practices are only observed at one point in time – strictly speaking the practices are observed in two years but five such practices are unique to the 1994 survey while six are asked of both surveys – but a long panel(s) is constructed by matching establishments in the cross-section data to the 1977 LRD. Here the assumption is that the innovative work practices did not exist in 1997, so that the innovative practices in levels can be used in fixed effect specifications as all observations necessarily represent changes. The authors present results for two panels, 1977–1993 and 1977–1996, as well as for two cross sections, 1993 and 1996.

As in the row 1 and 2 studies, Cappelli and Neumark also look for synergies in innovative work practices. They first enter the practices individually and then in groupings suggested by design (e.g., teamwork training is designed to improve performance in teams and should be more productive with employee meetings and self-managed teams) or prior research. As shown in the table, the results are mixed, pointing to few synergies and much evidence of heterogeneity bias in cross section. In terms of the three outcome indicators, the main results are that high performance work practices raise labor costs per worker, that they may raise productivity (sales per employee) and that they have no effects on unit labor costs. The last result may mean that, at a pinch, productivity and labor costs may be offsetting.

The study in row 4 of the table is notable for being the first to do more than simply include a union variable in the control set. Cooke (1994) examines whether unionism positively or negatively influences the effectiveness of employee-participation programs and group-based incentives in his sample of Michigan manufacturing firms. Cooke's measure of performance is value added net of labor cost per employee. To calculate this magnitude he estimates three equations: value added per employee, wage rates and labor cost/total cost. That is to say, he subtracts the estimated wage differential, adjusted by the labor cost share differential, from the estimated value added per employee differential to derive the performance measure.

His measure of employee involvement is a dummy variable indicating the presence or otherwise of team working and his group incentives variable is another dichotomous variable capturing the presence or otherwise of either profit-sharing or gainsharing plans. These dichotomous variables are separately and jointly interacted with the union status of the firm; the omitted category being the compound "nonunion/no gainsharing/no group incentive pay." The other regressors include firm size, depreciable assets per employee (albeit only at the firm's 2-digit primary industry, so that the same capital intensity is assigned to union and nonunion workplaces alike) and proxies for the skill composition of the workforce, technology and market power.

Using the estimated differentials associated with each combination of employee involvement, group incentive pay and union status, Cooke estimates their performance effects. The basic result is that unions elevate pay less than they increase productivity. Alone among the studies, Cooke reports that firm performance is 13% higher in unionized plants without either employee involvement or incentive pay than in comparable nonunion firms. The introduction of team working raises this differential to around 35%. By contrast, its introduction in the nonunion sector does not improve the innovating (nonunion) firm's net performance. In the absence of teamwork, group incentive pay has a much larger effect on efficiency in nonunion firms (+18.5%) than in union firms (+6.5%). In combination, the two measures also have a much bigger performance payoff in nonunion (+21%) than union (−0.7%) firms. While suggesting that the payoff to employee involvement and incentive pay may sharply differ in union and nonunion regimes, this study clearly paints a much rosier picture of union impact than the material reviewed earlier. Unfortunately, we are not told the statistical significance of the effects on performance. The cross section nature of the study, the deficiencies of the capital measure and issues of representativeness are additional sources of concern.

The study in row 5 of the table attempts to deal with each concern. Black and Lynch (2001) estimate production functions for a large, nationally representative sample of manufacturing establishments. The authors fit augmented Cobb-Douglas production functions to both cross-section and panel data for 1987–1993, having much richer data than Cooke.[9] The authors identify six *high performance work*

[9]For example, their capital stock measure is constructed from information on the book value of assets in the base year (1987) and new investments in each year, net of an estimate of the fraction of capital that is used up each year.

systems (TQM, benchmarking, number of managerial levels, number of employees per supervisor, the proportion of workers in self-managed teams and the {log} number of employees in training), two *employee voice* measures (unionization and the proportion of employees meeting regularly in groups) and two types of *profit sharing* (for management/supervisors and production/clerical/technical). Panel techniques are used in an attempt to tackle a potential omitted variables problem due to unobserved establishment characteristics (but see below). Specifically, Black and Lynch use a two stage procedure that involves first estimating a fixed, time-invariant firm effect for each establishment using data for the time-varying factors – namely, capital, labor and raw materials – and then regressing these fixed effects (or firm-level efficiency parameters) on all the time-invariant factors.[10]

The cross-sectional estimates indicate that, although most of the high performance work practices are positively associated with labor productivity, only benchmarking is statistically significant at conventional levels. Of the separate voice and profit sharing arguments the proportion of workers meeting regularly in groups – although not unionization – and nonmanagerial profit sharing are positively and significantly associated with productivity in most specifications. Contrary to the studies in the first two rows of Table 3.1, Black and Lynch report that there is no evidence of a synergistic bundling of the high performance work systems, although there is a positive and statistically significant association between unionization and nonmanagerial profit sharing.[11] Finally, the results for the panel data second-step estimation are much the same as the cross-sectional results, although in the specification containing interaction terms the own effect of unionism becomes negative, much larger in absolute magnitude and marginally statistically significant (even if this effect is more than counteracted by the positive interaction between unionism and nonmanagerial profit sharing).

Black and Lynch use their estimates to show how unionized establishments that embrace what are termed "transformed" industrial relations practices can have higher productivity than nonunion plants. Specifically, a hypothetical union plant practicing benchmarking and total quality management, with 50% of its workers meeting on a regular basis and operating profit sharing for its nonmanagerial employees is reported to have 13.5% higher labor productivity than a nonunion plant with none of these practices. By contrast, the corresponding differential for a high performance nonunion establishment is just 4.5%. In contrast to the row 4 study, however, if union and nonunion plants possess none of these workplace practices, the nonunion establishments have 10% higher labor productivity than their union counterparts. Note there is no attempt in this study to discover whether the workplace practices in question are positively related to *average costs per worker* (for evidence of which, see Cappelli & Neumark, 2001, in row 3).

---

[10]In addition to the within estimator, the authors use a GMM estimator in the first step to deal with biases stemming from the endogeneity of capital, labor, and materials.

[11]The authors note that, although individually only one interaction term was statistically significant, they rejected at the 0.05 level the joint null that all four interactions were zero.

Black and Lynch (2001, p. 443) themselves note that, despite the use of panel methods, the regressors used in the second stage procedure referred to earlier may still be correlated with unobserved time-invariant plant-level characteristics, thereby biasing the coefficient estimates on high performance workplace practices. One such variable is of course managerial quality. If more able managers are found in establishments that have greater recourse to such practices then the effects attributed to them may simply reflect good management. In the last row entry of Table 3.1, in their new study Black and Lynch (2000) present results from using a second wave of the dataset containing information on work practices and estimate a model in first differences, 1993–1996 (as well in cross section for 1996). As can be seen from the table, compared with the row 5 study there is some redefinition of high performance work systems, including the use of a new variable "reengineering." Another change is that unionization is now interacted with four of the work practices, including the proportion of workers meeting regularly in groups.

The results differ somewhat from Black and Lynch (2001). Commenting here on just the fixed effects findings, we see that reengineering is the only work practice with a statistically significant pro-productive own effect. For its part, the own effect of unionization is poorly determined throughout, although when interacted with the worker meeting variable the effect is positive and statistically significant. The own effect of worker meetings is strongly negative.

A final study not included in Table 3.1 nevertheless merits attention because it addresses the vexed issue of contingency in a union and high performance work practice context. The setting is the course of productivity at a single commercial aircraft manufacturer, 1974–1991. During the sample period there were three strikes and one work to rule, one major production change, three (five) changes in top management (union presidents) and the introduction of a TQM program towards the end of the sample period that included high levels of employee involvement and formal assurances of job security. Kleiner, Leonard, and Pilarski (2002) provide a detailed history of these industrial relations developments based on interviews with production managers and union leaders and estimate an engineering production function to investigate the effect of these changes. As expected, withdrawals of labor are associated with reduced productivity; also as expected, it only takes between one to four months to return to pre-strike productivity. Union leadership also seems important, while productivity appears highest when hawks on both sides negotiate. The effects of TQM are if anything perverse, which result is held to reflect the manner of its introduction (a top-down process that never fully won the embrace of the line workers or their front-line supervisors) (Kleiner et al., 2002, p. 203). It is argued that TQM was bearing fruit at exactly the time management's patience ran out (although the shift back to an authoritarian structure produced improved results in the short run). No less interesting, the authors observe that the effects of TQM may be upwardly biased because they fail to account for the impact of firms that are unsuccessful in implementing it – an inference that might be generalized to other practices pending improved datasets containing information on EI/HPWP dissolutions as well as adoptions.

The results of the more recent US literature on innovative work practices are, then, decidedly mixed both as regards the impact of particular practices and synergies between them. Abstracting from causation issues, positive effects on productivity where observed are only one-half the story, impact on the bottom line is the other. Contingencies may also cast a long shadow in reaching a judgment on the role of such practices. This setting is not altogether propitious for assessing the contribution of the union institution and it is unsurprising that the mainstream EI/HPWP literature has practiced benign neglect here. Yet recent US evidence does point to some positive interaction effects between individual work practices and unionism. Indeed, even though the nature of some of the associations is at times opaque, it has sometimes been claimed that there is a hierarchy for productivity performance with unionized plants having innovative practices at the top – above those of nonunionized plants with the same set of practices – and at the bottom in traditional workplaces. This conclusion and the use of synthetic workplaces to identify hierarchies are premature but the literature has demonstrated potential offsets to the union wage differential.

3.4 German Evidence

Reflecting the facts of workplace representation, most German studies have focused on the impact of works councils rather than unions, although interest in the union role – strictly, the collective agreement external to the establishment – has increased in the wake of Freeman and Lazear's (1995) extension of the voice model.[12]

3.4.1 Works Councils and Establishment Performance

In a reversal of the pattern of the US literature on worker representation, the early German literature pointed to mostly negative effects of works councils on firm performance. These studies cover a wide range of performance outcomes – total factor productivity, profits, product innovation and R&D, investment in physical capital and (excessive) quits – do not lack rigor,[13] and contain detailed information on

[12] Indeed, since this chapter was written, there has been a mini-explosion of interest in the effect of German collective bargaining proper on wages. See, inter al., Addison, Teixeira, and Zwick (2010), Gerlach and Stephan (2005, 2006), Gürtzgen (2005, 2006), Fitzenberger, Kohn, and Lembcke (2008).

[13] Among the best examples are the papers by Fitzroy and Kraft (1985, 1987, 1990) that deploy systems of equations in examining profitability, total factor productivity, and innovation, respectively, in a small sample of a little over 60 metal working firms. The unifying theme of all three papers is a managerial pressure/managerial competence model: efficient managers are able to elicit greater effort from their workforces without interference from works councils and are also able to institute adequate systems of communication and decision-making and avoid (the delays associated with)

establishment characteristics and sometimes union density (on the interaction of union density and works councils, see FitzRoy & Kraft, 1990; Schnabel & Wagner, 1994). But they have the disadvantage of small sample size – typically around 100 firms/plants – raising obvious issues of external validity.

More recently, analysts have been able to work with much larger data sets, namely, the *Hannover Firm Panel*, the *NIFA-Panel* and the nationally representative Establishment Panel of the Institute for Labor Market Research, or *IAB Establishment Panel*. The Hannover Panel is a stratified random sample of all manufacturing plants with at least five employees in the German state of Lower Saxony, 1994–1997. It comprises around 1000 establishments in 1994, declining to a little over 700 establishments in 1997 because of sample attrition. The NIFA-Panel is a survey of all establishments in the German machine-tool industry, 1991–1998. The sample base is around 6,000 firms and the realized sample approximates 1500 per wave. Finally, the IAB Establishment Panel has been conducted each year from 1993 (for eastern Germany from 1996 onward). It is based on a stratified random sample – strata for 17 industries and 10 size classes – from the population of all establishments with at least one employee covered by social insurance. The panel is created to serve the needs of the Federal Labor Agency and so its focus is on employment-related matters. The two other surveys contain somewhat richer establishment information and, in the case of (one wave of) the NIFA-Panel, subjective data on works council "type" and "degree of works council involvement" from a management perspective (see Dilger, 2002). That said, the IAB Establishment Panel contains the only nationally representative longitudinal sample of establishments and unlike the other two data sets it is possible to proxy the capital stock.[14]

Table 3.2 reports results from a selection of studies using two of the new data sets. It is not meant to be fully inclusive of the literature for two reasons. First of all, it is deliberately "light" on the more descriptive labor turnover literature.[15] Second of all, it is only one-half the research that we review: Table 3.4 contains studies that also consider EI/HPWP (to include training).

The most notable feature of the studies using these large datasets is their more favorable evaluation of works council impact. For the two studies using the Hannover Firm Panel (rows 1 and 2 of the table), the pro-productive effect of the works council depends on establishment size and collective bargaining regime, respectively. Addison, Schnabel, and Wagner (2001) report that works council presence is associated with higher value-added per employee only for the all-firm sample; for smaller establishments with between 21 and 100 employees the association is statistically insignificant. This result might reflect a theme of the earlier literature

works councils, partly by paying higher wages. As can be seen in Table 3.4, below, this model is qualified in a more recent study by the authors (Fitzroy & Kraft, 1995).

[14]Most recently, analysts have used data from the IAB Linked Employer-Employee Data set (LIAB), combining data from the IAB Establishment Panel and the Employment Statistics Register (e.g. Addison, Teixeira, & Zwick, 2010)

[15]But for an interesting study of quits, hires, and labor fluctuation using the NIFA-Panel that exploits works council type, see Dilger (2002)

Table 3.2 Recent estimates of the effects of works councils on performance

Study	Data	Dependent variable	Methodology	Works council measure/controls	Findings
1. Addison, Schnabel, and Wagner (2001)	Hannover Firm Panel. 900 establishments from the 1994 wave	Value added per worker; subjective measures of financial performance; wages and salaries per employee (and the percentage "wage gap"); three labor turnover measures (hires, separations and gross turnover); and two innovation measures (introduction of new processes/products)	Single equation estimates. Separate results for all - establishment sample and subset of plants with 21–100 employees	Works council presence. Controls include establishment size/age; measures of workforce composition and skill; product market competition; capacity utilization; profit sharing; state of technology; "excessive" wage costs; and industry dummies	Works council presence is associated with higher productivity overall but not for plants with 21–100 employees. Profitability systematically lower in works council regimes. Wages higher in the presence of works councils, but sources of these higher earnings are not transparent. All labor turnover measures are reduced in works council settings, other than for the subset of smaller establishments. Innovation measure unaffected by works control presence

(continued)

Table 3.2 (continued)

Study	Data	Dependent variable	Methodology	Works council measure/controls	Findings
2. Hübler and Jirjahn (2003)	Hannover Firm Panel. Pooled data from the 1994 and 1996 waves	Value-added per employee and wages and salaries per employee	Bivariate probit maximum likelihood estimates of works council presence and coverage by collective agreement to form selection arguments in wage and productivity equations	Works council presence and coverage by a collective agreement. Controls include establishment age/size/legal status; product market competition; measures of workforce composition and skill; profit sharing; state of technology; team work; and industry dummies	In separate productivity regressions by collective bargaining status, the positive effect of works councils on performance is statistically significant only where the plant is covered by a collective agreement. Wages are higher under both collective bargaining regimes, although the effect is better determined where there is no collective agreement
3. Frick and Möller (2003)	IAB Establishment Panel. Cross sections for 1998 and 2000	Log value added	OLS estimation of Cobb-Douglas, CES and translog production functions. Separate estimates for West and East Germany; summary results for manufacturing and services	Works council presence. In addition to log capital (proxied by replacement investment) and log employment, the regressors comprise: profit sharing for employees; process innovation;	Plants with works councils have sharply higher productivity of 25% (30%) in West (East) Germany. Disaggregation by sector confirms results other than for West German manufacturing in both 1998 and 2000. Works council

Table 3.2 (continued)

Study	Data	Dependent variable	Methodology	Works council measure/controls	Findings
				product innovation; employment of apprentices; coverage by a collective agreement; percentage of qualified employees; percentage of sales exported; and 14 sector dummies	interaction with profit-sharing is statistically insignificant
4. Schank, Schnabel, and Wagner (2004)	IAB Establishment Panel, 1993–2000. Unbalanced and balanced panels of plants with 21–100 employees	Log sales	Fixed effects stochastic frontier production function estimated separately for plants with and without works councils and for balanced and unbalanced panels. The comparison is between the technical efficiency estimates – and the 95% confidence intervals – of the median works council plant and its non-works	Works council presence. Regressors include log employment; shares of part-time, skilled and female employees; 41 sector dummies; and 7 year dummies	There are no statistically significant differences in efficiency between establishments with and without works councils. Results robust to disaggregation by broad sector and to time interval (use of two different five-year balanced panels)

(continued)

Table 3.2 (continued)

Study	Data	Dependent variable	Methodology	Works council measure/controls	Findings
			council counterpart		
5. Addison, Bellmann, Schnabel, and Wagner (2004)	IAB Establishment Panel. Initial sample of 1544 establishments all without works councils in 1996	Changes in quits, sales per employee, employment and the profit-situation	Nonparametric propensity score matching model. "Treated" group comprises all plants in which a works council was set up between 1996 and 1998. Matched plants derived from 1513 controls	Introduction of a works council	Mean values for the performance indicators in establishments that experienced the formation of a works council are not statistically different from those obtaining in plants that remained free of works councils. Results are robust to outliers
6. Addison, Bellmann, and Kölling (2004)	IAB Establishment Panel, 1996–2000	Plant closings	Probit estimates for the following samples: all-plants; plants stratified by whether or not they are covered by a collective agreement; plants stratified by size	Works council presence. Apart from coverage by a collective agreement and establishment size, the covariates include plant age; workforce composition and skills; recent layoff	For the aggregate sample, works councils associated with significantly higher closings. Works council effect is sharper for the uncovered sector and for smaller plants but the difference is not statistically

Table 3.2 (continued)

Study	Data	Dependent variable	Methodology	Works council measure/controls	Findings
			(less than or more that 50 employees); and plants stratified by both collective agreement and size	and industry dummies experience; state of technology; regional unemployment rate; and industry dummies	significant. Only for small plants that are covered by a collective agreement is there any suggestion that collective bargaining coverage can lower closure rate

to the effect that for smaller firms the advantages of codetermination may be lower and the costs higher than for larger firms. But the authors' own justification for focusing on this sub-sample is two-fold: first, lacking information on works council type, it makes sense to estimate the effect of the entity for a sample in which the rights/powers of the works council are a datum; second, only in this employment range are establishments with and without works councils present in reasonable numbers, as opposed to being dominant (see above). Apart from there being no evidence of higher productivity within this sample, the authors also report that works councils are not associated with reduced labor turnover in this sub-sample either, although they obtain the conventional effect for the all-establishment sample.

For their part, Hübler and Jirjahn (row 2) emphasize not establishment size but instead whether or not the establishment is covered by a collective agreement. They seek to test the argument of Freeman and Lazear (1995) that where a works council is embedded in a (external) collective bargaining arrangement, the control exerted by the union and the employers' federation can serve to dissipate distributional squabbles at the workplace allowing the voice effects to realized. As can be seen, they report that pro-productive works council effects are only found where the establishment follows a collective agreement.

The Hannover Firm Panel does not contain information on the capital stock, raising the potential problem of omitted variables bias. However, the capital stock may be inferred from data on replacement investment in the IAB Establishment Panel. Using this information, the study by Frick and Möller (2003) in row 3 of the table provides one of the first estimates of a works-council-augmented production function for Germany (see also Addison, Kraft, & Wagner, 1993; FitzRoy & Kraft, 1987). The magnitude of the works council "effect" is highly reminiscent of the union effects reported by Brown and Medoff (1978) some 25 years earlier for the United States. Plants with works councils have 25–30% higher productivity than their counterparts without works councils. Drawing on secondary evidence that wages are higher in works council plants, Frick (2003, p. 448) nevertheless cautions that these effects may not compensate firms for the additional costs.[16]

The next two studies summarized in Table 3.2 that also use the IAB Establishment Panel question this large productivity effect of works councils (as indirectly does the final study in row 6). Substituting sales for value added on data grounds – for example, respondents often fail to answer the materials cost question in the panel survey[17] – and restricting the sample to plants with between 21 and 100 employees, Schank Schnabel, and Wagner (2004) estimate frontier productions for separate samples of firms with and without works councils, 1993–2000. They report no

[16]Most German studies report that the simple association between works councils and wages is positive and statistically significant. But there is the suggestion that the excess of wages over those agreed to in collective agreements does not point in the same direction (see Addison, Schnabel, and Wagner 2001). Investigation of the works council-pay nexus is currently the subject of considerable research interest.

[17]No less important, the materials costs – measured as the percentage share of sales represented by materials costs – often seem to be little more than informed guesses.

statistically significant differences in the technical efficiencies of the median plant in the two samples. Using a very different approach, the study by Addison, Bellmann, Schnabel, and Wagner (2004) in row 5 of the table seeks to carefully match plants that witnessed the formation of works councils in 1998 and those that remained free of works councils for the entire sample period (1996–2000). For their four outcome indicators, including a proxy for productivity growth, they report no statistically significant differences in mean performance among the two types of firms. This study represents one attempt to deal with selection problems that dog the other estimates in Table 3.2.

The final study in row 6 of the table is relevant from the perspective of another type of potential bias attaching to both cross section and (balanced) panel estimates alike, namely, the problem that we are observing survivors. Underperforming works council plants may have been evolved out of the system, leaving an unrepresentative sample of survivors. The study by Addison, Bellmann and Kölling (2004) does find that other things being equal works council firms do fail at higher rates than their non-works council counterparts. Although this hints at survival bias, we would caution that this evidence is indirect (i.e., survivability should be modelled simultaneously with the performance equation), that the failure rates in question are low and that the number of industry controls is for statistical reasons quite limited.

By the same token, one should continue to be wary of the ambitious pro-productive works council effects provided in row 3 of the table. In Table 3.3, we

Table 3.3 Works council coefficient estimates from a translog production function fitted to IAB establishment panel data, 1997–2000

Sector			
Sample	All	Manufacturing	Services
Firms with 21–100 employees			
Pooled	0.118***	0.046	0.183***
	(3.22)	(1.03)	(3.05)
1997	0.123*	0.119*	0.098
	(1.86)	(1.73)	(0.82)
1998	0.107*	0.063	0.145
	(1.94)	(0.94)	(0.61)
1999	0.056	−0.011	0.139*
	(1.16)	(0.18)	(1.72)
2000	0.172***	0.070	0.269***
	(3.60)	(1.22)	(3.41)
Firms with ≥5 employees			
Pooled	0.232***	0.177***	0.275***
	(8.88)	(5.71)	(3.05)

Notes: The dependent variable is log total sales. In addition to the log capital stock, log number of employees and works council presence, the regressors are: investment in ICT, state of technology, workforce composition (share of part timers, apprentices and skilled workers), coverage by a collective agreement, a dummy for eastern Germany and, for the relevant pooled/cross section regressions, sector and/or year dummies.
$|t|$ - values in parentheses
***,* denote significance at the 0.01 and 0.10 levels, respectively.
Source: Addison, Schank, Schnabel, and Wagner (2006).

present summary results from fitting a translog production function to pooled data for 1997–2000 from the IAB Establishment Panel. The dependent variable is sales (rather than value added) and the sample is restricted to firms with 21–100 employees, other than in the last row of the table where pooled estimates for the entire sample of establishments with five or more employees are provided for purposes of comparison. Although works council "effects" are in our view still uncomfortably large,[18] note first the differences between the coefficient estimates in the pooled regressions for the sample comprising plants of all sizes and the sub-sample of establishments with 21–100 employees. Next observe the generally insignificant coefficients for the all-important German manufacturing sector in both pooled and cross-section data. Then observe the variation in the coefficient estimates in cross section. Finally consider that there are no a priori grounds for expecting major differences in works council effects as between manufacturing and services. Taken in conjunction with the evidence in Table 3.2, we would argue that although the early German literature most likely encouraged an overly negative view of works council operation some of the more recent estimates are likely to do exactly the opposite.

3.4.2 EI/HPWP, Works Councils and Firm Performance

As with the US evidence on unions, however, the issue of works council impact does not end here. Rather more than in the United States perhaps, discussion of the effects of worker representation in Germany has become linked with those of innovative work practices. Table 3.4. The themes of this literature, mostly based on the three large-scale datasets introduced earlier, include works council type, incentivization issues, complementarities and training. Together the studies point to circumstances in which works councils may be associated with favorable productivity outcomes.

We begin with the study by Frick (2002) (row 1) which, although descriptive, is notable for its attempts to investigate directly the impact of works council type (see also Dilger, 2002). The only large-scale data set to contain such information is the (sixth wave of the) NIFA-Panel. As can be seen, Frick's results using this information are mixed. Thus, HPWPs are not significantly more numerous in works council plants than others and while "more involved" councils tend to have more HPWPs those works councils dubbed "antagonistic" by management have the most. Similarly, while HPWPs are reported in the author's multiple classification analysis to benefit firm performance, they are also said to be associated with reduced labor demand.

[18]Quantile regression analysis suggests that the large pro-productive effects for the all-establishment sample in the Hannover Firm Panel are driven by highly productive plants; that is, they are detected only in plants at the top end of the conditional productivity distribution (see Wagner, Schank, Schnabel, & Addison 2006). The suggestion may be that only highly competent managers can cooperate with a works council in a way that materially advances productivity.

Table 3.4 Selected German studies examining the effect of innovative work practices and works councils on performance

Study	Data	Methodology	EI/HPWP measure	Works council variable	Findings
1. Frick (2002)	1700 establishments from the sixth wave of the NIFA-Panel. Dataset identifies five types of HPWP, as well as five types of works council and a measure of works council involvement	Descriptive analysis of HPWPs by works council presence, involvement and type. Multiple classification analysis for same categories as in descriptive treatment and five covariates (number of employees, sales per-employee, stock of orders and capacity/manpower utilization)	The five measures are: reduction in hierarchies; delegation of decision making; work groups with independent budgets; group or team work; flexible working time	Five works council types range from "antagonistic" to "excluded." The measure of works council involvement signifies greater engagement of works council in processes of technological or organizational change than laid down under law or by collective agreements	Plants with works councils use more HPWPs than do plants without works councils, but difference is not statistically significant. Plants with more involved works councils have more HPWPs than their less involved counterparts. But the number of HPWPs is highest in circumstances where works council is rated "antagonistic." HPWPs are reported to have a positive effect on establishment performance but a negative effect on labor demand[a]
2. FitzRoy and Kraft (1995)	Pooled data for 1977 and 1979 for 65 metal working firms in West Germany	Value added production function estimated separately for profit-sharing	Profit sharing for the workforce	Presence of works council	The effect of works councils on productivity measure is negative and statistically significant in non-sharing firms but

(continued)

Table 3.4 (continued)

Study	Data	Methodology	EI/HPWP measure	Works council variable	Findings
		firms and non profit-sharing firms, with correction for selection into profit sharing status. Also estimation of full interaction equation for the whole sample			is positive and statistically significant in profit-sharing firms. Interaction effect between works councils and profit-sharing in full interaction equation is statistically insignificant
3. Jirjahn (2003)	Pooled data from the 1994 and 1996 waves of the Hannover Firm Panel	Single-equation OLS pooled productivity estimates. Dependent variable: value added per employee. Auxiliary probit model of works council presence provided but not used to instrument works council presence. Separate estimates for all plants and subset of plants with	Managerial profit sharing. Some innovative practices are also included in the probit equation (e.g. further training and flexible working hours) but not in the productivity equation	Works council presence	Executive profit sharing schemes are productive for both samples. The interaction between executive profit sharing schemes and works council presence is negative and statistically significant for the all-establishment sample. Works council effect is positive and statistically significant throughout

Table 3.4 (continued)

Study	Data	Methodology	EI/HPWP measure	Works council variable	Findings
		plants with 21–100 employees			
4. Wolf and Zwick (2002)	IAB Establishment Panel, for a 1999 cross section and a 1996–1999 panel	Cobb-Douglas production function with capital approximated by replacement investments. Dependent variable: value added and average value added. Cross-section estimates with and without correction for endogeniety of HPWP. For the panel, the estimation follows the Black-Lynch (2001) two-step procedure (see Table 3.1, row entry 5), with the second stage also controlling for the endogeneity of HPWPs	Six HPWPs: shifting of responsibilities to lower level in hierarchy; team work; work groups with independent budgets; employee share ownership; profit sharing; training to support organizational change; and incentive training. These HPWPs are aggregated into two independent factors ("organizational change" and "incentives") using principal components factor analysis	Works council presence	In cross section, the positive effect of "incentives" (share ownership, profit sharing, supportive training and incentive training) results from their being introduced by firms when they are prospering. In the panel (second- step) estimates and after correcting for selection, "organizational changes" (shifting responsibilities, team work and independent work groups) have significantly positive effect on productivity and are also introduced to deal with structural problems – structural productivity gaps. No interaction between "incentives" and "organizational changes." Separate works council

(continued)

Table 3.4 (continued)

Study	Data	Methodology	EI/HPWP measure	Works council variable	Findings
					effect not robust in cross section, but in panel estimates work council presence has significantly positive impact on the establishment-specific fixed effect
5. Zwick (2004)	IAB Establishment Panel, 1997–2000	Cobb-Douglas production analysis function. Dependent variable: value-added and average value-added. In other words, the same two-step procedure used by Wolf and Zwick (2002) but with GMM (GMM-SYS) estimator rather than within estimator alone to estimate the production function	"Participative work forms" based on delegation of responsibility and decision making to lower-levels in the hierarchy, team work and work groups with independent budgets. These three measures are aggregated to form a single participation variable set equal to one if at least one of the three measures had been introduced	Works council presence and estimated works council presence in a switching regression model	After accounting for the endogeneity of participative work forms, the productivity effect of such practices is around 25%. In the switching regression, (the instrumented value for) participation is only (marginally) statistically significant in the sample of works council firms. Selection into works council status is well determined in both equations

Table 3.4 (continued)

Study	Data	Methodology	EI/HPWP measure	Works council variable	Findings
		and the time-invariant establishment fixed effect	by 1997. Participation is endogenized using instrumental variable regressions		
6. Zwick (2002)	IAB Establishment Panel. Cross sections for 1998 and 1999 and panel data for 1997–1999	Cobb-Douglas production function approach. Dependent variable: value added. Cross-section estimates of effect of continuous training in 1997 on labor productivity levels in 1998 and 1999. Selection into training status handled through a probit regression used to calculate the probability that a firm offers training and a Tobit equation to predict the intensity of training.[b]	Training incidence (i.e. presence); training intensity (number of trainees divided by number of employees); and training type (formal external courses, formal internal courses, training on the job, seminars and talks, job rotation, self-induced learning and quality circles)	Works council presence	In cross section, (lagged) effects of training intensity – actual and predicted – are positively associated with labor productivity. Formal external training courses also positive and statistically significant throughout. Negative effect of training on-the-job. Selection term is negative. Panel estimates indicate that training intensity, formal external and internal courses and self-induced learning have a positive and significant effect on productivity; after selection is accounted for, the coefficient estimates for training intensity and

(continued)

Table 3.4 (continued)

Study	Data	Methodology	EI/HPWP measure	Works council variable	Findings
		Fixed effects panel estimation using two-step procedure as in row 4 and 5 studies, also with and without selectivity correction			formal training courses increase in size. Coefficient estimate for the works council dummy is positive and statistically significant in cross-section estimates but not in second-step regression estimates

[a]See also Addison and Teixeira (2006).
[b]See also Gerlach and Jirjahn (2001).

Unlike the other studies reviewed in Table 3.4, the analysis in row 2 is based on an early sample of just 65 metal working firms. This dataset was first used by FitzRoy and Kraft (1985, 1987, 1990) in a series of studies charting the impact of works councils on profitability, total factor productivity and innovation. Adverse works council effects were reported for all outcome indicators in systems of equations that typically endogenized works council presence. But FitzRoy and Kraft (1995) report that works council effects on labor productivity are actually positive and statistically significant in regimes practicing profit sharing, even if they are negative and well determined in other regimes. The authors attribute the positive works council effect to cooperative labor relations.

FitzRoy and Kraft find that profit sharing is associated with sharply higher productivity – in both single and simultaneous equation treatments. This result is not generally found in the large-scale data sets; for example, neither Frick (2002) (row 3, Table 3.2), using the IAB Establishment Panel, nor Hübler and Jirjahn (2003) (see row 2, Table 3.2), using the Hannover Firm Panel, observe discern this outcome.[19] Abstracting from measurement issues (on which more below), an interesting extension of the incentives discussion is profit sharing for executives. The study by Jirjahn in row 3 of Table 3.4 finds that managerial profit sharing strongly improves productivity, suggesting that it attenuates agency problems. But this issue is only part of the story; the other is works council presence and its interaction with executive profit sharing. As can be seen from the table, the coefficient estimate for works council presence is positive and statistically significant while that for the interaction term is negative and again well determined. Focusing on the latter result, Jirjahn argues it is consistent with two hypotheses: either profit sharing reduces the commitment value of agency in circumstances where works councils cannot foster trust and loyalty absent the cooperation of management, or management rent seeking is curbed by profit sharing and the works council is not important for building cooperation in situations of reduced opportunism on the part of management. Although these empirical findings have yet to be replicated in other data sets, central to Jirjahn's approach is the notion that the management environment of the firm matters in evaluating works council impact. His approach is also very much in the spirit of collective voice model, even if the contribution of the works council is not transparent. This latter remark in turn reflects some negative correlations noted in the literature such as that between team work and works councils and most of all the simple fact that the magnitude of the negative interaction term between works councils and executive profit sharing exceeds that of the positive point estimate for works council presence.

The remaining three studies in Table 3.4 follow a common methodology suggested in part by Black and Lynch (2001) (see row 5, Table 3.1) and use the IAB Establishment Panel. Especial emphasis is placed upon on selection issues and unobserved plant heterogeneity. The study by Wolf and Zwick (2002) in row 4 of

---

[19] The exception is that the latter do find a positive effect of profit sharing on labor productivity for the uncovered sector which they speculate reflects *union* opposition to profit sharing.

the table provides the template for the other two studies. Wolf and Zwick focus on the productivity of HPWPs. They identify seven such practices, which are reduced to two independent factors – termed "organizational changes" (participatory practices such as team work) and "incentives" (e.g., profit sharing) – using principal components analysis. The authors fit an augmented Cobb-Douglas production function to cross-section data for 1999, including a selectivity-adjusted specification. They also fit a Cobb Douglas function to data for 1996–1999 using panel estimation methods and derive a fixed time invariant establishment-specific effect for each plant after Black and Lynch, which values are then regressed on the time invariant covariates – including organizational changes and incentives – again allowing for selection. Note that the dataset only enables the authors to identify whether or not the innovative practices were present at a point in time, not when they had been introduced.

As shown in Table 3.4, both selection and accounting for structural differences matter. The positive effect on productivity of incentive mechanisms, observed in both cross section and in the panel, seem to result from such schemes being introduced in times of plenty, when firms are doing well. That is, after correction for the endogeneity of such measures, the variable is no longer statistically insignificant. For their part, the effect of organizational changes is statistically insignificant in cross section, with and without correction for selection. However, the coefficient estimate is both positive and well determined in the panel estimates and after controlling for selection strengthens somewhat. The main message of this study, therefore, is that firms which introduce organizational changes seemingly have unobserved time invariant characteristics that decrease their productivity. Expressed differently, participatory work practices raise productivity.

The results pertaining to incentive mechanisms clearly qualify the results of FitzRoy and Kraft (1995) (row 2 of Table 3.4) who control for the endogeneity of profit sharing but not unobserved firm heterogeneity. Wolf and Zwick nevertheless do argue that works councils work – the coefficient estimate for the variable is positive and statistically significant in the second step panel procedure – even if they do not seem to influence organizational change. Finally, two other results from the cross section part of this study might be noted. Not only is there no evidence of synergies between the two sets of innovative practices but also very few instances of complementarities between their components.

The role of the work council is secondary in Wolf and Zwick. In the next study in the table, it shares equal billing with innovative work practices. Zwick (2004) (row 5) considers just those practices found to be statistically significant in Wolf and Zwick – namely, organizational changes, now termed "participation" – and considers their impact on productivity over the same interval, albeit using a different panel estimator for the (first stage) production function and now controlling for the endogeneity of works council presence (as well as that of the participation variable). Other differences reside in the measurement of capital and the construction of the participation measure.

Zwick's second-step regression results for a specification in which works councils are assumed exogenous indicates that the innovative work practice(s) is positive and well determined, elevating productivity by 25%. However, accounting for the nonrandom distribution of works councils in an endogenous switching regression model shows that the pro-productive effect of participatory work practices (or one such practice) only obtains in works council firms. Although the story is similar to that told by Wolf and Zwick – in the sense that innovatory practices are conceived to rectify structural productivity deficits – the main result is very different: innovations only bear fruit in works council regimes. This result is analogous to that of Cooke (1994) for the United States (see Table 3.1, row 4). No other German study reaches so stark a conclusion. Unfortunately, the coefficient estimate for the participation variable in works council plants is only marginally significant and it is not clear that it differs from the statistically insignificant coefficient estimate for participation in the sample of plants without works councils.

One of the routes through which works councils and all voice institutions are supposed to influence productivity is by encouraging investments by the firm in training. Although the German literature has dwelt at some length on the labor turnover issue, there has been little direct analysis of further training. A rare exception is the final study in row 6 of Table 3.4, in which Zwick (2002) examines the determinants of such training and its impact on labor productivity. As in the two preceding studies, the basic framework is again a mix of cross-section and two-stage panel estimation with endogenous training. In addition to detailed establishment controls, the equations include three other innovative work practices. These are identical to those previously grouped under "participation" in the row 5 study, namely, a shifting of responsibilities to lower levels in the hierarchy (termed "employee participation"), the presence of teamwork and self-directed groups and work groups/units with their own costs and results accounting. These three practices are now treated as exogenous, as is works council presence.

Zwick finds that training intensity (share of trained employees in the workforce) as well as several types of training (especially formal external training) have positive effects on productivity in cross section and in the second-step panel estimation procedure. In both cases, accounting for the endogeneity of training either by predicting training intensity or by including an inverse Mills ratio term (the probability of a training presence) in the regression tends to increases the magnitude of well determined coefficient estimates, suggesting that there is a greater tendency to train when productivity is low. The effect of the panel estimation is also to increase the measured impact of training. But of the other innovative work practices none is statistically significant in cross section and just one (shifting of responsibilities to lower levels in the hierarchy) is significant in most of the second-step panel estimates. A positive association between works council presence and productivity is discerned in cross section but in the second-step panel estimates the coefficient estimates for this variable are statistically insignificant throughout. Unlike the work practices, however, work councils are included as a determinant of training (i.e. training intensity) and their influence is positive and statistically significant, a result that is ascribed to the absence or attenuation of hold up – this time on the part of employers – in works council regimes.

The increasing sophistication of the German studies in the attempt to control for selection into either innovative working practices, including training, or works council status while taking account of unobserved plant heterogeneity has to be measured against the limitations of the data. As far as selection into works council status is concerned, we do know that works council presence is not random but is related to establishment size and the structure of the work force, among other things. However, the introduction or presence of a works council is not the result of a rational choice made by the owners or managers of an establishment based on comparing costs and benefits; rather, it is the results of actions taken by the employees. Employee action might well be related to the past performance of the establishment, but whether an establishment has a works council at a point in time should not be viewed as caused by contemporaneous productivity. This might suggest that one can treat works council status as exogenous to the establishment, not least given the difficulties of identifying the works council equation in a convincing manner. Identification issues also loom large when it comes to the identifying equations for innovative work practices, including training. Understandably, analysts have switched between treating works councils and innovative work practices as endogenous according to the emphasis of the particular study, but this approach has produced some inconsistencies.

More might be expected from panel estimates that control for unobserved plant heterogeneity. Unfortunately, there is the problem that that works council status tends not to change over time – for example, in balanced panels from the IAB Establishment Panel, 1997–2000, around 2% of plants change their works council status. And as regards the innovative works practices, we lack longitudinal data altogether. Given the lack of temporal variation in works councils status and HPWPs for either reason, the Black-Lynch procedure has found favor. The problem is course that the estimated coefficients for works councils and innovative practices in the second step may be biased by reason of correlation between these variables and unobserved time-invariant establishment characteristics. For example, as noted earlier, better managers are more likely to be observed in plants with HPWPs so that what appears to be a productivity effect of innovative practices is good management. A further problem is that if we lack longitudinal information on HRWPs, we cannot assume that their frequency did not change materially over the sample period in question.

Despite these and other concerns (not least the lack of information on the numbers of workers covered by HPWPs and on "type" of works council), the German literature suggests that works councils and innovative practices may have favorable effects on productivity. Indeed, we have reviewed material that attributes major productivity gains to each. But average wage council effects are unlikely to be strong. We mainly base this conclusion on estimates obtained for samples of plants employing between 21 and 100 workers where there is a "balanced" representation of works council and non-works council establishments and where the powers of works councils are a datum. As far as HPWPs are concerned, it would be idle to claim that there is agreement as to which measures work. The disputation over profit sharing is the main case in point. Not surprisingly, therefore, when we come to consider the interaction between works councils and innovative work practices the situation is not transparent. But we think the evidence supports the notion that innovative work practices allow works council firms to achieve improved productivity if not necessarily higher productivity than non-works council firms.

One further caveat is the near exclusive focus in recent German work on the productivity outcome. We have no information on possible increases in labor costs occasioned by operating HPWPs, as suggested by some US research. We do know that wages are higher in works council regimes (see Table 3.2, row 1), but we can infer little from this because there is no indication that works council presence has any effect on the excess of wages over those laid down under collective agreements. It remains a possibility that works councils may cause wage or incentive pay drift, or that they inflate nonwage costs, but there is as yet no real evidence to substantiate this.

Two final issues are whether works councils and HPWPs are substitutes or complements and the status of the often-encountered argument that German employers would have to invent works councils were they not legislated. There is evidence to favor both substitutability and complementarity. Examples of the former include the finding of a negative association between team working and works council presence and the greater frequency of HPWPs in the absence of works councils (or, in their presence, where works councils are antagonistic). Examples of the latter are the positive association between works councils and group incentives and training.

The idea that works councils would have to be invented were they not legislated is certainly not a finding that leaps from the pages of the early works council literature or even from the more recent literature without qualification. One such qualification may be plant size. Unfortunately, it is difficult to test the argument that works councils are beneficial in large establishments. This is because nearly all such plants are covered by works councils and conversely for small plants. Another qualification is that establishments may need to be covered by a collective agreement to blunt the bargaining or hold-up implications of codetermination. A final set of qualifications would presumably involve the contribution of innovative working practices and cooperative relations.

3.5 Conclusions

We have examined the effects of worker representation in unions and works councils and of employee involvement/innovative work practices – and their interaction – on firm performance. In view of the doubts concerning the performance effects of high performance work practices and controversies surrounding the effects of worker representation – both of which aspects are addressed in our analysis – the results of the exercise might have been predicted to be both compartmentalized and pessimistic. But we have largely interpreted the worker representation literature as indicating that the effects of *unions* on productivity are likely to be small on average so that we should look to factors such as innovative work practices in explaining the diversity in the effects of worker representation in different settings. Expressed in terms of the theoretical conjectures on unions, the interaction of worker representation and HPWPs provides a practical means of peering inside the union black box. We would incline to the same interpretation of the works council institution, even if unobserved works council type may turn out to be a more important source of variation in overall performance.

We reported evidence of positive HPWP effects in both countries and also of positive interactions with worker representation. The suggestion is, then, that combinations of innovative practices and worker representation can yield substantial productivity gains. In Germany, HPWPs may be an important means of restoring competitiveness and works councils may assist in their introduction. Yet there should be no pretence to precision in all of this in the sense of the literature having uncovered a well determined hierarchy for productivity performance, or blueprint for the future of unions in particular. This is not unexpected and reflects the limitations of both literatures – namely, the notion of collective voice and the high performance paradigm – underscored throughout by pronounced measurement difficulties. Here the cautionary remarks of Wood and De Menezes (1998) based on the British experience (see below) are also apropos for the United States and Germany.

3.6 Addendum: The British Evidence on Unionism and Firm Performance

3.6.1 Introduction

There is no doubt that unions declined in Britain after 1979, after a period of period of substantial growth. At that time some 53% of workers were union members, but by 1999 this had fallen to 28%. Correspondingly, there also occurred a sharp fall in the share of employees whose wages are set by collective bargaining: from 70% in 1980 to around 45% in the mid-1990s. Moreover, all indicators of union presence point in the same direction and for all sectors other than the public sector (Machin, 2000).

And since 1980 there have also occurred some profound changes in observed union effects, some of which are more controversial than others. We should perhaps preface our presentation of establishment performance outcomes – along the dimensions of financial performance, productivity, investment, employment and plant closings – with some brief remarks on the course of the union-nonunion differential, one of the more controversial areas. It has been conventional to argue that the union premium remained more or less stable during the 1980s but declined fairly precipitously during the first half of the following decade (see Stewart, 1995; Hildreth, 1999). That said, studies based on individual rather than workplace data pointed to persistence in the union premium of around 10% (e.g. Blanchflower, 1997). Nevertheless, the most recent work using individual-level data now seems to confirm the workplace findings (e.g. Machin, 2001), so that wage evidence does after all tend to favor a marked decline in the union differential. Today, the premium seems to have all but disappeared (e.g. Bryson, 2002a).

3.6.2 Establishment Performance Outcomes

3.6.2.1 Profitability

Almost all of the early British studies pointed to negative effects of various indicators of union presence on financial performance/profitability. Some of the more recent evidence is reported in Table 3.5. The starting point is the study by Machin and Stewart (1996) that identifies a sharp decline in the union effect on productivity over the first three WIRS. Note that by 1990 any negative effect was confined to closed shop or analogous situations in conjunction with some degree of market power. The second row contains the results of a replication of the previous study by Addison and Belfield (2001) using the 1998 WERS which, if anything, points to a sharper retardation in the negative effect of union recognition.

The study in the third row is of interest because it attempts to capture the effects of financial participation and employee involvement practices on financial performance. In particular, it is argued that employee involvement will be more capable of yielding a dividend where it is associated with financial participation – and, further, that different types of employee involvement and financial participation will vary in their impact on financial performance. Mixed effects are duly reported for union recognition: unions have positive impact where the organization practices downward communication and upward communication, but in the absence of such schemes the union effect is negative and statistically insignificant. But Addison and Belfield's (2000) replication of this study (row 4) finds no statistically significant effect of union recognition for any concatenation of employee involvement and financial participation.

The study in row 5 allows for a wider variety of union measures. Booth and Mcculloch (1999) find no evidence of a hierachy of (nonmanual) union effects on aggregate. However, disaggregation by union type – manual versus nonmanual – reveals such effects for nonmanual unions: the effect of "recognition only" is negative and statistically significant and the presence of a closed shop increases the absolute value of the union effect. In contrast, manual unions have no discernible impact on financial performance.

The principal concern in the study by Conyon and Freeman (2001) is the role of (four types of) financial participation on financial performance (row 6), so that the union effects for the 1998 WERS cited in the table are secondary – and, interestingly, never commented on in the paper. As can be seen, the association between union recognition and financial performance is negative and statistically significant throughout (i.e. irrespective of the form of financial participation). Unlike the row 3 study, it is argued that employee involvement and financial participation are complementary.

Table 3.5 Union effects on profitability

Study	Data/ Methodology	EI/HPWP measure	Works council variable	Findings
1. Machin and Stewart (1996)	WIRS 80, WIRS 86; WIRS 90. Ordered Probit	Financial performance	Union recognition (manual), closed shop/management recommends unionism (manual)	Negative effect of union recognition halved during the 1950s. By 1990 a significantly negative effect was confined to the closed shop. Moreover, that effect was conditional on presence of market power (as proxied by relative size of the establishment)
2. Addison and Belfield (2001)	WERS 98. Ordered probit	Financial performance	As above	Coefficient estimates for union recognition and closed shop/management recommends unionism variables statistically insignificant throughout
3. McNabb and Whitfield (1998)	WIRS 90. Probit	Financial performance better than average	Union recognition	Union effect hinges on interaction with financial participation and employee involvement mechanisms. Union effect statistically insignificant in conjunction with financial participation. Union effect positive and statistically significant in the presence of employee involvement schemes and is negative and statistically significant in their absence
4. Addison and Belfield (2000)	WERS 98. Probit	As above	As above	Coefficient estimate for union recognition statistically insignificant irrespective of employee involvement and financial participation mechanisms
5. Booth and McCulloch (1999)	WIRS 90. Ordered probit	Financial performance	Union recognition, closed shop; manual and nonmanual union recognition, manual closed	Positive and statistically significant effect of union recognition; negative and statistically significant effect of closed shop. But these overall effects hinge on union type. Coefficient estimates for recognition and closed shop are

Table 3.5 (continued)

Study	Data/ Methodology	EI/HPWP measure	Works council variable	Findings
			shop, nonmanual closed shop	only statistically significant for nonmanual unions
6. Conyon and Freeman (2001)	WERS 98. Ordered probit	Financial performance	Union recognition	Coefficient estimates for union recognition negative and statistically significant throughout
7. Menezes-Filho (1997)	Sample of 494 firms, 1984–1990. Pooled regressions and fixed effect specification	Rate of return on sales	Union recognition/derecognition; bargaining structure	Coefficient estimates for union recognition negative and statistically significant but declining through time. Strongest negative effects observed in single establishment firms and where different unions bargain jointly with the firm at the industry level. Fixed effects specifications show that derecognized firms have faster increases in profitability. Also fragmentation of bargaining structure associated with higher profitability

Note: Unless otherwise indicated, financial performance is based on a five-element categorical measure derived from the manager respondent's assessment of the financial performance of the establishment relative to others in the same industry.

The final study by Menezes-Filho (1999) is notable for its use of financial data per se and incorporation of bargaining structure (and theory) (row 7). The main result is that although union firms have lower profitability overall this effect narrowed between 1984 and 1990 and had all but disappeared by the end of the sample period. That being said, consistent with the author's theoretical priors, there is evidence that profit retardation remains strongest in areas where unions have greater bargaining power. Bargaining power is shown to be greatest in single establishments (vis-à-vis multi-establishment undertakings) and among multi-establishment firms where the firm bargains jointly and an industry agreement applies.

3.6.2.2 Productivity

The sparse early British evidence points to negative effects of unionism on firm and establishment productivity, despite contemporaneous estimates of the union wage premium of around 10% (e.g. Machin, 1991). The dominant theme of the newer literature reviewed in Table 3.6 is that unionized firms/plants increased their productivity most at the end of the 1980s (and arguably in 1979–1984, too) and/or that there is no longer evidence of a productivity shortfall in union firms/establishments. The evidence is not overwhelming, however, so that it would be premature to claim more than what has been observed is a reduction in the disadvantages of unionism. The relevant factors here include the possibility that least productive unionized undertakings may have been evolved out of the system (for this reason the empirical suggestion that the union effects were most positive in situations where competitive pressures were more acute is not altogether compelling), the fact that the strongest productivity gains are actually reported for union de-recognitions and to some degree the odd timing of the observed spurts of improvement.

As in the case of parallel developments in profitability, reviewed earlier, it is conventional to ascribe the observed changes to the impact of the Thatcher reforms in conjunction with heightened competitive pressures (e.g. Gregg, Machin, & Metcalf, 1993, p. 895). At issue of course is whether the observed changes will persist. Also worrying is the apparent shift in the effects of other covariates – not just the union argument – on the productivity outcome indicators in the WIRS/WERS.

3.6.2.3 Investment

Consistent with the US evidence reviewed in the main paper, the early British research also provides evidence of some negative effects of unions on investment in physical capital (Denny & Nickell, 1992). However, this does not carry over to investments in intangible capital. While they are able to replicate Hirsch's (1991) results for the United States, Menezes-Filho, Ulph, and Van Reenen (1998) do not obtain similar results for the United Kingdom. Indeed, in their recent review, Menezes-Filho and Van Reenen (2003) contrast the very different results for the United States on the one hand and Britain and continental Europe on the other. Although the association between unionism and R&D is negative in this bloc as

Table 3.6 Union effects on productivity and changes in productivity

Study	Data/methodology	EI/HPWP measure	Works council variable	Findings
1. Gregg, Machin, and Metcalf (1993)	328 trading firms from EXSTAT, 1984–1989. Production function estimated using panel regression methods	Growth in log real sales	Union recognition, changes in union status	For union recognition along, the union effect is negative and statistically insignificant for 1984–1997, but positive and statistically significant for 1988–1989 (+3–4%). For changes in union status, repudiation of the closed shop has no incremental impact over (favorable) effect of union presence (1988–1989) but derecognitions considerably elevate differential productivity growth of union sector
2. Conyon and Freeman (2001)	(i) 284 firms, 1995–1998. Fixed effects production function estimates	Log real sales	Union recognition	Union effect negative but statistically insignificant throughout
	(ii) WERS 98. Ordered probit	Relative labor productivity	Union recognition	Union effect negative but statistically insignificant throughout
3. Moreton (1999)	WIRS 98. Ordered probit	Relative labor productivity and union density	Union bargaining power, proxied by separate multi-unionism and firm endorsement of union membership, inter al	Multiunionism associated with significantly lower productivity. Where management recommends unionism there is a positive effect on labor productivity. Also some suggestion that union effect might be positive where labor demand elasticity is higher
4. Fernie and Metcalf (1995)	WIRS 90	(i) Relative labor productivity	Union recognition alone. Pre-entry closed shop, post-entry closed shop,	For the one-dimension measure, the union effect is negative but only marginally significant. For the fuller representation, no hierarchy of effect beyond union recognition

(continued)

Table 3.6 (continued)

Study	Data/methodology	EI/HPWP measure	Works council variable	Findings
			management recommends union member-ship, union recognition only	only, which is negatively signed and highly significant
		(ii) labor productivity improve-ment, 1987–1990	As above	For the one-dimension dimension measure, the union effect is negative but only marginally significant. For the fuller representation, only the pre-entry closed shop is associated with significantly lower productivity growth
5. Addison and Belfield (2001)	WERS 98	(i) As above	As above	Coefficient estimates for each union measure are negative but statistically insignificant throughout
		(ii) As above	As above	For union recognition alone, the union effect is positive and statistically significant. For the fuller representation, no hierarchy of effect beyond union recognition only, which is again positively signed and highly significant

Note: The relative labor productivity measure is a categorical variable based on management perceptions of labor productivity of the workplace compared with other similar workplaces (WIRS 90) or workplaces in the same industry (WERS 98). The labor productivity improvement measure is a categorical variable based on management perceptions of the level of labor productivity at the survey date relative to 3 years earlier (WIRS 90) or 5 years earlier (WERS 98).

well, it is seemingly driven by unions being concentrated in older, low-tech industries. In interpreting these results, the authors critique the notion that unions will hold up firms by expropriating sunk R&D investments through demanding higher wages while also observing that the hold-up problem may be mitigated by strategic incentives to compete in R&D races. The latter such considerations imply that the union effect on R&D might exhibit nonlinearities – being positive at lower levels of union density.

3.6.2.4 Employment

We turn next to the evidence of union effects on employment, broadly interpreted. There is reasonable agreement in the WIRS/WERS on the seeming role of unions in retarding employment growth or exaggerating employment reduction. As can be seen from Table 3.7, there is the suggestion that unionized establishments in the 1980s tended to grow by 3% less per year than their nonunionized counterparts. There was some early disputation over the results reported by Blanchflower, Millward, and Oswald (1991) for 1980–1984 because of the concern that unionized firms at this time were more likely to see an erosion of restrictive practices (partly related to overmanning arrangements and demarcation) so that union plants were more likely to be shedding labor. But this criticism does not seem to be unduly damaging. First, the organizational change measure in the WIRS used to proxy such reform of working conditions does not overturn the negative impact of the union density measure on employment for the specific time period under consideration. Second, Booth and Mcculloch (1999) (row 2) report for later intervals that the union effect on employment is robust to the inclusion of organizational change measures. Subject to obvious limitations – the data-driven failure to model the dynamics of the employment adjustment process – there is a large measure of accord in repeated cross sections of the workplace surveys that union slow job growth.

That said, there is some lingering ambiguity as to some other relationships in the employment change equations. Thus, for example, Addison and Belfield (2001) (row 4) report very different findings for financial participation arguments than do Fernie and Metcalf (1995) (row 3). On the other hand, neither finds evidence of a hierarchy of union effects (i.e. greater retardation of employment growth/heavier employment decline in circumstances where unions are more powerful). In this connection, Bryson (2002b) has recently argued that stronger unions are more likely to bargain over employment so that one should not expect to see declines in employment for this subset of unions.

Not shown in the table is union impact on quits and absenteeism rates. But there is no disagreement in successive WIRS/WERS of the union role in reducing quits, even if the data are less than ideal. Rather, disagreement centres on absenteeism rates: Addison and Belfield (2001) report sharply higher absenteeism rates in union regimes using WERS 98 whereas Fernie and Metcalf (1995) report no such association for WIRS 90. There has been no investigation in Britain of the effect of quits in improving productivity and of absenteeism in reducing it.

Table 3.7 Union effects on employment

Study	Dataset/ methodology	Outcome measure	Union variable	Findings
1. Blanchflower, Millward, and Oswald (1991)	WIRS 84. OLS	Log employment (effectively an employment change equation as coefficient estimate on lagged dependent variable approximates 1)	Union recognition; union density; post-entry closed shop, pre-entry closed shop, membership	Coefficient estimate for union recognition is negative and statistically significant. Effect of density is better determined . Together the recognition and density results imply that union establishments contract 3 percentage points more per year than their nonunion counterparts. Allowing for different union types, union membership and pre-entry closed shop are associated with lower employment growth (though the latter variable is poorly determined) while the effect of the post-entry closed shop is positive and statistically significant
2. Booth and McCulloch (1999)	WIRS 90. OLS	Change in log employment, 1989–1990 and 1987–1990	Union recognition	Coefficient estimate for union recognition is negative and statistically significant throughout. The union-induced reduction in employment growth is 2.6% (5.7%) for 1989–1990 (1987–1990)
3. Fernie and Metcalf (1995)	WERS 90. OLS	Change in employment 1984–1990	Union recognition only; pre-entry closed shop, post-entry closed shop, management recommends unionism, recognition alone	For union recognition only, the association between employment change and union recognition is negative and statistically significant. For the fuller characterization of unionism, the coefficient estimates for all but the post-entry closed shop are negative and highly statistically significant. No hierarchy of effect
4. Addison and Belfield (2001)	WERS 98. OLS	Change in employment, 1993–1998	As above	For union recognition only, the association between employment change and union recognition is negative and highly statistically significant. For the fuller characterization of unionism, only the two weaker measures of union presence are associated with a material reduction in quits

3.6.2.5 Plant Closings

The last outcome indicator we examine is plant closings. The evidence summarized in Table 3.8 is again mixed. Studies based on WIRS 84 reveal scant evidence of any association between unionism and plant closings, irrespective of the union measure (see row 1). Note also the absence of a hierarchy of union effect: more powerful unions, as proxied by the magnitude of the wage premium or presence of the closed shop, have no discernible incremental effect on closings.

The plot thickens when we come to consider the most recent evidence. Broadly speaking, the sign of the coefficient estimate for the union variable is positive and statistically significant in the two studies using WERS 98 (rows 2–3). But this broad result hides as much as it reveals. Although reporting a material and robust positive association between either of two measures of unionism – recognition for collective bargaining purposes and union coverage – Addison, HeyWood, and Wei (2003) find that this relation holds only for establishments that are part of larger (i.e. multi-establishment) undertakings. For single-plant entities (here firms), the direction of the association is reversed. (All studies support the more general result that single independent plants are less likely to close than their counterparts that are part of multi-establishment undertakings). The authors interpret the former result as consistent with a decline in union bargaining power in the wake of a decade of anti-union legislation, either by emboldening employers in multi-plant enterprises to close unionized establishments, or by weakening union influence over employment in such settings (see Machin, 1993). The single-plant result, on the other hand, is rationalized in terms of (differential) union concessions in conjunction with rents.

While not contesting these findings, the recent study by Bryson (2004) (row 4) offers a very different interpretation of the positive association between plant closings and unionism. Rather, it is now union weakness – accentuated by the legislation – that is said to explain the sea change in union effect detected in the more recent workplace survey. Bryson reports that where unions are strong the coefficient estimate for unionism is no longer statistically significant. Stronger unions are identified by the closed shop and a combination of high density, bargaining coverage and on-site representatives, inter al. The "weak union" result is taken to be consistent with such unions being an ineffective voice for workers – and an inefficient agent for management as well – and conversely for strong unions. However, while downplaying rent seeking and emphasizing the ineffective voice/weakened agency function of weak unions, Bryson reports that where unions bargain over physical working conditions, the likelihood of closure is increased.

3.6.3 Unions, Employee Involvement and Other Workplace Practices

As we have seen, two of the studies considered earlier examine the association between unionism and employee involvement/financial participation and financial

Table 3.8 Union effects on plant closings

Study	Dataset/Methodology	Outcome measure	Union variable	Findings
1. Machin (1995)	1984 WIRS, using data on plants that subsequently closed from the WIRS 1984–1990 Panel. Probit model	Union recognition	Log number of employees, proportion nonmanual workers, single plant, manufacturing dummies, below average financial performance, operating well below capacity	Union recognition effect statistically insignificant both overall and by type of union (manual and nonmanual) and in the presence or otherwise of the closed shop. Result robust to inclusion of one-digit industry dummies
2. Addison, Heywood, and Wei (2003)	1990 WIRS, using data on plants that subsequently closed from WERS 1990–1998 Panel. Probit mode	Union recognition; union coverage	Establishment size, establishment age, proportion female, proportion manual, proportion professional/technical, proportion short-term contracts, wide range of employee-involvement and participation mechanisms, industrial relations climate, technology variables, flexibility at workplace, change in ownership, market power, layoff experience, export exposure, regional unemployment rate, one-digit and more detailed (three or four digit) industry controls	Robust positive and statistically significant association between union measures and probability of plant closure. But the result is driven by plants that are part of multi-establishment undertakings. For single-establishment firms, the union effect(s) is negative and generally statistically insignificant
3. Bryson (2004)	As above. Probit model	Union recognition; union strength (3 measures); union type; number of unions; bargaining arrangements (e.g. single vs. joint bargaining); and bargaining scope	Industry-level union density, log number of employees, proportion non- manual, single plant, (10) regional dummies, (18) two-digit industry controls, degree of competition, use of flexible contracts, financial performance better than average, operating considerably below capacity, increase in employment	Union measure(s) positively associated with plant closure. But magnitude and significance of the effect is sensitive to form of measure. Statistically significant effects where union is weak, for manual worker unions, single unions and where union bargains over physical working conditions

performance. But the key study is Wood and De Menezes (1998), exploiting data from the 1990 WIRS and the Employers' Manpower and Skills Practices Survey. The authors first attempt to test whether the range of employee involvement and participative mechanisms used in the literature form a unity and can be used as indicators of a high commitment orientation on the part of management. Wood and de Menezes use latent variable analysis to search for identifiable patterns in the use of twenty-three such practices. They are unable to identify high commitment management as a well-defined continuous variable. But they are able to fit a latent class model to the data, that is, identify a progression of types of high commitment management. There are four such types: high HCM, medium-low HCM, low-medium HCM and low HCM.

As far as unionism is concerned, Wood and de Menezes first examine the association between union recognition and high commitment management. Neither high HCM nor low HCM workplaces emerge as distinctive with respect to unionism. This suggests among other things that the tendency of the (British) industrial relations literature to treat nonunion workplaces as "bleak authoritarian houses" is erroneous. Second, the authors include the establishment's HCM class as an argument in conventional performance equations alongside unionism and controls for workplace characteristics and industry affiliation. They examine seven such performance outcomes: labor productivity, change in labor productivity, financial performance, job creation, employee relations climate, quits and absenteeism. High HCM establishments are not found to be more effective than others. That is, in no case do they perform better than all the others on any performance criteria. For example, although high HCM plants do have better employment growth and better financial performance that the two medium HCM categories, this does not carry over to the low HCM plants. Evidently different types of plants can perform differently according to the outcome measure. The plot only thickens when it comes to the effect of union recognition since five out of seven coefficient estimates are negative, of which four are statistically significant. (For a more optimistic interpretation, see Metcalf, 2003.)

3.6.4 Interpretation

We have discussed changes in the impact of British unionism on firm performance, examining en passant the modern notion that the decline in union influence has downside efficiency consequences. Our findings may be summarized as follows. First, there is evidence of a diminution of union effects on wages, financial performance and productivity through time. As the measures are not commensurate, we cannot quantify the degree of efficiency improvement stemming from these changes in unionized regimes. Arguably, the economic impact could have been small, but we incline to the view that more than redistribution (from workers) has been involved. But by the same token, certain unfavorable effects of unionism persist (e.g. slower employment growth and higher absenteeism) and so it is also appropriate to conclude that there has been a reduction in the disadvantages of unionism, not a reversal.

Larger efficiency gains are likely to have accrued from the decline in union density and the ability of newly formed enterprises to avoid union organization.

Second, we have found little direct support for the revisionist notion that the reduction in union power is responsible for worse outcomes. The argument that unions have to be strong to be an effective vehicle of pro-productive voice and to act as an authoritative agent of the employer principal has been found to be pushing things too far (Addison & Belfield, 2004). In the specific context of plant closings given that there is no evidence of a union effect in the WIRS 84, why should the reduction in union power have yielded a significantly positive association between union recognition/density and plant closings in the WERS 90? The Bryson argument would be that the legislation led to weaker unions that could neither deliver voice nor act as a responsible agent of the employer. Unfortunately, the test hinges on differences in effect between strong and weak unions rather changes in union power. We favor the alternative efficiency argument that there were too few closings in earlier years. In other words, we would argue from the finding for multi-establishments in the literature (see Table 3.8, row 3) that management in such undertakings have taken the opportunity to rid themselves of inefficient plants. This contention gels with the reasoning of Manning (1993), who argues that the requirement for pre-strike ballots (under the Conservative administration's Trade Union Act of 1984) may have led to a loss of union influence over employment. Unions, so the argument goes, had hitherto kept open unprofitable plants by threatening to strike profitable ones. The new need to ballot members destroyed the credibility of this mechanism because workers whose jobs were not in jeopardy would not vote for a strike.

Caveats attach to the interpretation that the workplace surveys unambiguously reveal evidence of a decline in union influence (see, in particular, the analysis of the WERS 98 panel by Addison & Belfield, 2002). A final issue is the interaction between unionism and high performance work practices. Here the British evidence has nicely exposed the ambiguities in the literature rather than advanced that literature.

References

Acemoglu, D., & Pischke, J-S. (1999). The structure of wages and investment in general training. *Journal of Political Economy, 107*, 539–572.

Addison, J. T. (2005). The determinants of firm performance: Unions, works councils, and employee involvement/high performance work practices. *Scottish Journal of Political Economy, 52*, 406–450.

Addison, J. T., & Belfield, C. R. (2004). Union voice. *Journal of Labor Research*, 25, 563–596.

Addison, J. T., & Belfield, C. R. (2000). The impact of financial participation and employee involvement on financial performance: A re-estimation using the 1998 WERS. *Scottish Journal of Political Economy, 47*, 571–583.

Addison, J. T., & Belfield, C. R. (2001). Updating the determinants of firm performance: Estimation using the 1998 WERS. *British Journal of Industrial Relations, 39*, 341–66.

Addison, J. T., & Belfield, C. R. (2002). *The impact of unions and employee involvement on workplace performance: Panel data evidence from the United Kingdom*. Unpublished Paper, Department of Economics, University of South Carolina.

Addison, J. T., & Belfield, C. R. (2004). Unions and establishment performance: Evidence from the British Workplace Industrial/Employee Relations Surveys. In Phanindra V. Wunnava (Ed.), *The changing role of unions – New forms of representation* (pp. 281–319). Armonk, NY: M.E. Sharpe.

Addison, J. T., & Belfield, C. R. (2007). Training, unions and firm performance. *Zeitschrift für ArbeitsmarktForschung, 40*, 361–381.

Addison, J. T., Bellmann, L., & Kölling, A. (2004). Works councils and plant closings in Germany. *British Journal of Industrial Relations, 42*, 125–148.

Addison, J. T., Bellmann, L., Schnabel, C., & Wagner, J. (2004). The reform of the German Works Constitution Act: A critical approach. *Industrial Relations, 43*, 392–420.

Addison, J. T., Heywood, J. S., & Wei, X. (2003). New evidence on unions and plant closings: Britain in the 1990s. *Southern Economic Journal, 69*, 822–841.

Addison, J. T., & Hirsch, B. T. (1989). Union effects on productivity, profits and growth: Has the long run arrived? *Journal of Labor Economics, 7*,. 72–105.

Addison, J. T., Kraft, K., & Wagner, J. (1993). German works councils and firm performance. In B. E. Kaufman & M. M. Kleiner (Eds.), *Employee representation: Alternatives and future directions* (pp. 305–338). Madison, WI: Industrial Relations Research Association.

Addison, J. T., Schank, T., Schnabel, C., & Wagner, J. (2003). *German works councils in the production process*. (Discussion Paper No. 812). Bonn: Institute for the Study of Labor/IZA.

Addison, J. T., Schnabel, C., & Wagner, J. (2001). Works councils in Germany: Their effects on establishment perfomance. *Oxford Economic Papers, 53*, 659–694.

Addison, J. T., Schnabel, C., & Wagner. (2004). The course of research into the economic consequences of German works councils. *British Journal of Industrial Relations, 42*, 255–281.

Addison, J.T. & Teixeira, P. (2006). The effect of works councils on employment change: *Industrial Relations, 45*, 1–25.

Addison, J. T., Teixeira, P., & Zwick, T. (2010). *Works councils and the anatomy of wages. Industrial and Labor Relations Review* (forthcoming).

Applebaum, E., & Batt, R. (1984). *The new American workplace: Transforming work systems in the United States*. Ithaca, NY: ILR Press.

Baldwin, C. Y. (1983). Productivity and labor unions: An application of the theory of self-enforcing contracts. *Journal of Business, 56*, 155–185.

Black, S., & Lynch, L. M. (2000). *What's driving the new economy: The benefits of workplace innovation*. (NBER Working Paper No. 7479, Revised October 2000).

Black, S., & Lynch, L. M. (2001). How to compete: The impact of workplace practices and information technology on productivity. *Review of Economics and Statistics, 83*, 434–445.

Blanchflower, D. G. (1997). Changes over time in union relative wage effects in Great Britain and the United States. Unpublished paper, Dartmouth College.

Blanchflower, D. G., Millward, N., & Oswald, A. J. (1991). Unionism and employment behavior. *Economic Journal, 101*, 815–34.

Booth, A. L., & Mcculloch, A. (1999). Redundancy pay, unions and employment. *Manchester School, 67*, 346–366.

Brown, C., & Medoff, J. L. (1978). Trade unions in the production process. *Journal of Political Economy, 86*, 355–378.

Bryson, A. (2002a, January). *The union membership premium: An analysis using propensity score matching*. Unpublished Paper, Policy Studies Institute.

Bryson, A. (2002b). Unions and workplace performance: What is going on? Unpublished Paper, Policy Studies Institute.

Bryson, A. (2004). Unions and workplace closure in Britain, 1900–1998. *British Journal of Industrial Relations, 42*, 283–302.

Cappelli, P., & Neumark, D. (2001). Do high-performance work practices improve establishment-level outcomes? *Industrial and Labor Relations Review, 54,* 737–775.
Cavanaugh, J. K. (1998). Asset-specific investment and unionized labor. *Industrial Relations, 37,* 35–50.
Clarke, K. B. (1984). Unionization and firm performance: The impact on profits, growth and productivity. *American Economic Review, 74,* 893–919.
Conyon, M. J., Freeman, R. B. (2001, August). *Shared modes of compensation and firm performance: UK evidence* (Working Paper 8448). National Bureau of Economic Research.
Cooke, W. N. (1992). Product quality improvement through employee participation: The effects of unionization and joint union-management administration. *Industrial and Labor Relations Review, 46,* 119–134.
Cooke, W. N. (1994). Employee participation programmes, group-based incentives and company performance: A union-nonunion comparison. *Industrial and Labor Relations Review, 47,* 594–609.
Denny, K., & Nickell, S. J. (1992). Unions and investment in British industry." *Economic Journal, 102,* 874–887.
Dilger, A. (2002). *Ökonomik betrieblicher Mitbestimmung.* Munich and Mering: Rainer Hampp Verlag.
Dinardo, J., & Lee, D. S. (2002). *The impact of unionization on establishment closure: A regression discontinuity analysis of representation elections* (Working Paper No. 8993). National Bureau of Economic Research.
Dustmann, C., & Schönberg, U. (2004). *Training and union wages.* (Discussion Paper No. 1435). Bonn: Institute for the Study of Labor/IZA.
Fallick, B. C., & Hassett, K. H. (1999). Investment and union certification. *Journal of Labor Economics, 17,* 570–582.
Fernie, S., & Metcalf, D. (1995). Participation, contingent pay, representation and workplace performance: Evidence from Great Britain. *British Journal of Industrial Relations, 33,* 379–415.
Fitzenberger, B., Kohn, K., & Lembcke, A. C. (2008). *Union density and varieties of coverage: The anatomy of union wage effects in Germany* (Discussion Paper No. 3356). Bonn: Institute for the Study of Labor/IZA.
FitzRoy, F., & Kraft, K. (1985). Unionization wages and efficiency: Theories and evidence from the US and West Germany. *Kyklos, 38,* 537–554.
FitzRoy, F., & Kraft, K. (1987). Efficiency and internal organization: Works councils in West German firms. *Economica, 54,* 493–504.
FitzRoy, F., & Kraft, K. (1990). Innovation, rent-sharing and the organization of labor in the Federal Republic of Germany, *Small Business Economics, 2,* 95–103.
FitzRoy, F., & Kraft, K. (1995). On the choice of incentives in firms. *Journal of Economic Behavior and Organization, 26,* 145–160.
Forth, J., & Millward, N. (2004). High involvement management and pay in Britain. *Industrial Relations, 43,* 98–119.
Freeman, R. B. (1976). Individual mobility and union voice in the labor market. *American Economic Review, Papers and Proceedings, 66,* 361–368.
Freeman, R. B., & Lazear, E. P. (1995). An economic analysis of works councils. In J. Rogers & W. Streeck (Eds.), *Works councils, consultation, representation and cooperation in industrial relations.* Chicago, IL: University of Chicago Press.
Freeman, R. B., & Medoff, J. L. (1984). *What do unions do?* New York: Basic Books.
Frick, B. (2002). High performance work practices und betriebliche Mitbestimmung: Komplementär oder substantive? *Industrielle Beziehungen, 9,* 79–102.
Frick, B., & Möller, I. (2003). Mandated works councils and firm performance: Labor productivity and personnel turnover in German establishments. *Schmollers Jahrbuch, 123,* 435–454.
Gerlach, K., & Jirjahn, U. (2001). Employer provided further training: Evidence from German establishment data. *Schmollers Jahrbuch, 121,* 139–164.

Gerlach, K., & Stephan, G. (2005). Wage settlements and wage setting; results from a multi-level model. *Applied Economics, 37*, 2297–2306.

Gerlach, K., & Stephan, G. (2006). Pay policies of firms and collective wage contracts- an uneasy partnership. *Industrial Relations, 45*, 47–67.

Godard, J. (2004). A critical assessment of the high-performance paradigm. *British Journal of Industrial Relations, 42*, 349–378.

Gürtzgen, N. (2005). *Rent-sharing and collective bargaining coverage – Evidence from linked employer-employee data* (Discussion Paper No. 05—90). Mannheim: Zentrum für Europäische Wirtschaftsforschung/Centre for European Economic Research.

Gürtzgen, N. (2006). *The effect of firm- and industry-level contracts on wages – Evidence from longitudinal linked employer-employee data* (Discussion Paper 06—82). Mannheim: Zentrum für Europäische Wirtschaftsforschung/Centre for European Economic Research.

Handel, M. J., & Levine, D. I. (2004). Editor's introduction: The effects of new work practices on workers. *Industrial Relations, 43*, 1–43.

Hildreth, A. (1999). What has happened to the union wage differential in Britain in the 1990s? *Oxford Bulletin of Economics and Statistics, 61*, 5–31.

Hirsch, B. T. (1991). *Labor unions and the economic performance of US firms*. Kalamazoo, MI: Upjohn Institute for Employment Research.

Hirsch, B. T. (2004). What do unions do for economic performance? *Journal of Labor Research, 25*, 415–455.

Hirsch, B. T., & Morgan, B. A. (1994). Shareholder risk and returns in union and nonunion firms. *Industrial and Labor Relations Review, 47*, 302–318.

Hübler, O., & Jirjahn, U. (2003). Works councils and collective bargaining in Germany: The impact on productivity and wages. *Scottish Journal of Political Economy, 50*, 471–491.

Huselid, M. A. (1995). The impact of human resource management practices on turnover, productivity and corporate financial performance. *Academy of Management Journal, 38*, 635–670.

Huselid, M. A., & Becker, B. E. (1996). Methodological issues in cross-sectional and panel estimation of the human resource-firm performance link. *Industrial Relations, 35*, 400–422.

Ichniowski, C. (1990). *Human resource management systems and the performance of US manufacturing businesses* (Working Paper No. 3349). National Bureau of Economic Research.

Ichniowski, C., Kochan, T. A., Levine, D., Olson, C., & Strauss, G. (1996). What works at work: Overview and assessment. *Industrial Relations, 35*, 299–333

Ichniowski, C., Shaw, K., & Prennushi, G. (1997). The effect of human resource management practices on productivity: A study of steel finishing lines. *American Economic Review, 87*, 291–313.

Jirjahn, U. (2003). Executive incentives, works councils and firm performance. *Schmollers Jahrbuch, 123*, 397–421.

Kleiner, M. M., Leonard, J. S., & Pilarski, A. M. (2002). How industrial relations affects plant performance: The case of commercial aircraft manufacturing. *Industrial and Labor Relations Review, 55*, 195–218.

Lalonde, R. J., Marschke, G., & Troske, K. (1996). Using longitudinal data on establishments to analyze the effects of organizing campaigns in the United States. *Annales d'Economie et de Statistique, 41/42*, 155–185.

Levine, D. I., & Tyson, L. D. (1990). Participation, productivity and the firm's environment. In A. S. Blinder (Ed.), *Paying for productivity*. Washington, DC: The Brookings Institution.

Machin, S. (1991). The productivity effects of unionism and firm size in British engineering firms. *Economica, 58*, 479–490.

Machin, S. (2000). Union decline in Britain. *British Journal of Industrial Relations, 38*, 631–645.

Machin, S. (2001). Does it still pay to be in a union or join a Union? Unpublished Paper, University College London.

Machin, S., & Stewart, M. (1996). Trade unions and financial performance. *Oxford Economic Papers, 48*, 213–241.

Machin, S., & Wood, S. (2005). HRM as a substitute for trade unions in British workplaces. *Industrial and Labor Relations Review, 58*, 201–218.

Malcomson, J. M. (1983). Trade unions and economic efficiency. *Economic Journal, 93*, 50–65.
Manning, A. (1993). Pre-strike ballots and wage-employment bargaining. *Oxford Economic Papers, 45*, 422–429.
Mcnabb, R., & Whitfield, K. (1998). The impact of financial participation and employee involvement on financial performance. *Scottish Journal of Political Economy, 45*, 171–187.
Menezes-Filho, N. (1997). Unions and profitability over the 1980s: Some evidence on union-firm bargaining in the United Kingdom. *Economic Journal, 107*, 651–670.
Menezes-Filho, N., Ulph, D., & Van Reenen, J. (1998). R&D and union bargaining power: Evidence from union companies and establishments. *Industrial and Labor Relations Review, 52*, 45–63.
Menezes-Filho, N., & Van Reenen, J. (2003) Unions and innovation: A survey of the theory and empirical evidence. In J. T. Addison & C. Schnabel (Eds.), *International handbook of trade unions* (pp. 293–334). Cheltenham, England and Northampton, MA: Edward Elgar.
Metcalf, D. (2003). Unions and productivity, financial performance and investment: International evidence. In J. T. Addison & C. Schnabel (Eds.), *International handbook of trade unions* (pp. 118–171). Cheltenham, England and Northampton, MA: Edward Elgar.
Milgrom, P., & Roberts, J. (1995). Complementarities and fit: Strategy, structure and organizational change in manufacturing. *Journal of Accounting and Economics, 19*, 179–208.
Moreton, D. (1999). A model of labor productivity and union density in British private sector unionized establishments. *Oxford Economic Papers, 51*, 322–344.
Schank, T., Schnabel, C., & Wagner, J. (2004) Works councils: Sand or grease in the operation of German firms. *Applied Economic Letters, 11*, 159–161.
Schnabel, C., & Wagner, J. (1994). Industrial relations and trade union effects on innovation in Germany. *Labor, 8*, 489–503.
Stewart, M. (1995). Union wage differentials in an era of declining unionization. *Oxford Bulletin of Economics and Statistics, 57*, 143–166.
TUC. (1999). *Partners for progress: New unionism at the workplace*. London, England: Trade Union Congress.
Wagner, J., Schank, T., Schnabel, C., & Addison, J. T. (2006). Works councils, labor productivity and plant heterogeneity: First evidence from quantile regressions. *Jahrbücher für Nationalökonomie und Statistik, 226*, 505–518.
Wood, S., & De Menezes, L. (1998). High commitment management in the UK: Evidence from the Workplace Industrial Relations Survey and Employers' Manpower and Skills Practices Survey. *Human Relations, 51*, 485–515.
Wolf, E., & Zwick, T. (2002). *Reassessing the impact of high performance workplaces* (Discussion Paper No. 02-07). Mannheim: Zentrum für Europäische Wirtschaftsforschung/Centre for European Economic Research.
Zwick, T. (2002). *Continuous training and firm productivity in Germany* (Discussion Paper No. 02-50). Mannheim: Zentrum für Europäische Wirtschaftsforschung/Centre for European Economic Research.
Zwick, T. (2004). Employee participation and productivity. *Labour Economics, 11*, 715–740.

Chapter 4
Ageing and Pension Reforms: Simulations and Policy Options

Joachim Winter

4.1 Introduction and Overview

The aim of this paper is to discuss the macroeconomic and capital market effects of population ageing and of fundamental pension reforms. The focus is on those European countries that are most affected by population ageing (France, Germany and Italy); the effects in other European countries and the United States are somewhat smaller.[1]

Population ageing is just about to hit the industrialized countries in full force. The younger generations in these countries are becoming aware of the need to provide for an increasing share of their retirement income through private savings. In many countries, these effects of demographic change are accentuated by fundamental pension reforms that aim at more pre-funding. Population ageing changes households' saving behavior not only directly, but also indirectly via pension reforms. To a varying degree, additional assets will be invested in the capital market. As a consequence, demographic change will have an important effect on capital markets and interest rates are set to decline once strong cohorts start to dissave upon entering retirement. A small glimpse of hope is, however, that there are differences in the age structure across countries. These differences result in differences in the domestic rates of return, which will be equalized by capital flows to the extent that capital is internationally mobile. These capital flows should have a dampening influence on the macroeconomics effects of demographic change in fast-ageing countries. While these mechanisms are intuitive, the quantitative effects are unclear and can only be predicted with a fully specified macroeconomic simulation model.

The simulations presented in this paper are taken from Börsch-Supan et al. (2006). The main insights from their multi-country overlapping generations (OLG)

[1] We do not attempt a review of the existing literature on the macroeconomic effects of population ageing here; the interested reader is referred to the above-cited paper by Börsch-Supan, Ludwig, and Winter (2006).

J. Winter (✉)
University of Munich, Munich, Germany

P.J.J. Welfens, J.T. Addison (eds.), *Innovation, Employment and Growth Policy Issues in the EU and the US*, DOI 10.1007/978-3-642-00631-9_4,

model are as follows. Demographic change alters the time path of aggregate savings within each country. This process may be amplified when a (fundamental) pension reform shifts old-age provision from pure pay-as-you-go towards more pre-funding. Finally, while the patterns of population ageing are similar in most countries, timing and initial conditions differ substantially. Hence, to the extent that capital is internationally mobile, population ageing will induce capital flows between countries. All three effects influence the rate of return to capital and interact with the demand for capital in production and with labor supply.

The simulations by Börsch-Supan et al. (2006) predict substantial capital flows due to population ageing. The countries most affected by ageing such as France, Germany and Italy will initially export capital, while countries less affected by ageing like the United States und other OECD regions will import capital. This pattern will reverse around the year 2020 when baby boomers decumulate their assets and the fast ageing economies therefore become capital import regions. Pension reforms with higher degrees of pre-funding are likely to induce more capital exports. They also increase labor supply considerably, but the effects on the rate of return to capital are small. Overall, while the rate of return to capital declines in response to population ageing, simulations suggest there will not be a devastating "asset market meltdown". Another important insight is that the timing of these adjustments is complex.

These results have several implications for the current public policy debate. In countries such as Germany, the objectives of fundamental pension reforms are twofold. First, these reforms seek to strengthen private old-age provision and second, they aim at increasing the average retirement age. The simulations presented by Börsch-Supan et al. (2006) suggest that there are no major macroeconomic obstacles for such reforms. Moreover, these simulations show that international capital mobility is important since it allows to diversify away some of the demographic burden. Finally, fundamental pension reform should be augmented by measures that increase returns to capital (see also Börsch-Supan, Köke, and Winter, 2005).

The remainder of this paper is structured as follows. In Sect. 4.2, we review some well-known facts about population ageing very briefly. Section 4.3 describes the overlapping generations model and Sect. 4.4 contains the simulation results. Section 4.5 concludes.

4.2 Some Facts About Population Ageing

Throughout the world, demographic processes are characterized by falling mortality rates and a decline in birth rates, resulting in population ageing and reducing the population growth rate (in some countries, even turning it negative). These two mechanisms, increasing longevity and falling fertility, have many reasons, but essentially they have to be accepted by economists as external factors which cannot be influenced by public policy – at least not in the short run (in the medium and long run, it may be possible to induce a reversal in the fertility rate). While demographic change occurs in almost all countries, the extent and timing thereof differ substantially. Europe and some Asian countries have almost passed the closing

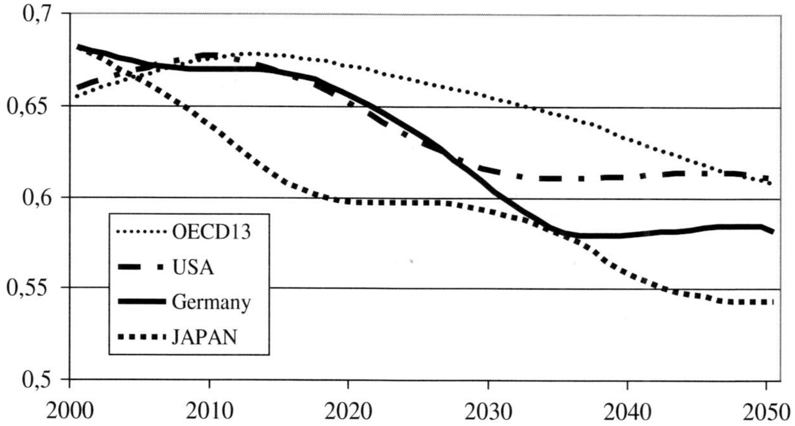

Fig. 4.1 Working-age population ratio
Source: Demographic projections based on United Nations (2002)

stages of the demographic transition while Latin America and Africa are only at the beginning.

Figure 4.1 shows projections of the ratio of the working-age population (age 15 through 65) to the total population in various countries. It is obvious from this figure that population ageing affects all countries. However, the timing and extent of demographic change differ substantially between countries. Ageing in Japan and Germany is particularly fast. Non-OECD countries age much more slowly. Ageing in the two other large continental European countries with large public pensions systems, France and Italy, is similar to Germany. In our subsequent analysis, we therefore treat France, Germany and Italy as a single "country" with fast population ageing and a generous public pension system.

From a macroeconomic point of view, population ageing will change the balance between capital and labor, particularly in the industrialized countries. Labor will be scarce whereas capital will be relatively abundant. This will drive up wages relative to the rate of return on capital, reducing households' incentive to save (if the interest elasticity of saving is positive). In addition, decreasing labor supply reduces demand for investment goods since less capital is used in the production sector. When we allow for capital mobility, differences in the speed and extent of population ageing that result in differential national rates of return to capital will result in international capital flows which equalize return differentials (to the extent that capital is internationally mobile). To quantify these effects, a macroeconomic simulation model is needed.

4.3 The Overlapping Generations Model

In order to analyze the effects of demographic change and pension reform, Börsch-Supan et al. (2006) construct a dynamic macroeconomic simulation model. Their model is based on a version of the overlapping generations (OLG) model

(Samuelson, 1958; Diamond, 1965) introduced by Auerbach and Kotlikoff (1987, Chap. 3). OLG models are well suited to study the effects of population ageing on social security systems, because they are based on households' and firms' optimal reactions to variations in the demographic structure and to public policy measures.

Traditionally, OLG models have been constructed for closed economics. The first open-economy version of the OLG model that allows for an analysis of the effects of demographic change on international capital flows is attributed to Börsch-Supan, et al. (2002) and was elaborated and refined by Börsch-Supan et al. (2006). A full description of this model is beyond the scope of the present paper. We summarize only the main features of the model here and present results in the next section.

The first building block is the demographic background. The demographic structure is taken as exogenous in the simulation, since it cannot be influenced by public policy in the short run. The demographic projections are based on United Nations data (2002). Crucial parameters such as the fertility rate and life expectancy as well as their entire time paths are set to conventional values. Migration rates also affect the demographic structure; while they can be influenced by public policy, we fix them at their current levels since the effects of migration and national immigration policies are not the focus of this paper. The forecasting horizon extends until 2200, well beyond UN projections, for reasons related to the simulation horizon of the OLG model which will become clear below.

The second building block of the model is the public pension system which is initially taken as fully pay-as-you-go in the countries of interest (France, Germany and Italy). Under the assumption of a balanced budget of the public pension system, country and year-specific contribution and replacement rates can be projected given the demographic data and labor supply (which is endogenous, see below). The simulations are based on two extreme cases: Either the 2006 benefit level (i.e., the replacement rate) is maintained by increasing contribution rates (the "no reform" scenario), or 2006 contribution rates are fixed which results in (dramatically) declining benefit levels (the "fundamental reform" scenario). Neither of these scenarios is likely to happen in its pure form in any of the countries of interest, but these scenarios can nevertheless serve as useful benchmarks. In particular, they allow us to separate the pure effect of population ageing and the additional effect of a fundamental reform.

The third building block is the OLG model itself. The macroeconomy consists of a production sector with a Cobb-Douglas production function, convex capital adjustment costs and age-specific labor productivity and of a household sector in which households maximize intertemporal utility with CRRA within-period utility functions and exponentially discounting. The choice variables are consumption (which of course also determines saving) and labor supply. There is uncertainty about the length of life and accidental bequests are absorbed by an implicit annuity market.

Any private saving that occurs in this model reflects the fact that retirement income is below labor income while working, in combination with households' desire to smooth utility and thus consumption over time. Specifically, private retirement saving arises endogenously and when future pension benefits are reduced by a fundamental pension reform, households react optimally by decreasing consumption and saving more. Therefore, we only need to fix contribution rates in our "fundamental reform" scenario – under the assumption of a balanced budget, resulting thereby in lower future pension benefits to which households react by saving more. Thus, a funded pillar of the pension system is introduced automatically.

The final building block of the simulation model is the international capital market. An important issue is the degree of capital mobility. We know, of course, that capital is not fully mobile for many reasons (for instance, private investors do not fully diversify their portfolios due to a "home bias" as well as capital restrictions, political risk, effects of exchange rate uncertainty and so on). It is virtually impossible to address these issues structurally in a macroeconomic simulation model. We therefore take a different approach. We assume that capital is fully mobile and that rates of return are fully equalized, within a certain group of countries. In relation to the rest of the world, this group of countries is, however, a closed economy. By increasing the number of countries in this group, we increase capital mobility and allow capital to flow to countries where population ageing is less severe and returns on capital are higher. In the results presented in the next section, we begin by considering France, Germany and Italy as a single closed economy. In the subsequent steps, we allow capital to be mobile within the EU, the OECD and finally the whole world.

Once these four building blocks are in place and their equations are fully specified (see Börsch-Supan et al., 2006, for details), the model can be simulated. The values of all technology and preference parameters are set to conventional values (and shown in a sensitivity analysis so as not to affect the results in a substantive way). The dynamic solution of the model is obtained by numerical methods (specifically, by a suitably modified version of the Gauss-Seidel algorithm). We calibrate and test the model using historic data for 1960 through 2001, the simulation period is 2002 through 2100 and we allow for a phase-out period which extends through 2300 to ensure that the simulations end in a steady state. This also explains why we had to extend the demographic projections beyond the limits of the UN's official forecasts.

Below, we present our simulation results for the period 2000 through 2070. As discussed before, we distinguish between two scenarios – one with no pension reform and one with a fundamental reform that freezes the contribution rate at their 2006 level. The figures presented below are for the "fundamental reform" scenario, since it generates larger capital market effects. The effects in the "no reform" scenario are generally somewhat smaller.

4.4 Results

We begin by considering the saving rate. As Fig. 4.2 shows, the combination of population ageing with the implied strongly declining ratio of working age population to total population and fundamental pension reforms results in declining saving rates until about 2050. The different lines shown in this figure refer to the capital

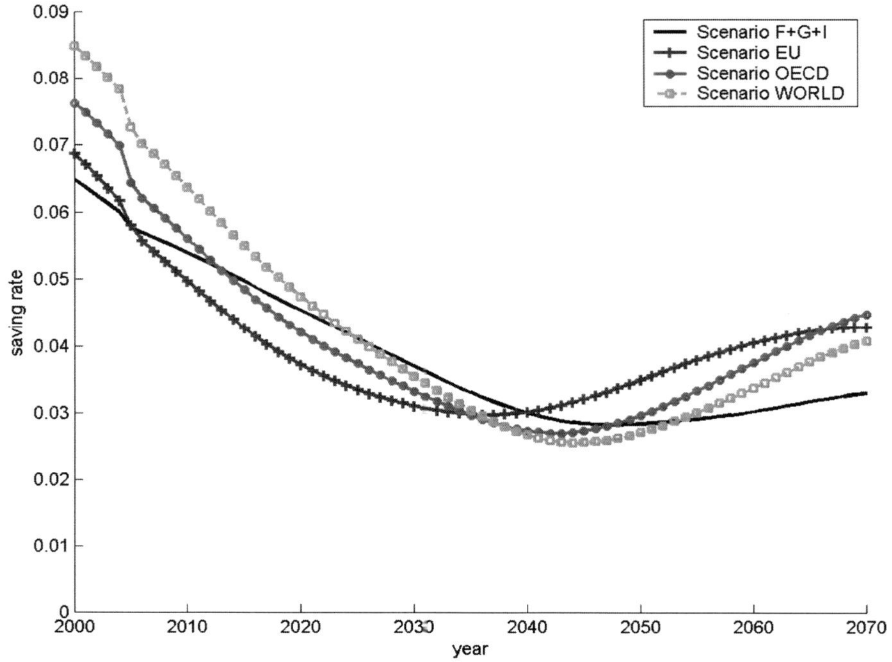

Fig. 4.2 Saving rates in the "fundamental reform" scenario
Source: Simulations based on the model by Börsch-Supan et al. (2006).

mobility scenarios discussed above. These capital mobility scenarios affect the level of the saving rate due to technical details of our historic calibration approach, but the shape of time path is similar in all scenarios: Saving rates decline until the strong baby-boom cohorts have retired and then pick up again, but they do not reach their current high levels (which are mostly due to the currently already quite high private retirement savings).

This trend in the saving rates translates into an increasing capital-to-output ratio (not shown), but the differences in the level and shape for the various capital mobility scenarios are somewhat larger. This is not surprising given that under increased capital mobility, a larger fraction of domestic savings is initially invested abroad, which in turn reduces initial capital formation.

In the third step of the analysis, we can now look at the dynamics of the rate of return to capital (Fig. 4.3). Again, we present our different capital mobility scenarios in the same graph. As with the savings rate, our calibration and simulation procedure implies differences in levels between the scenarios, but this is not the main object of interest anyway. The important finding is that the rate of return to capital declines in all scenarios by about one percentage point, but its dynamics are less pronounced when capital is fully mobile. Intuitively, this makes sense because more capital mobility (i.e., adding countries with more favorable demographic structures) reduces the pressure on the domestic capital market. At the same time, since all

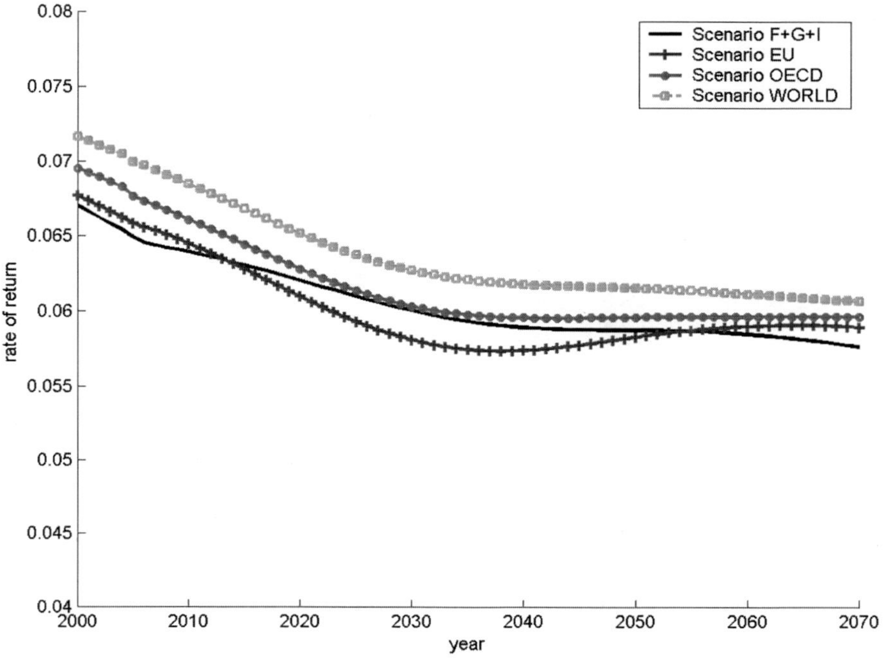

Fig. 4.3 Rates of return in the "fundamental reform" scenario
Source: Simulations based on the model by Börsch-Supan et al. (2006).

countries experience population ageing, international capital mobility cannot undo the negative trend in rates of return. As an aside, these results suggest that the assumption of a constant world interest rate that is sometimes made in other macroeconomic models of pension reform is certainly wrong.

In a final step, we also analyze the direction and size of capital flows implied by our model. Not surprisingly, the model predicts a U-shaped time path of current accounts for the group of fast-ageing countries (France, Germany and Italy) – depending on the capital mobility scenario, these countries will remain capital exporters for the next one or two decades and then start to import capital as the strong cohorts enter retirement and consume their savings.

4.5 Concluding Remarks

The most important result of the research reviewed in this paper is that while demographic change will lead to decreasing saving rates and also to declining interest rates, the effects on interest rates are smaller than many researchers as well as members of the public and media fear. In the projections of the macroeconomic simulation model presented here, the rate of return to capital declines by about one percentage point. The overall effect on the interest rate should be of a similar magnitude. This is not what we would consider an "asset market meltdown."

Another important result is that a fundamental pension reform involving partial transition to a funded pension system has an additional negative effect on capital returns (because it implies initially some extra saving and later corresponding amounts of dissaving), but this additional effect is smaller than the direct effect of demographic change. Finally, international capital mobility can – at least partially – mitigate these effects by allowing such fast-ageing countries as France, Germany and Italy to invest some of their retirement savings in slower-ageing countries that offer higher returns.

From a methodological perspective, the research program summarized in the paper by Börsch-Supan et al. (2006) offers some important insights as well. The size of the effects of demographic effects depends non-trivially on many features of a simulation model – this is of course neither new nor surprising. Traditionally, researchers have focused their "sensitivity analysis" of OLG and other macroeconomic simulation models on the values of technology and preference parameters that enter production and utility functions. However, this is only part of the story. It is also important and arguably more important, to investigate the sensitivity of simulation results to much more fundamental modelling decisions related to such model features as the existence of capital stock adjustment costs, the existence of perfect annuity markets, the endogeneity of labor supply and so on. Such a "fundamental" sensitivity analysis has been conducted by Börsch-Supan et al. (2006), but this is not yet standard in published scientific papers. This sensitivity analysis shows that endogenous labor supply has the strongest effects on the simulation results (since it gives households another degree of freedom in the reaction to demographic change and pension reforms).

Future research on the macroeconomic effects of demographic change is likely to focus on other extensions of the basic OLG model presented here, most importantly, the role of risky assets (the OLG model presented in this paper had only one form of capital and no return uncertainty). Other important issues for future research on the capital market consequences of demography change are capital market frictions and home bias, exchange rate effects and the effects of international migration, all of which have not been addressed in the present analysis.

From a substantive viewpoint, this paper suggests that the recent pension reform proposals introduced by the "Rürup commission" for Germany (which by now have for the most part been implemented) were well-founded. The main elements of this and similar reforms in other countries – increasing the retirement age and strengthening private pension provision – do not have major adverse macroeconomic effects. So the good news is that at least for the time being, the German public pension system has been stabilized (which is not to say that there will never be the need for an additional reform, of course).

Looking ahead, while there are still many open issues related to pension reforms, the next big challenge is already on the horizon – the health insurance system. Awareness for the problem of financing future health costs, in particular of prescription drug and nursing home expenses, is much stronger in the United States, but just as with ageing and pension reform, these issues will have to be addressed by economists, politicians and the public in Europe as well, and the sooner the better. The interactions between demographic change, labor markets, and health care costs as well as the future of health care financing are quite daunting.

Acknowledgments The author would like to thank the participants of the EIIW workshop "EU-USA: Structural Change, Employment, and Growth Issues", Wuppertal, July 2005, for helpful comments. Financial support for the research project on demographic change on which this paper is based was provided by the VW Foundation, the Deutsche Forschungsgemeinschaft, the Land of Baden Württemberg, the Gesamtverband der deutschen Versicherungswirtschaft, and the US Social Security Administration via grants to the Mannheim Research Institute for the Economics of Ageing (MEA).

References

Auerbach, A., & Kotlikoff, L. (1987). *Dynamic fiscal policy*. Cambridge, MA: Cambridge University Press.

Börsch-Supan, A., Köke, J., & Winter, J. (2005). Pension reform, savings behavior, and capital market performance. *Journal of Pension Economics and Finance, 4*, 87–107.

Börsch-Supan, A., Ludwig, A., & Winter, J. (2002). Ageing, pension reform, and capital flows. In A. Auerbach & H. Hermann (Eds.), *Ageing, financial markets and monetary policy* (pp. 55–83). Heidelberg: Springer.

Börsch-Supan, A., Ludwig, A., & Winter, J. (2006). Ageing, pension reform, and capital flows: A multi-country simulation model. *Economica, 73*, 625–658.

Diamond, P. (1965), National debt in a neoclassical growth model. *American Economic Review, 55*, 1126–1150.

Samuelson, P. A. (1958). An exact consumption-loan model of interest with or without social contrivance of money. Journal of Political Economy, *66*, 467–482.

United Nations (2002). *World population prospects*. New York: United Nations.

Chapter 5
Ageing and Global Capital Flows

Volker Clausen

The world economy, and in particular the major industrial economies, will experience a considerable ageing of their populations over the next decades. It is generally accepted that this demographic transition will have important effects on macroeconomic developments and capital markets. This panel contribution concentrates on the role of international capital flows in an ageing world economy. It addresses the question to which extent international capital flows may soften the demographic burden on national economies and facilitate the adjustment to the global demographic transition. Particular attention is paid to the case of Germany as a country which ages relative to the world economy.

It is well known from the system of national accounts that (net) international capital flows have to equal the difference between aggregate savings and investment in the national economies. The following two sections therefore briefly discuss the impact of demographic developments on both variables and the respective implications for international capital flows. The final section discusses to which extent these developments can already be identified in the case of Germany.

5.1 Ageing and Savings

Aggregate savings consist of private and public sector savings. For private savings, the simple life-cycle theory of consumption predicts that savings rates typically rise over the first course of the working life, then start to fall and eventually turn negative during retirement. This broad direction and timing of effects is generally confirmed in microeconomic studies, which implies that population ageing will lead to

The first version of this panel contribution was written during a research visit at the American Institute of Contemporary German Studies (AICGS) in Washington, DC, in 2005. Financial support from the German Academic Exchange Service (DAAD) is gratefully acknowledged.

V. Clausen (✉)
Department of Economics, University of Duisburg-Essen, Germany

P.J.J. Welfens, J.T. Addison (eds.), *Innovation, Employment and Growth Policy Issues in the EU and the US*, DOI 10.1007/978-3-642-00631-9_5,

an overall fall in the private savings ratio. While the qualitative pattern is more or less undisputed, the size of the macroeconomic effects is somewhat hard to assess. First, empirical analyses often find that the amount of dissaving during retirement is generally much smaller than predicted by the life-cycle model, which dampens the negative impact of population ageing on private savings (Deutsche Bundesbank, 2004). Second, and more importantly, it may be misleading to project the results from microeconomic studies to the macroeconomic level, given that income and wealth distributions as well as savings are considerably skewed in the economy and may not be adequately represented in microeconomic studies (Bosworth, Bryant, & Burtless, 2004). Third, the impact of ageing on private savings will critically depend on the reforms of the social security systems and the design of tax incentives.

The amount of public savings is determined by the financing decisions at the various levels of government and the financial balance of the social security system. The qualitative impact of population ageing on public savings is also largely undisputed. Spending on pensions and social security will increase relative to tax revenues and social security contributions such that public sector budget deficits are likely to increase. In sum, population ageing is likely to result in a fall in aggregate savings. This relationship is confirmed by macroeconometric cross-country evidence (see for example IMF, 2004).

5.2 Ageing and Investment

Standard neoclassical theory predicts that a contraction of the labor force will lead to wage increases and a lower return on capital. The change in relative factor prices will trigger some capital deepening. Nevertheless, due to the lower return on capital, population ageing will be accompanied by a slump in investment. A simple projection based on the evolution of the labor force clearly overstates the effects. There are many ways to mobilize the labor force – such as an increase in female labor force participation, a lengthening of working hours, an increase in the retirement age, an improvement of the functioning of the labor market, extended immigration and so on (see Deutsche Bundesbank, 2004). Empirical evidence on the role of demographic variables on aggregate investment is somewhat scarce but tends to confirm that population ageing lowers investment (see again IMF, 2004).[1]

Taken together, the overall impact of ageing on international capital flows is ambiguous. Both, aggregate savings and investment are likely to fall. As the difference between aggregate savings and investment determines the direction of *net* international capital flows, it depends on the relative impact of ageing on aggregate savings and investment. However, econometric cross-country evidence tends to

---

[1] In the context of investment, the impact of ageing on the productivity the labor force plays an important role as it affects the return on capital over time and investment activity. However, empirical evidence on the implications of population ageing for labor productivity does not yet allow for a reliable quantification of the effects.

support the conclusion that – holding other factors constant – ageing economies tend to be characterized by net capital outflows (IMF, 2004).

5.3 Ageing and Capital Flows: The Case of Germany

Germany will clearly age relative to the rest of the world over the coming decades. As savers and investors already anticipate this demographic trend *current* savings ought to be relatively high compared with future savings and investment to be comparatively low under otherwise unchanged circumstances. This means that Germany needs a current account surplus or net capital exports in the anticipation of ageing and will therefore accumulate net foreign assets over time. The higher stock of net foreign assets in the future will then allow Germany to finance a part of the future demographic burden via the returns on the international investment position.

As a matter of fact, Germany has seen in recent years an increasing savings ratio and a fairly weak domestic investment activity as measured by the corresponding ratios from the national accounts (see Fig. 5.1). Since the turn of the millennium private households seem to have somewhat lost their faith in the support by the public pay-as-you-go pension system and raised their private savings ratio. This tendency was reinforced by pension reforms and strong tax incentives for savings targeted at supplementing the public pensions. In recent years more than half of these savings were channelled not into domestic investment but into international financial markets. These net capital exports mirror the current account surplus.

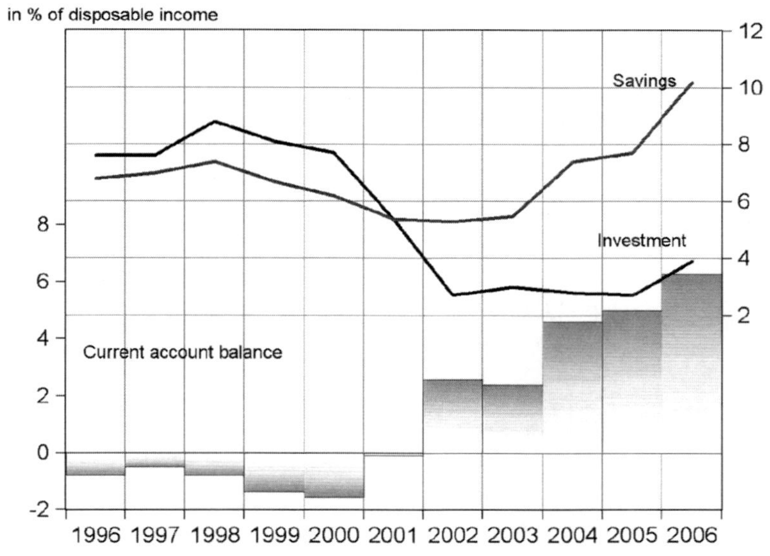

Fig. 5.1 Germany – savings, investment and net capital flows

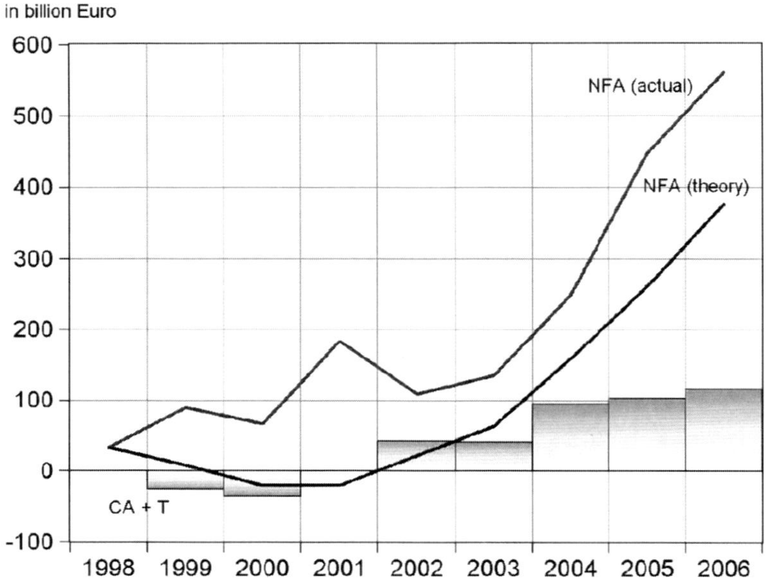

Fig. 5.2 Germany – international investment position

In theory, this international adjustment mechanism buffers the impact of ageing on the German economy. The sequence of current account surpluses in recent years has already led to a considerable improvement in the net international investment position of Germany (Fig. 5.2). The current account switched into surplus at the start of this decade. Taking the stock of net foreign assets (NFA) at the beginning of EMU (end 1998) as the point of departure, the sequence of current account surpluses would have led in theory to an NFA position amounting to about 380 bn Euro (labelled NFA theory) which would be about 16% in relation to German GDP in 2006.

As a consequence of a substantial increase in the gross cross-border asset positions in the last decade, valuation effects gained considerable importance in the external adjustment process. Starting again at the end of 1998 with the same level, Fig. 5.2 compares the evolution of the NFA position based only on the current account (including one-time transfers, CA + T) and the evolution including current account and valuation effects mainly arising from financial market and exchange rate developments. The actual value of NFA went up to 560 bn Euro at the end of 2006, which is about 25% of German GDP.

The ultimate benefits of the net asset accumulation towards the rest of the world generally depend not only on savings and investment decisions but also on the composition of international balance sheets. International differences in the rate of return across different types of assets and liabilities as well as differences in the composition of international balance sheets affect the degree to which the international investment position generates a future net income stream. Taking the example of Germany, the net foreign asset position at the end of 2005 of almost 450 bn Euro generated a net factor income stream in 2006 of about 24.3 bn, which to a first approximation implies a rate of return of slightly more than 5%.

Lane and Milesi-Ferreti (2005) present for a number of countries a systematic decomposition of the change of the NFA position into the contributions of the cumulative current account and of capital gains over the period 1995–2000. Our case, Germany, started at the end of 1994 with an initial NFA position divided by GDP of 9.5% and experienced over this relatively short period of six years a decline of this share by 8.1 percentage points. The main contributors to this deterioration were the cumulative current account deficit including cumulative investment income (−5.1) and an important capital gains component (−3.1). In a more detailed disaggregation, they also find that the average real return on international assets is *lower* than the average real return on liabilities (6.8%). Furthermore, changes in the real effective exchange rate have a considerable impact on the evolution of the NFA position. Here, as in the case of the US, a depreciation typically leads to an improvement in NFA as most liabilities are denominated in domestic currency, while most international assets are denominated in foreign currency. This effect can also be identified in Fig. 5.2, where the strong depreciation of the euro in the first three years of EMU led to a significant improvement in the German NFA position despite a deficit on the current account.

Overall, to a first approximation, the qualitative pattern implied by the standard theory of savings and investment seems to work in the case of Germany.[2] However, the quantitative importance judged on the basis of current income flows is quite small. The role of net international factor payments (including cross border labor income) as measured by the percentage difference between gross domestic product and gross national income is currently about 0.5%. So even if most net international factor payments come from net foreign assets, it does not represent a large component of national income. Furthermore, the important lesson from the recent experience in Germany is that return and valuation effects are very important relative to the standard effects arising from differences between aggregate savings and investment. The direction and the strength of the valuation effects depend on the overall return differentials between cross-border assets and liabilities as well the composition of the international balance sheets. International capital markets will therefore play an important part in the global adjustment to the demographic transition in the 21st century. The global allocation in assets and liabilities will increasingly matter and it will ultimately pay to be a "smart" investor. Time may be better spent to decide on the appropriate asset allocation of savings rather than to work, to earn and to save!

---

[2] For a more detailed assessment see Deutsche Bundesbank (2004). There are some other country-specific examples where intertemporal smoothing in response to demographic developments appears to happen. For example, starting at the beginning of the 1980s Japan continues to run considerable current account surpluses in the anticipation of an ageing population.

References

Bosworth, B. P., Bryant, R. C., & Burtless, G. (2004). *The impact of ageing on financial markets and the economy: A survey*. Washington, DC: Brookings Institution, Mimeo.

Deutsche Bundesbank. (2004, December). Demographic burdens on growth and wealth in Germany. *Monthly Report*, 15–29.

International Monetary Fund (IMF). (2004, September). *World economic outlook: The global demographic transition* (Chapter 3). Washington, DC: IMF.

Lane, P. R., & Milesi-Ferreti, G. M. (2005). *A global perspective on external positions* (NBER Working Paper No. 11589). Washington, DC: National Bureau of Economic Research.

Chapter 6
Effects of the Ageing Population on Capital Markets

Freddy Van den Spiegel

6.1 Introduction

The ageing of the population in the developed world, and especially in Europe, will have far-reaching consequences for the economic, financial, social and political environment over the coming decades. Capital markets in general are considered an important instrument in coping with this challenging situation. While they cannot solve all the issues linked to ageing, well functioning capital markets could at least provide a channel for the transmission of purchasing power between generations. Capital accumulation over one's working years would be followed by disinvestments during retirement, which would facilitate the financing of ageing. This mechanism could be used at the individual level and at the macro level.

However, massive intergenerational capital flows, linked to the rapid ageing of a substantial proportion of the world economy, could also change the behaviour of capital markets themselves, leading to some unpleasant surprises. It would therefore be interesting to look at the trends which could fundamentally influence the functioning of markets over the coming decades.

6.2 Supply of Savings

Saving and investing are nothing more than means of exchanging actual purchasing power for a promise of future purchasing power. Investors are constantly looking for vehicles capable of carrying their purchasing power over time. These vehicles may be real or financial assets. At any point in time, there is an exchange rate between current purchasing power and future (expected) purchasing power – the interest rate. Because of the existence of money and its potential for, inflation this purchasing power exchange rate is often blurred. Economic theory tells us that saving behaviour depends on the level of the interest rate, which reflects the prospects for economic growth. However, in the event of massively ageing populations, the level of savings

F. Van den Spiegel (✉)
Fortis Bank, Brussels, Belgium

P.J.J. Welfens, J.T. Addison (eds.), *Innovation, Employment and Growth Policy Issues in the EU and the US*, DOI 10.1007/978-3-642-00631-9_6,

may not depend on interest rates or potential productivity gains, but merely on the need to accumulate future purchasing power. This is a rather frightening scenario: during the capital accumulation phase, the prices of all investment vehicles (real estate, stocks, bonds, etc.) start to rise. At the same time, risk premiums fall, not because the risks have disappeared, but because of increased demand and TINA (There is no alternative). From time to time, these excesses become all too apparent and lead to a temporary crash on certain markets. However, these crashes do not change fundamental behaviour. Only when ageing becomes a fact is a massive preference displayed to exchange future purchasing power for present purchasing power. The massive selling of assets then leads to a fundamental downturn on all asset markets. Increases in interest rates will not stop this process, once again because the old investors have to sell their assets at any price in order to pay for their retirement. This boom/bust scenario is quite devastating as it means that those who are saving today pay high prices for assets, which they will have to sell at low prices in the future. Saving enough to finance ageing becomes a "mission impossible" in this case. Economically however, this process can be explained. Partly, it comes down to the "saving paradox", which states that if we all try to save more, in reality we will collectively save less. Partly, the negative interest rate in this scenario reflects the changing preferences of economic operators.

Are there any signs that this scenario is occurring at the moment? Unfortunately, some indicators seem to point to excessive savings today. The crash of the stock markets in 2000–2002 can be explained as a symptom of this process. "TINA" can explain, among others, the continuing rise in real estate prices worldwide and the growth of leveraged investments such as hedge funds.

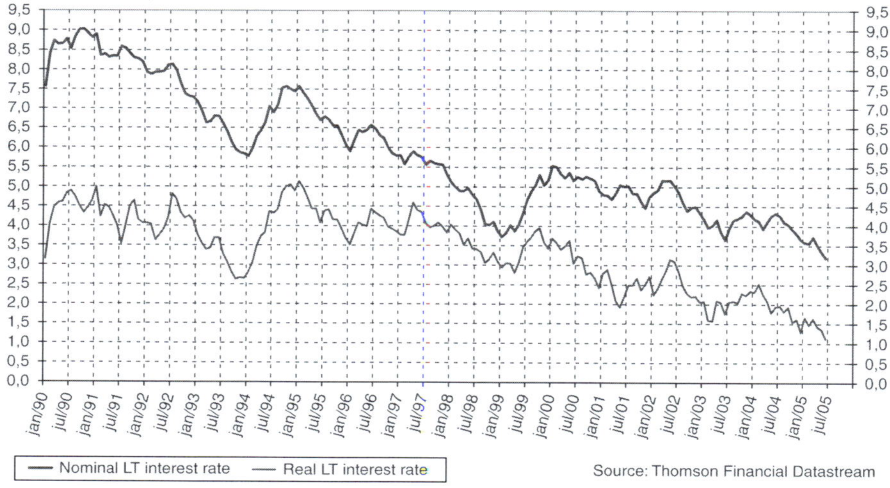

Fig. 6.1 Nominal and real long-term interest rates in EU (1990–2005)

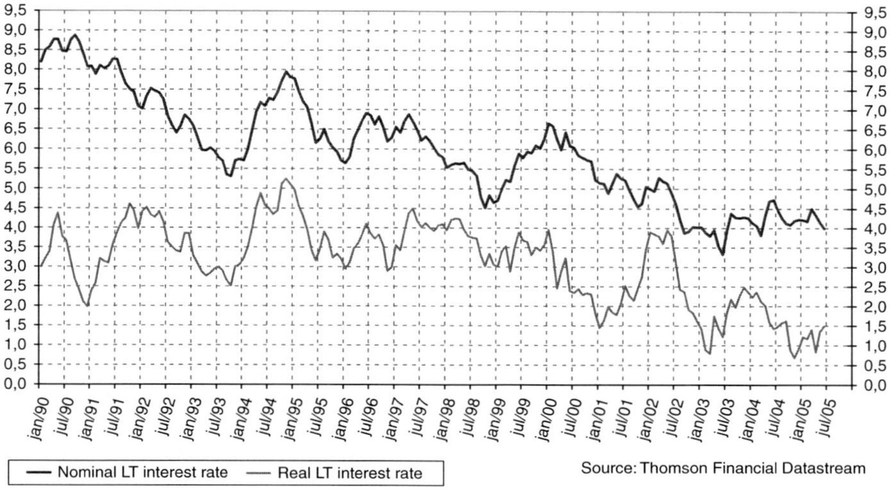

Fig. 6.2 Nominal and real long-term interest rates in US (1990–2005)

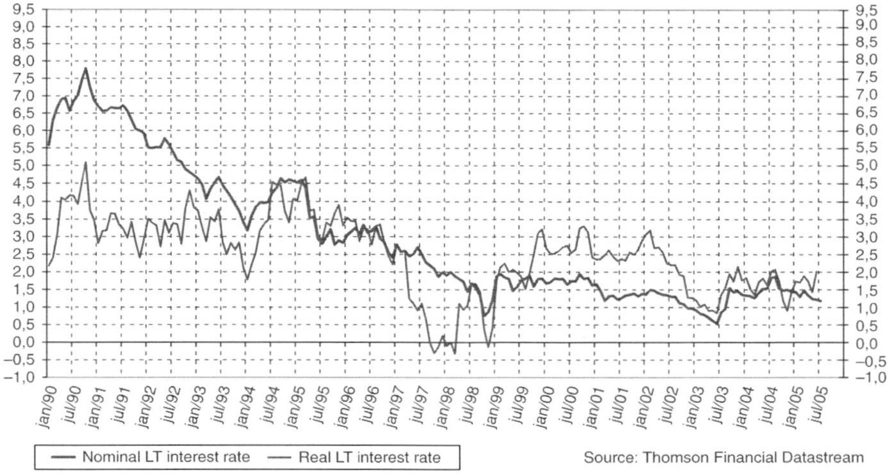

Fig. 6.3 Nominal and real long-term interest rates in Japan (1990–2005)

More worrying is the fact that real long-term interest rates have declined all over the developed world, from about 5% in the early 1990s to a mere 1% today. This decline has been rather gradual, notwithstanding periods of high and low economic growth (Figs. 6.1, 6.2, and 6.3).

Moreover, figures relating to the level of worldwide savings, produced by IMF, indicate that savings increased from 23% of worldwide product in 1999 to about 25% in 2005. Surprisingly, however, this sharp rise does not come mainly from the "old" developed world, but from the high levels of macro-economic savings in the newly industrialised countries such as China. This fact can also be considered encouraging, as it points to the reality that the globalised world is increasingly influenced by these new economic powers. Supposing a fully globalised economy in the decades to come, the effects on capital markets of the ageing of Europe will not necessarily have a boom-bust effect, as they will be offset by opposite behavioural patterns in other parts of the world, where the timing of the ageing phenomenon is different. In other words, the assets which Europeans should now be accumulating will be sold to foreign investors somewhere in the world, in economies which are starting to accumulate at that time.

6.3 Availability of Investment Vehicles

At first sight, investment opportunities seem to decline with the ageing of populations. Ageing is expected to have a negative influence on the level of productivity and on the potential for growth in productivity, which reduces competitiveness and the potential return on investments. Ageing populations will lead to more capital intensive investments, which could also undermine potential returns. In general, ageing reduces dynamism, innovation and demand, all of which are essential ingredients for the creation of attractive investment opportunities.

However, ageing could have a number of positive consequences which are not recognised today. Gradually, the need to work longer will become apparent and politically acceptable in ageing societies. Working at least until the age of 65 and perhaps longer will undoubtedly necessitate the reinvention of the concept of "working". More flexibility will be needed on the part of both employers and employees. This flexibility in our economies, dominated by services, will have to be based on optimal and innovative use of IT. This could well provoke the development of a new economy for ageing populations, with many attractive investment opportunities.

Ageing economies will have to develop a policy of "targeted" immigration. This development will also bring its own investment opportunities and will compensate for some of the deficiencies of ageing economies.

Ageing populations will also see an increasing need for new types of services, linked to the organisation of daily life and health care. Being the first economies to adapt to this situation in practice could, if we are very optimistic, become a competitive advantage in comparison with other economies where the ageing phenomenon occurs at a later date. As a matter of fact, if world population growth is to be contained, sooner or later the whole world will experience the process of ageing, as human life is constantly expanding.

Finally, in the search for investment opportunities, as with regard to the levels of savings discussed earlier, the globalization of the world economy makes investment opportunities less dependent on the local economic situation. If the globalization process can develop smoothly and peacefully, based on the concept of free capital markets, the needs of ageing and newly industrialised countries could well be complementary.

6.4 Financial Equilibrium of Individual Countries

The development of financial markets during the ageing process will depend on the equilibrium worldwide, both in countries (or economically integrated areas) and at the level of the individual. All these balances are interconnected. A major imbalance at the individual level will lead to political chaos at the country level, which could in turn influence the worldwide equilibrium.

Looking at the equilibrium at the country level, it is clear that ageing populations will have an impact on government budgets. Any significantly increasing imbalance at that level for the EU will undermine confidence in the monetary system, which will lead to changing monetary policies and potentially to monetary chaos, with all its consequences for the financial markets. To avoid this scenario, it is essential that a politically acceptable and affordable balance be found between the responsibilities of the government and the responsibilities of the individuals in financing retirement. In theory, the EU, and more specifically the EMU, should be well protected against this risk of imbalance, given the Growth and Stability Pact. Experience has shown, however, that governments are just beginning to become politically aware of the "ageing time bomb" under their budgets. The Growth and Stability pact is essentially concerned with short-term budgetary discipline and far less with the need to achieve long-term equilibrium. The discipline needed to comply with the GSP is limited, and major member states have succeeded in changing the GSP in order to hide their lack of courage in implementing it. This is not very encouraging, but at least some debate at the national political level has started in most countries.

Besides the actual effect of increasing retirement costs, a far more difficult issue has to be tackled at country level: the increasing and even unlimited costs of health care. Current technological improvements in the health-care industry are gradually leading to a situation in which technological solutions are available to extend almost without limits individual lives. However, the price of this technology means that it is not available to all and is unaffordable for government budgets. Health care, which is essentially a "public good" in most developed economies, has an increasing impact on government budgets. Covering all health problems at all ages, using every technology available, is no longer an option. However, there is no easy alternative: privatisation of health care seems difficult in EU democracies, and refusing to finance certain available technologies using public funds is an impossible message as well. What is at stake here is not just a problem of budget equilibrium, but a series of fundamentally ethical problems, including euthanasia and many others. This extremely complicated political debate has not yet seriously begun.

Looking at equilibrium at the individual level, there are indications that the level of private savings in the EU is high enough to finance ageing, at least the retirement side of it. However, these savings are not equally distributed. Political pressure regarding redistribution of available savings through taxation on capital will increase in the years to come, and it is unclear how this debate will evolve and what will be the outcome. Creating uncertainty could have disastrous effects on financial and political behaviour.

6.5 Allocation of Savings

The balance between individual and collective responsibilities in financing the ageing process is clearly evolving towards a less collective and a more individual sphere. This is not only the case in the debate about public versus private pensions. Even within the private pension sector, the shift towards more responsibility for the individual is clear: pension funds and life insurance are increasingly changing from Defined benefit systems to Defined contribution systems. While this trend may be discerned in many sociological processes, there is a more fundamental problem linked to financing retirement. Retirement financing is essentially about financing the longevity risk. The only way to efficiently manage the longevity risk is to manage it collectively, which is just the opposite of what is happening.

It has already been clearly demonstrated that individuals perform extremely poorly in managing investments based on an uncertain life cycle. Even when optimising simple Asset Allocation problems, individuals fully demonstrate the irrationality described by behavioural finance.

Even with rational investment behaviour, it is clear that an individual, facing longevity risks, cannot afford to take the same level of risk which a collective portfolio should be able to take. However, if we believe in more or less efficient and rational financial markets, taking lower risk levels will collectively reduce performance. The individualisation of retirement financing leads in this case to a collective loss of future purchasing power. It also leads to a collective reduction in the supply of risk-taking capital, which undermines the growth potential of the economy.

6.6 Some Conclusions

Trying to assess the effects of ageing populations on capital markets is an extremely ambitious exercise. Ageing is not occurring within an isolated world where all the other elements remain stable. Ageing is occurring within a globalising world, in which all relations are fundamentally changing and which is becoming increasingly interdependent. On top of that, ageing itself will change human behaviour and the organisation of society.

In this changing world, it is, for example, becoming almost impossible to decide whether collectively saving more or saving less will provide a solution. It could be argued that in terms of the globalization of the world economy, ageing in Europe is only a minor issue.

The final solution towards ageing will probably be a policy mix, which reduces the strategy risks. Working longer, saving more, and making the individual more responsible while collective responsibilities are reduced seem to be the ingredients for the future. Financial markets can play an important role in helping to solve the problem, but relying solely on building financial reserves and creating efficient financial markets could prove a dramatic mistake.

Building economic models in this complicated and totally new world is necessary and interesting if we are to gain an idea of what the future may look like. However, we have to be aware that any model is in practice simply a scenario analysis of what could happen, while many other possibilities are not taken into consideration. Building a model which creates a false feeling of certainty is probably the most dangerous way of looking at the challenge of ageing populations in Europe.

Chapter 7
US Economic Outlook: Beyond the Credit Crunch

Holger Schmieding

Rarely have opinions on the US economic outlook differed so much. Some observers interpret the sharp slowdown in US growth in late 2007 and early 2008 as the beginning of a long and nasty recession during which the US needs to purge itself of its excessive addiction to credit. Many of these pessimists had maintained for years that the comparatively strong US economic performance for the last 5, 10 or even 15 years was built on unsustainable foundations which have now – belatedly – started to crumble. However, others see the current calamity as a temporary and comparatively shallow interruption in a remarkable uptrend.

We agree that the pessimists have a point or two. As always in life, linear trends cannot last forever. Consumption cannot forever expand faster than overall GDP, external deficits cannot rise forever as the capacity to attract capital from abroad at favorable conditions will eventually be impaired by doubts about the ability to serve and repay a mounting burden of debt and by doubts about the stability of the currency. Thus, some rebound in the US national savings rate, some retrenchment in the expansion of domestic absorption relative to domestic production and, reflecting such changes, some correction in the US external deficit is inevitable. Also, risk appetite and the eagerness of financial institutions to use ever more complex instruments to offer ever more credit at ever more favorable rates has to hit limits eventually as the economy runs out of good borrowers. But by and large, we count ourselves among the optimists. In our view, the US can, in the context of a healthy global economy, get through these corrections with only limited upheaval.

7.1 International Background: Globalization

Before we lay out the optimistic case for the US, we first need to sketch the global environment. The opening up of post-communist Asia and Europe in the last 20 years has unleashed an unprecedented surge in global growth. With China and India

Source for all data: Haver Analytics

H. Schmieding (✉)
Bank of America, London, UK

P.J.J. Welfens, J.T. Addison (eds.), *Innovation, Employment and Growth Policy Issues in the EU and the US*, DOI 10.1007/978-3-642-00631-9_7,

making rapid progress, more people are rapidly moving out of abject poverty than ever before, escaping the risk of dying of hunger or of suffering debilitating malnutrition. The recent inclusion of almost half of mankind into the global economy has also benefited the traditionally advanced countries of Europe and North America more than even optimists had hoped some 10 years ago. Corporate profits are close to record highs (13.5% of GDP in the US in 2007 vs. a post-1960 average of 8.8%) despite a dent in late 2007 caused by the recent write-downs in the financial industry. Also, unemployment is either very low (5.0% in the US, 5.2% in the UK, 2.0% in Japan and between 2 and 3% in Switzerland and some other small European countries). Or where unemployment is still high, it has at least fallen rapidly so far this decade, hitting for instance a 15-year low of 7.2% in the Eurozone at the end of 2007. In this sense, the starting situation for the global and the US economy upon the onset of the serious financial market turmoil in August 2007 was excellent. This starting situation provides economies some resilience to shocks.

7.2 A Long US Upswing

Since the end of the 1991 recession, the US economy has outperformed most other major Western economies by a wide margin. From the end of 1991 to late 2006, the US expanded at an average pace of 3.0%, well ahead of the 2.2% growth in the Eurozone. Since the mid-1990s, the US benefited from a revival in productivity growth driven by the comparatively fast diffusion of new information and communication technologies through the flexible US economy. Better logistics supported by such new technologies also helped. US productivity growth held up astonishingly well even throughout the mild 2001 recession.

Fast growth went along with an unexpected decline in US core consumer price inflation (excluding energy and food) from an average of 5% in 1990/1991 to an average of 2% in the years since 2002, with some cyclical variations. Beyond the better productivity performance and a more credible monetary policy, the dampening impact of globalization on the domestic wage-setting power of workers and the price-setting power of companies served to contain inflation despite strong rates of trend growth.

Subdued core inflation, in turn, helped to reduce inflation expectations and the inflation risk premium in real interest rates. Two other factors also contributed to a long period of low interest rates, especially in the years since the 2001 recession.

1. The US Federal Reserve mistook the decline in underlying inflationary pressures caused largely by an increase in global and US supply as a potentially deflationary problem reflecting insufficient demand. Afraid of falling prices, the US Fed steered rates down to 1% in 2003 and raised them only haltingly in the following years. With an aggressive monetary and fiscal stimulus, US consumers never experienced a recession early this decade although the economy as a whole had contracted mildly in two out of four quarters in 2001.

2. Looking for a safe haven to invest their rapidly rising foreign exchange reserves, China and other current account surplus countries in Asia and the oil-rich Middle East invested rising sums into US Treasuries, pushing yields well below normal historic valuations.

The prolonged period of low interest rates, ranging from central bank rates to bond yields at the long end of the maturity spectrum, encouraged a greater appetite for risk and a spate of financial innovations which, in effect, made it easier to borrow and lend.

7.3 From Low Interest Rates to High House Prices

In many regions of the world, including the US, globalization led to a major rise in the relative price of houses. In general, the phenomenon makes sense. The relative price for scarce local factors of production typically rises in periods of fast growth. Land is relatively more valuable in rich than in poor countries. As globalization (and faster productivity growth in the US) raised the global rate of output growth, prices for land and real estate had to rise relative to the prices for tradable goods and services. Low interest rates, both in nominal and real terms, added momentum to the process, making it cheaper to take out a mortgage. In technical terms, the factor to discount the expected future benefits from very durable real estate relative to the present value of the benefits derived from less durable goods and services decreased with low real interest rates. In financial markets, low interest rates supported an increasing degree of leverage and securitization.

As usual in largely unfettered markets with their endogenous trial-and-error search for a new equilibrium, benign and fundamentally supported trends can carry on until they have obviously gone too far. In retrospect, it seems clear that the volume of US residential construction and the level of US house prices overshot its sustainable level at some time around 2005.

Since 2Q 2006, residential construction in the US has contracted at an average annualised rate of almost 20%, in mid-2007, house prices started to fall. In August 2007, fears of a major rise in mortgage defaults finally caused a near-seizure in money and credit markets. As most US mortgage loans are securitised, repackaged and sold on in various guises, the risk from mortgage defaults is now spread to the global financial community, in ways that are not fully transparent. More precisely, the complex nature of the products mean that the distribution of losses can differ substantially in response to comparatively small changes in the underlying valuation assumptions.

7.4 The US Sub-prime Crisis

For many countries, corrections in the real estate market are a normal feature of the business cycle. But the current US situation is different in a way which accentuates the short-term risks as well as the long-term recovery prospects.

In times of yore, mortgage lenders took deposits from customers and lent the money on to homebuyers. Nowadays, mortgage lenders and other financial institutions higher up the value chain often take bundles of mortgage loans, re-package them and sell them on as tradable securities to investors. If homeowners default on their mortgage, the resulting loss mostly hits those investors who have bought the mortgage-backed bonds, and the financial institutions which had agreed to back up such mortgage bonds if need be.

The securities are tradable. Expected future mortgage defaults change their market value. As a result, the mortgage-related losses are now more front-loaded, visible almost immediately in the prices for these securities. In the non-securitised past, such losses typically hit balance sheets only over a longer time.

Let us take a look a rough magnitudes. The major concern focus on the market for US subprime mortgages, that is on mortgages for borrowers with a weak income position, an impaired credit history or inadequate documentation. Financial innovation, low interest rates and the ability for lenders to redistribute risk via asset-backed securities helped people to get onto the property ladder who, before, would not have been able to get a mortgage. The volume of US subprime mortgages repackaged and sold on as mortgage-backed securities amounts to roughly $1200 billion. If one third of all borrowers default in the end, a historically extreme default rate even for weak borrowers, the direct losses would total $400 billion, plus some extra hit due to the costs of dealing with defaults.

However, even in the case of default, the loans are not worthless. They are backed by houses and land. At the moment, the real estate is difficulty to sell. But the US has a rapidly growing population. Most of the mortgage problems are in the sunshine states of the US such as California and Nevada which look set to attract further immigration. In other words, even with some sizeable correction in house prices and including the costs of foreclosure, lenders should be able to eventually recoup, say, 60% of the value of a loan on which the borrower has defaulted. As the houses may only be sold some 2–3 years into the future, we also have to discount the possible future proceeds. But even after that, the net present value of subprime mortgage loans on which borrowers are defaulting should be at least 50% of the face value. As a result, the total final losses from the US subprime crisis may not exceed $200 billion. Even adding other costs, namely rising defaults fro some mortgage categories which are not sub-prime and rising losses on credit cards, which some borrowers may be using to pay their mortgages for a few more months until they hit their credit limit, the total damage to the financial industry from the US mortgage crisis should not exceed $400 billion in our view.

7.5 A Major Credit Crunch?

According to our back-of-the-envelope guesstimates, the potential losses from the US crisis are thus huge, around 2.8% of US GDP. But until January 2008, major US financial banks and brokers had already announced write-downs of more than $120 billion. Many of the final losses are likely to reside in non-US institutions and with asset managers and pension funds in the US who may, for instance, be indirectly engaged in sub-prime mortgages via their credit portfolios and money market funds. In addition, the recapitalisation of the US banking system has proceeded swiftly so far, with either major new rights issues or capital infusions from sovereign wealth funds out of Asia or Arabia. We see a good chance that, at least for major US financial institutions, the period of ever bigger write-downs could soon be over. If so, the credit-granting business in the US need not have to tighten its credit conditions severely for much longer.

We also need to put the magnitudes of losses into context. First, US financial institutions had been highly profitable until the start of the US mortgage crisis. In 2006, US domestic financial institutions made profits of $482 billion. In 2Q 2007, that is just before the advent of the recent subprime write-downs, profits were running at a quarterly annualised rate of $521 billion. By and large, the US financial industry seems well placed to digest its part of the global fallout from the subprime crisis.

In addition, the US Fed has responded aggressively to the financial turmoil, cutting its funds rate from a mildly restrictive 5.25% in September 2007 to 3.0% in late January 2008, with the promise of more cuts to come if need be. Courtesy of the Fed, the yield curve for US government bonds has steepened substantially, with the yield differential between 2-year and 10-year Treasuries widening from a mild inversion (−10 basis points) in May 2007 to 180 basis points in mid-February 2008. For many banks, this matters. Term transformation (playing the yield curve) is one of their bread-and-butter businesses. By engineering a steep yield curve, the US Fed is de facto helping to raise the profit opportunities for US banks.

Pessimists now maintain that the implosion of financial leverage and the repricing of risk will cause a severe credit crunch, that is a tightening of credit supply to end-users (households and non-financial businesses) pronounced enough to trigger a protracted fall in household consumption and business investment. Optimists maintain that the financial turmoil will largely remain within the financial sphere. The tightening of credit conditions due to the reduced risk appetite and the diminished opportunities for financial institutions to re-allocate risks to willing bearers of risk could be largely offset by lower interest rates reflecting the aggressive rate cuts from the US Fed.

7.6 Cyclical Outlook: Rebound in Late 2008

Our optimism that the US economy will hit bottom soon and start to recover before the end of 2008 thus rests on three major pillars:

1. While we expect some further tightening of credit conditions, the impact will likely be mitigated by low interest rates. Although banks are raising the margin between their refunding costs and the loan rates they offer, this is offset by a lower level of overall interest rates. For most consumers, that is those outside

the subprime section, loans should be available at rates that are not significantly more unfavorable than they were in mid-2007. Data on US mortgage loan rates showing a small decrease over the 6 months to January 2008 bear this out.
2. The decline in residential construction, which has dragged down real GDP growth by an average of almost 1.1% per year since 2Q 2006, looks set to peter out over the course of the year, following a total peak-to bottom fall in residential construction of roughly 40%. In terms of GDP arithmetic, the end of a major negative is a positive.
3. Although the US still has a major external deficit, export volumes (up 7.7% year-on-year in 4Q 2007) are rising much faster than import volumes (up 1.4%). We expect net exports to contribute at least 0.6% to GDP growth in 2008 after an average contribution of 1% in the last three quarters of 2007.

Put differently, the positive contribution from net exports and the gradual end of the major drag from residential construction will probably counterbalance the temporary loss of momentum in business investment and private consumption in the wake of the credit crisis.

7.7 Longer-term Outlook: Global Rebalancing

Beyond the cyclical outlook for 2008, the key question is whether the US is heading for a prolonged period of comparatively soft growth, a pay-back for the period of astonishingly rapid expansion since the mid-1990s. The case for slow growth typically rests on the view that overextended US consumers have to revert to more prudent spending patterns. Slower trend growth in consumption, in turn, will show up in a more subdued pace of overall GDP expansion.

In general terms, we doubt that arguments that focus on demand are very relevant for the expected average growth rate of US GDP for, say, the 5 years after the 2008 stagnation. But even in terms of demand dynamics, we consider such pessimism far overdone.

There is no compelling evidence that the US consumer is overindebted. Pessimists typically point to a higher level of debt or to rising debt service obligations. However, we need to look at both sides of the consumer balance sheet, comparing debt to assets and debt service costs to asset incomes. Although debt levels have risen, asset values have surged much more. Data from the US Fed show that the real net worth of US households, that is assets minus liabilities adjusted for inflation, has risen from \$33.2 trillion in late 1997 to \$58.6 trillion 10 years later. Although more households are defaulting on their debt, especially but not only on subprime mortgages, there is little evidence that the US household sector is heading for major debt service problems beyond the usual cyclical fluctuations in loan default rates (outside subprime). Admittedly, the debt service of households has risen from 12% of disposable income 10 years ago to 14.2% in late 2007. However, asset incomes (interest and dividends) are still much higher, at 19.3% of disposable income. While this share had fallen until 2003, partly because paying dividends had become unfashionable and partly because bond yields were trending down, it has now rebounded from the 16% reached in 2003.

Globally, the period of low interest rates is likely to come to an end incoming years. Although increased global competition will continue to prevent major sustained spikes in labor costs, export prices of major emerging markets are starting to rise gradually as labor becomes less abundant in China and India. In addition, the major global buyers of government bonds, the reserve managers of Asia and the oil countries, are likely to diversify ever more into other assets such as equities and away from bonds. The role which sovereign wealth funds are playing in the recapitalisation of major US financial institutions exemplifies this trend. More normal interest rates will constrain the potential for any rebound in US house prices after the current correction. They will also, in a neoclassical sense, enhance the incentive to save for US and other consumers.

We thus agree with the pessimist that gains in US consumption will likely be more restrained incoming years than they were in the recent period of falling and/or exceptionally low interest rates. But this need not show up in a major and protracted slackening of US GDP growth. With continuing immigration, a proven ability to integrate immigrants comparatively well and a birthrate at roughly the replacement level of 2.1 children per woman, the US population will likely increase by 0.8% per year for the foreseeable future. Even if productivity growth slows down from its 1996 to 2007 average of 2.5%, US growth trend could remain close to 3%.

In terms of demand, we expect the US to be a prime beneficiary of the next stage of globalization. So far, China and some other fast-growing emerging markets have embarked on growth driven by mostly be investment and exports. These countries supply world markets with cheap and mostly labor-intensive manufactures while they import capital goods and intermediate products, running up huge trade surpluses in the process. But due to the success of this strategy, formal employment and wage levels are now rising rapidly in coastal China. The next major stage of development will likely be the rise of an urban class of consumers with enough means to buy imported quality products. For China and other export-surplus countries among the emerging markets, we look for a gradual transition away from the export focus to stronger domestic demand. Better developed financial and social systems in these countries will probably lead to a secular decline in the personal savings rates.

This transition will open up new opportunities for Western exporters of consumer goods. On a global scale, we thus expect a major rebalancing over the next 5–10 years. As consumers in China and some other emerging markets are gradually emerging as major drivers of world demand, US consumption growth can trail behind the overall advance in US GDP, with stronger net exports filling the gap. In the process, the US trade deficit and the Chinese trade surplus will probably decline to much more sustainable levels over the next 5 years. Many of the recent European concerns about the US, be it worries about the external deficits or the allegedly overindebted consumers, will likely turn out to be overdone. Relative to major parts of the European continent, with low birthrates, rising difficulties to integrate immigrants into the labor force and flagging productivity growth, the US still looks in fairly decent shape.

Chapter 8
Selected Innovation Factors: An International Comparison

Lothar Funk and Axel Plünnecke

8.1 Introduction

Contrary to the post-war period where growth and catching-up with the United States could largely be achieved through accumulation of production factors and from assimilating existing technologies, once European countries had moved closer to the technology frontier,[1] innovation has become the main engine of growth. In other words, the balance between imitation and innovation has shifted in favor of the second. At the heart of this ability to innovate lie all those factors that lead either to the introduction of new products (product innovation) or to the introduction of new production processes (process innovation). Additionally, a greater proportion of that innovation is radical rather than incremental (OECD, 2004c, p. 82). According to the Sapir-report, "growth becomes driven by innovation at the frontier and fast adaptation to technical progress" (Sapir et al., 2004, p. 38).

Different studies have attempted to assess the innovative capabilities of national economies by using a set of indicators. One such leading set of indicators is the European Innovation Scoreboard (EIS). This Scoreboard takes into consideration the following areas: human resources, the generation of knowledge, the dissemination and use of knowledge, the system of financing innovations as well as the output from innovation. In a recent European Innovation Scoreboard (EIS, 2005), Germany is ranked in fourth place behind Sweden, Switzerland and Finland. Germany,

[1] Several European countries were able, above all in the immediate post-War decades, to successfully pursue a process of catching up. It was therefore possible to reach the technological frontier in important areas. However, the growth that is attributable to catch up came to an end, at the latest, at the beginning of the 1980s when the easy gains from imitating and incrementally improving existing technologies became exhausted and demand became saturated for the output of leading industries. A decrease in the importance both of imitating successful economies and of simple incremental innovations for economic growth are inextricably linked with the ending of the process of catching up (see Sapir et al., 2004, pp. 35–37).

L. Funk (✉)
University of Applied Sciences Düsseldorf, Germany
e-mail: lothar.funk@fh-duesseldorf.de

P.J.J. Welfens, J.T. Addison (eds.), *Innovation, Employment and Growth Policy Issues in the EU and the US*, DOI 10.1007/978-3-642-00631-9_8,

therefore, is placed well ahead of many other EU member states. According to this set of indicators, Germany's strengths lie, above all, in patents, which make up four of the 22 indicators. Germany also has above-average scores for employment rates in sectors that rely on cutting-edge technologies. On the other hand, Germany has, within the EU, below-average scores for the number of graduates in mathematics, engineering, the life sciences and technology. It also has below-average scores in the area of life-long learning. In terms of risk capital, the use of ICT and in the creation of value in cutting-edge technologies, Germany receives approximately average values within the EU.

In terms of macroeconomic growth in the EU member states, Jungmittag (2004) has shown not only the importance of specialising in leading-edge and research-intensive technologies, but also the significance of the dynamic growth effects of patents. "Innovation measured by the growth rates of the patent stocks of the EU countries foster economic growth. . . ., the level of relative technological specialisation in the area of R&D-intensive industries and especially in the area of leading-edge industries contributes significantly to economic growth within the EU." Jungmittag also shows that countries farthest away from the technological frontier profit more in terms of growth from transferable technological knowledge.

The current study puts forward a more broadly conceived notion of the innovative capabilities of national economies. Drawing upon a set of indicators, this study will attempt to show how Germany's innovative capabilities could be assessed. In doing so, a cross-sectional comparison will be undertaken. Within this set of indicators, patent measures are excluded as patents are often drawn upon within similar studies for OECD countries, to a greater of lesser extent, in a strategic way. These country-specific attributes can, to be sure, be controlled for within time-series analyses; they are, however, very problematic in comparatively primitive sets of indicators. Moreover, in contrast to the EIS, the specialisation of a national economy in leading-edge technologies is endogenised particularly by using the differences in framework conditions between the countries in our analysis. This study examines data from the last decade for 22 of the most important OECD countries in order to assess whether the economies that were both at the technological frontier and had the highest rates of economic growth in recent years did, indeed, have the best frameworks for innovation. Even if different international organisations – for example, the European Commission and the OECD – have conducted a range of very useful benchmarking exercises in the area of innovation[2] (European Commission, 2004a; OECD, 2004b), this question has not been analysed in conjunction with a liberal regulatory framework policy approach.

Our study uses the United States as the benchmark economy.[3] The position of the USA as the most technologically advanced country (European Commission, 2004b,

[2]Such benchmarks show the position of various national economies across a range of measures of innovation; they do not, however, provide an overall ranking based on those indicators that are most relevant for economic growth.

[3]Above all, we choose Germany as the base country, as it was the laggard among the countries analysed here. If growth rates over the ten years until 2004 are averaged out, then Germany, with

pp. 174–177) results from the fact that it, in common with many European countries, has high hourly productivity rates, whilst it has, at the same time and in contrast to many European states, higher work volumes (Sapir et al., 2004, pp. 34–35). Obviously, the trade-off between productivity and work volumes that is so apparent in much of Europe does not arise in the US.[4]

Our analysis demonstrates that for countries at the technological frontier, the existing conditions for innovation have a significant effect on current growth rates. Human resources, financing possibilities as well as the institutional regulation of product and labor markets have a prominent position amongst those factors that influence innovation. At the technological frontier, those countries with high rates of growth also distinguish themselves from those with low rates of growth on important measures of innovation. Significant differences between the two groups of countries exist in the numbers of university graduates (including those in particularly important subjects such as mathematics, sciences and technology – in short MST-subjects), the availability of venture capital, the amount invested in information and communication technologies (ICT), the general conditions in which firms operate as well as in their demographic developments. Even if other indicators of innovation are considered, it can be shown that the average ranking on such measures is correlated with economic growth rates after the end of the boom in the new economy.

8.2 Theoretical and Methodological Background

In contrast to traditional growth explanations, "modern theories emphasize research inputs and human capital as the key drivers for long-run growth. They stress not only the importance of 'own' innovation but also the capacity to imitate and to absorb externally available know-how. Institutional factors and framework conditions are seen as an important part of the 'innovative system' in which innovative firms operate" (European Commission, 2004b, p. 175). There appears to be an emerging consensus that a narrow view of innovation (science should somehow cause innovation) is inappropriate (Arnold and Thuriaux, 2003, p. 1) and needs to be replaced by a broader view of innovation.

> Innovation is a fundamentally economic process. Schumpeter called innovation 'a new combination of factors of production'. This can be the result of an invention. But it can equally

---

an annualized growth rate in per capita GDP of only 1.2% in real terms, is last amongst 22 OECD countries.

[4]The USA and other successful economies have managed to increase their productivity rates despite a simultaneous increase in their rates of employment. This also applies to low-skilled labor in the US. This implies that the USA has obviously been more successful at transforming inventions in basic science into growth-enhancing innovations. High rates of hourly productivity alone are no indication of being a technological frontrunner. Several European countries, of which Germany is a good example, have high levels of labor productivity per hour; at the same time, however, they exhibit low levels of employment. Other countries that have higher employment levels have lower hourly productivity rates.

> involve the exploitation of new natural resources, copying an idea from a distant market, or describing an old product in a new way. *Entrepreneurship – the act of making innovations – is not something related to science and research, but about changing the rules of the game in economic competition.* Exploiting an invention is, therefore, an important special case of innovation, but it is not the general case. Economists see innovative activity as a driver of economic development because it provokes imitation. Innovation forces competitors to react – often in creative ways involving improvement and 'innovating around' the first innovator's design to erode the 'supernormal' profit of the original innovator. Innovations give rise to changes in the economy, which may be several times larger than the effect of the original innovator. *The main driver of economic growth is therefore the process by which change diffuses through the economy* (Arnold and Thuriaux, 2003, p. 2).

In other words, an economy's "national innovation capacity" is appropriately defined "as the ability of a nation to not only produce new ideas but also to commercialize a flow of innovative technologies over the longer term" (European Commission, 2004b, p. 175). It is vital that conditions for innovation and diffusion must be approached as a whole (Arnold and Thuriaux, 2003, p. 8). In the words of Jan Fagerberg (2003, p. 17): "Innovation processes are inter-temporal by nature. Current innovations depend on past innovations and future innovations will depend on current innovations. This means that there may be a strong aspect of path dependency in innovation processes. Radical innovations open up new paths for future innovation activities and can dramatically influence what can be done profitably. Consequently, 'innovation and diffusion should be seen as an integrated process'. It does not make sense to separate innovation and diffusion policy because a number of innovations occur in the wake of radical innovation and these only achieve economic impact as part of the diffusion process. What is important for innovation policy is to approach innovation as an integrated whole...; it is important to get on the bandwagon early, because as time goes by requirements become increasingly difficult to meet and unless one can jump on early it may be difficult to catch up at all because of the accumulated advantages associated with path dependency."

The arguments mentioned above have important repercussions for analysing the preconditions and effects of innovation empirically, as the incentive to engage in innovative investments – which involve risky experimentation and learning in particular at the technological frontier – is itself strongly affected by the economic environment (Sapir et al., 2004, p. 38). At least three implications result from such a broad perspective of innovation: firstly, it appears reasonable to link the rates of economic growth of countries to the conditions for innovation and diffusion, as is quite common nowadays (see e.g. Acemoglu, Aghion, & Zilibotti, 2002; Fagerberg, 2005; Fehn, 2004). Secondly, although in each and every country there will be specific factors at work (see e.g. Boyer, 2004), these will not be in focus here. Rather we will attempt to single out some general factors related to innovation capacity that may be of interest when debating the differences across countries in economic performance. These important innovation dimensions are human resources, financing conditions and more general framework conditions, for example in product and labor markets. The factors of innovation in this study are clearly multidimensional and are therefore rather difficult to handle. Hence, we shall identify four reliable innovation input indicators for each of the three dimensions, express these in a comparable format and weigh them together, giving each indicator and dimension an equal weight in the calculation of the composite ranking indicator, which will be computed as an average rank of the single indicators in this article. Admittedly, there is an element of arbitrariness involved in such a calculation. This analysis may, for example, have a problem of omitted variable bias which is alleviated, however, by using several different indicators for each dimension. Moreover, it would of course have been preferable to have prior knowledge about the true weights to use. Missing such information, it appeared least arbitrary to give each variable an equal weight. Finally, it should be noted that the general approach adopted here will – in line with the broad definition of innovation we follow – not only measure the direct effects of innovation activity (understood in a narrow sense as technological progress) on output but also indirect effects on output such as reductions of technical inefficiency and improvements in allocative efficiency that cannot be separated empirically (see Van Ark, 2003, pp. 10–13).

8.2.1 Innovation at the Technological Frontier: Catch-up States, Growth Stars and Laggards

A recent study (Acemoglu et al., 2002) shows that the closer a country finds itself to the global cutting edge of technology, the more important radical innovations combined with fast adaptation to technical progress in all sectors become for economic growth. The basic conditions that promote the flow of radical innovations are of particular importance. By contrast, the importance of the ability to imitate others is severely reduced. According to this approach, the factors affecting innovation are of relatively little importance for economic growth in countries that are a long way from the technological frontier (catch-up countries); however, in national economies that are close to this technological frontier these factors are of decisive importance for growth. The following empirical analysis is based on this general empirical approach.

Per capita gross domestic product (GDP) will be a proxy for the technological advancement of a country. The USA will be defined as the benchmark for technological advancement. Therefore, the US-GDP per capita will be normalized to 1. In connection with their growth rates over the last ten years, the 22 OECD countries considered here can be grouped into three categories (Fig. 8.1).

- *States that are catching up:* in the first group, countries can be found that have a relatively low level of per capita GDP and, at the same time, exhibit real GDP growth rates of more than 2%. These countries, such as Ireland, South Korea, Greece and Spain, can be described as states that are catching up. What also distinguishes these countries from most other countries is the fact that their rates of per capita GDP were less than two thirds of the US's level in 1994. This group of countries can profit significantly from imitating more successful economies and by incremental innovation.

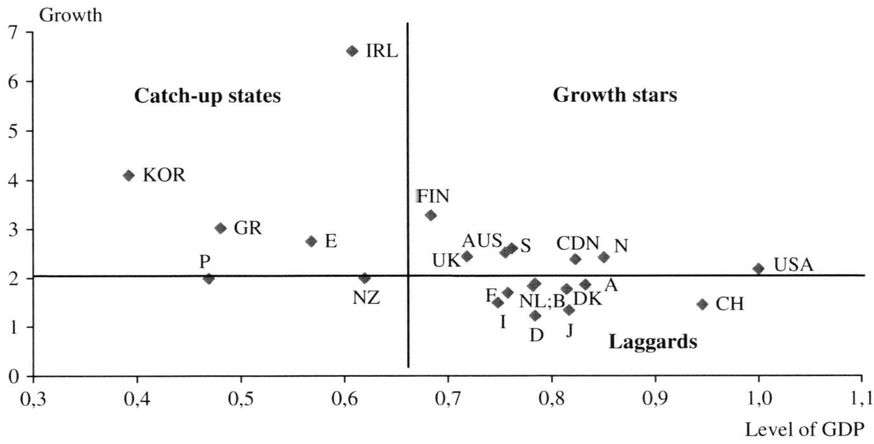

Fig. 8.1 Level of economic development and economic growth
Source: Average yearly growth rates in per capita gross domestic product over the period 1994–2004 in percent and per capita gross domestic product in 1994 (USA = 1) in 22 OECD countries

- *Growth stars:* a second group of countries is characterized, firstly, by real GDP growth rates of over 2% and, secondly, by already high levels of per capita GDP. This group can be described as growth stars. Over the last decade, Finland, Sweden, Australia, the UK, Norway, Canada and the USA have belonged to this group. These states have, despite a relatively high technological level, high rates of economic growth.
- *Laggards:* a third group of countries, which are characterized by growth rates of less than 2%, can be described as laggards. Belgium, Austria, the Netherlands, Denmark, France, Italy, Switzerland, Japan and Germany belong to this group. These countries are, because of their high levels of per capita GDP, undoubtedly highly developed economies. They, therefore, find themselves largely at the technological frontier. They have not, however, been able to take sufficient advantage of the opportunities for growth that have existed over the last decade.

8.2.2 Conditions for Innovation

The growth stars as well as the laggards are, as groups of countries operating at the technological frontier, more dependent on radical innovations and fast adaptation to technical progress in all sectors for economic growth. This raises a couple of main questions. Firstly, what are the conditions that have the most favorable effects on growth? Secondly, did those economies having the highest rates of growth between 1994 and 2004 have the best frameworks for innovation and did they also, in the latter years of that period, offer better conditions for innovation? Moreover, did they subsequently exhibit higher rates of growth? By considering more recent years, influences will be excluded that may have promoted innovation during the boom in the area of information and communication technologies (ICT) (the so-called new economy boom), but may nevertheless distort evaluations of the innovation factors in the medium term.

From a theoretical perspective, variations in conditions for radical innovation in particular play a central role in explaining the differences in growth rates between countries. These conditions cover:

1. *Human Resources*. In the case of human capital, a paper by Aghion, Meghier, and Vandenrusche (2003) demonstrates that for 19 OECD countries between 1960 and 2000, the impact on economic growth of highly skilled workers in the population increases as the distance to the technological frontier declines.
2. *Financing Possibilities*. The role of finance and investment for innovation and growth is shown in papers by Levine (1997) and by Bassanini and Scarpetta (2002). The latter show that technological change – embodied by new ICT capital goods – has been a primary source of output and productivity growth in ICT-using sectors.
3. *General Framework Conditions*. The role of regulations for innovation and growth is shown in papers by Scarpetta and Tressel (2004) and by Nicoletti and Scarpetta (2003). There is evidence that high labor adjustment costs can have a strong negative impact on productivity. Such costs can, in particular, reduce incentives for innovation and the adoption of new technologies and lead to lower productivity performance when institutional settings do not allow wages or internal training to offset high hiring and firing costs. Reforms promoting private governance and competition tend to boost productivity. Both privatisation and market-entry liberalisation are estimated to have a positive impact on productivity.

As mentioned above, the performance of an innovation system cannot sensibly be explained by one solitary factor. Therefore, in order to evaluate the forces that promote innovation within a national economy, it is necessary to look at a bundle of factors as explained above. Due to the research design, it will not be possible, here, to cover the measures of the output of innovation that do not include the effects of research and development on the national economy. In other words, measures such as the number of new patents will not be discussed here. This paper aims at elaborating on the link between basic input factors within an innovation system and economic growth per capita.

8.2.2.1 Human Resources

As mentioned above, the share of the population that is highly qualified is of particular importance for the ability to innovate (Aghion et al., 2003). If the technological distance to the leading national economies is great, then the share of the population that is highly qualified hardly has any role – empirically – to play in influencing economic growth. This is because imitation and incremental innovation dominate. The share of the population that is highly qualified, the share of the population with a post-graduate research qualification, the numbers of successful candidates in mathematics, sciences and technology, and investments in education are important measures which can be used as indicators of a national economy's human resources.

If the three groups of countries are considered, it can be shown that the group of growth stars performs considerably better on these measures of human capital than the average for the group of laggards (Table 8.1). The share of the population aged between 25 and 64 that has a tertiary education is, at 34% in the former group of countries, nearly a third higher than the figure for the laggards. Similarly, the number of graduates in mathematics, sciences and technology per 100,000 employees aged between 25 and 34 is considerably higher amongst growth stars than it is amongst the laggards. Moreover, the former group of countries have a higher share of graduates who have completed post-graduate research degrees amongst those of typical graduating age. Indeed, the growth stars invest more, as a percentage of GDP, in education than do the laggards.

The performance of the growth stars and the laggards differ significantly on the indicator measuring "persons with a tertiary education" and on the "MST" indicator. However, there is no significant difference between growth stars and laggards on investments in education and the share of the population with a doctorate.[5]

8.2.2.2 Financing

Expenditure on R&D as a percentage of GDP and the investment in ICT represent important indicators of investment and, therefore, the financing of innovations (see for the theoretical background Aghion and Howitt 1992; Romer 1990 and 1996). This is particularly true for governments' promotion of research and knowledge in the area of basic science. This is, to a large extent, supported by governments and is carried out at universities and other research institutes (BMBF, 2004, p. 167). One reason for the role of government in this area might be that free market mechanisms do not generate enough of such "products" because of their public-goods character. Moreover, state subsidies can fundamentally improve the sub-optimal supply of innovations. However, for political-economic reasons and because of the risks of inefficient demands for subsidies, funding should be treated with care; this becomes increasingly important as the tasks of R&D become more application oriented (Farhauer & Henke, 2002). A large amount of venture capital (even if it represents only a relatively small percentage of GDP) can most readily enable radical innovations which, as a result of the current structural changes in economies, obviously have a greater effect on growth than incremental innovations do.

Moreover, venture capital can increase the rate at which new companies are founded. Such companies, amongst other things, make a significant contribution to innovation as they are able to choose the most productive combination of factors

[5]The p-value of the t-test that the means for the two groups are different is 0.1 and 0.375 for investments in education and for the share of the population with a doctorate, respectively.

Table 8.1 An international comparison of human resources in 2002

Country	Persons with a tertiary education as a percentage of the population aged between 25 and 64		Persons with post-graduate research qualifications[1] as a percentage of the population aged between 25 and 64		Graduates in MST[2] per 100,000 employees aged between 25 and 34		Expenditure on education as a percentage of GDP	
	in percent	Rank	in percent	Rank	Value	Rank	in percent	Rank
Growth stars								
Finland	33	4	1.9	4	1785	2	5.8	11
Australia	31	6	1.3	8	1659	4	6.0	8
Sweden	33	4	2.8	1	1267	8	6.5	4
Norway	31	7	1.1	11	703	16	6.4	5
UK	27	11	1.6	6	1727	3	5.5	13
Canada	43	1	n.a.	n.a.	855[3]	12	6.1	7
USA	38	2	1.3	8	928	11	7.3	2
Laggards								
Denmark	27	10	0.9	14	799	14	7.1	3
Austria	14	20	1.7	5	528	20	5.8	11
Belgium	28	9	1.1	11	674	18	6.4	5
Netherlands	24	15	1.3	8	653	19	4.9	18
France	24	15	1.4	7	1609	5	6.0	8
Italy	10	21	0.5	20	703	16	5.3	15
Germany	23	18	2.0	3	721	15	5.3	15
Japan	36	3	0.7	18	1074	9	4.6	20
Switzerland	25	13	2.6	2	838[3]	13	5.3	15
Catch-up states								
Ireland	25	13	0.8	17	1514	6	4.5	21

(continued)

Table 8.1 (continued)

Country	Persons with a tertiary education as a percentage of the population aged between 25 and 64		Persons with post-graduate research qualifications[1] as a percentage of the population aged between 25 and 64		Graduates in MST[2] per 100,000 employees aged between 25 and 34		Expenditure on education as a percentage of GDP	
	in percent	Rank	in percent	Rank	Value	Rank	in percent	Rank
Korea	26	12	0.9	14	1788	1	8.2	1
Greece	18	19	0.7	18	n.a.	n.a.	4.1	22
Spain	24	15	1.0	13	935	10	4.9	18
Portugal	9	22	n.a.	n.a.	n.a.	n.a.	5.9	10
New Zealand	30	8	0.9	14	1497	7	5.5	13
Average values on the measures of human resources for the three groups of countries								
Growth stars	34	1	1.7	1	1275	1	6.2	1
Laggards	23	2	1.4	2	844	2	5.6	2
Catch-up states	22	3	0.9	3	n.a.	n.a.	5.5	3

Source : OECD 2004e and OECD 2004f.
[1] Those who have successfully completed a post-graduate research degree.
[2] Graduates of mathematics, engineering, the life sciences and technology. Values for Canada are for 2000 and for Switzerland are for 1998.
[3] 2000.

of production. Unlike established companies, new firms do not have any costs associated with adaptation (for example, training, redundancies) when they introduce new technologies and production methods. The establishment of new companies also exerts competitive pressures on existing ones; the latter are then spurred on to create innovations (OECD, 2004c, p. 88).

If an examination is made of the extent to which the growth stars and the laggards differ on the measures of financing and investment, the means across all four indicators are better for the growth stars than for the laggards. The differences in average values for investments in ICT and venture capital are statistically significant.[6]

In the availability of venture capital in the early and expansionary phases, the growth stars dominate. Younger firms have a far better climate in which to fund their expansions from external sources. For the growth stars, venture capital amounted to, on average, 0.218% of GDP between 1998 and 2001. In the laggard countries, not even half that amount was available to establish new companies. These latter countries had to be satisfied with venture capital amounting only to approximately 0.1% of GDP. More recently, the group of growth stars have also invested in new ICT; these not only contribute to higher productivity and growth, but they also represent an important infrastructure for the wide diffusion of advanced technologies (OECD, 2004c, p. 77). Whilst for the growth stars investment in ICT amounted to, on average, 22.2% of their fixed investment, this figure was a mere 16.8% for the laggards (Table 8.2).

Despite the fact that the growth stars perform better on indicators that measure investments in R&D and tax relief on R&D on average, there are states with low rates of growth that perform well on such indicators, and there are also states with high rates of growth and relatively poor records in these areas. For instance, Switzerland, Japan, Germany, Denmark, and France, which are all laggards, invest more in R&D than Canada, the UK, Norway and Australia, which are growth stars. Similarly, Austria, Denmark and the Netherlands, which are laggards, offer far greater tax relief on investments in R&D than do Finland, Sweden and Norway, which belong to the group of growth stars.

8.2.2.3 General Framework Conditions

The general conditions for companies are of great importance for the potential success of new technologies as such investments are associated with great uncertainties. These risks can, in general, be managed better in a situation in which there are fewer labor-market regulations. Furthermore, flexible markets create incentives for highly qualified young people to use their knowledge to create new knowledge; this means that they will not use their knowledge in socially unproductive rent seeking (Murphy, Shleifer, & Vishny, 1991).

---

[6]The *p*-value of the *t*-test on mean equivalence is 0.4 and just under 0.6 for tax relief on R&D and investments in R&D, respectively; this indicates that there are no statistically significant differences between the two groups on these two measures.

Table 8.2 An international comparison of financing conditions

Land	Share of GDP spent on R&D[1] (2002)		Tax relief per R&D dollar spent in large firms[2] (2001)		Share of GDP that is available as venture capital (1998–2001)		Share of fixed-capital investments that is spent on ICT[3] (2000)	
	in percent	Rank	in US$	Rank	in percent	Rank	in percent	Rank
Growth stars								
Finland	3.46[4]	2	– 0.01	13	0.138	8	17.5	10
Australia	1.54	16	0.20	3	0.093	14	19.9	6
Sweden	4.27[5,6]	1	– 0.01	13	0.207	5	21.6	3
Norway	1.67	15	– 0.02	18	0.125	10	n.a.	n.a.
UK	1.88	14	0.10	8	0.219	4	22.8	2
Canada	1.91	12	0.17	4	0.251	2	21.4	4
USA	2.67[7]	5	0.07	10	0.492	1	30.1	1
Laggards								
Denmark	2.52	7	0.11	7	0.082	16	19.1	7
Austria	1.93	11	0.12	6	0.044	21	12.8	16
Belgium	2.17[5]	10	– 0.01	13	0.172	7	17.5	10
Netherlands	1.89[5]	13	0.10	8	0.241	3	20.9	5
France	2.20	9	0.06	11	0.120	11	12.6	17
Italy	1.11[5]	19	– 0.03	21	0.076	17	16.1	12
Germany	2.52	7	– 0.02	18	0.127	9	17.6	9
Japan	3.12	3	0.01	12	0.020	22	17.8	8
Switzerland	2.57[4]	6	– 0.01	13	0.085	15	n.a.	n.a.
Catch-up states								
Ireland	1.15[5]	18	n.a.	n.a.	0.114	12	14.6	15
Korea	2.91[8]	4	0.13	5	0.202	6	n.a.	n.a.
Greece	0.65[5]	22	– 0.01	13	0.059	20	15.7	13
Spain	1.03	20	0.44	1	0.095	13	15.5	14
Portugal	0.93	21	0.34	2	0.066	19	11.4	18
New Zealand	1.18[5]	17	– 0.02	18	0.069	18	n.a.	n.a.
Average values of the financing measures for the three different groups								
Growth stars	2.49	1	0.07	2	0.218	1	22.2	1
Laggards	2.22	2	0.04	3	0.107	2	16.8	2
Catch-up states	1.31	3	0.18	1	0.101	3	n.a.	n.a.

[1] Research and development.
[2] Equivalent to the tax relief on every dollar spent on R&D in large companies.
[3] Information and Communication Technologies.
[4] 2000.
[5] 2001.
[6] Underestimated values.
[7] So-called capital expenditures have been excluded.
[8] R&D expenditures in the humanities and the social sciences have been excluded.
Source: OECD (2004b), Institut der deutschen Wirtschaft Köln.

Nonetheless, the growth stars Finland, Sweden and Norway have higher levels of labor-market regulation than many other growth stars. These latter countries are, however, typically small open economies with particularly high trade openness and labor demand elasticities that may not directly be – due to lower insider power to set wages above market-clearing levels – comparable in this respect with larger economies (Brandt, Burniaux, & Duval, 2005, p. 65).

However, the other regulatory conditions for companies are better in the growth stars. A ranking of the conditions for firms (Matthes & Schröder, 2005), which is based on World Bank data (World Bank, 2004) and which does not take into consideration labor market flexibility, shows that the growth stars perform significantly better than the laggards and those states catching up. This measure covers data on the founding of companies, the registration of homes and property, the availability of credit, the disclosure requirements of public limited companies, the extent to which contracts are upheld, and the laws on insolvency.

Equally, the age structure of the population is of great importance for the willingness to innovate, as a younger population shifts the political majority in the direction of more innovative production methods (Gehrig, 2000, pp. 570–571). Of particular importance for the renewal of human capital is the ratio of pupils and students to the total number of employees. This indicator shows how strongly the basis of human capital as a share of total employment is being refreshed. It is, first and foremost, demographic developments that influence these factors. States whose populations are ageing quickly have a paucity of younger people who, as carriers of newly created knowledge, are particularly important in driving radical innovations. A demographically more favorable age structure creates more impetus for innovation when the education system is of high quality so that those starting work for the first time have a high level of competence. The results of the PISA studies for OECD countries in literacy, mathematics and the natural sciences are important measures of the quality of the education system (OECD, 2004d).

To sum up, the growth stars perform on net better in their general conditions for innovation than both the laggards and those states that are catching up (Table 8.3). The growth stars exhibit significantly better averages for indicators that measure the general product, capital market and bureaucracy environments in which firms operate and that capture demographic developments.[7]

8.2.3 Overall Evaluation of the Drivers of Innovation

If the average values for all the measures are considered, Canada performs the best, followed by the US, the UK, Sweden, Finland and Australia (Table 8.4). From the group of growth stars, Norway is the sole country not occupying a leading position.

[7] The *p*-value of the *t*-test on mean equivalence is 0.122 and 0.394 for labor-market regulation and the quality of the education system as measured by PISA, respectively. This means that the growth stars and laggards are not significantly different on these two measures.

Table 8.3 A comparison of the general framework conditions

	Labour market regulation (2003)		Conditions for companies[1] (2004)		Ratio of pupils and students to employees[2]		PISA results[3] (2003)	
Country	Index[4]	Rank	Index	Rank	in percent	Rank	Value	Rank
Growth stars								
Finland	2.1	11	52.3	7	63	9	545	1
Australia	1.5	6	55.8	5	80	1	525	5
Sweden	2.6	17	52.6	6	64	7	510	10
Norway	2.6	17	59.2	2	59	13	493	16
UK	1.1	2	58.0	3	74	3	n.a.	n.a.
Canada	1.1	2	50.2	9	n.a.	n.a.	526	4
USA	0.7	1	57.2	4	64	7	490	17
Laggards								
Denmark	1.8	8	45.9	15	56	14	494	15
Austria	2.2	12	44.6	17	52	19	496	14
Belgium	2.5	15	44.6	17	74	3	515	8
Netherlands	2.3	13	49.0	10	54	15	525	5
France	2.9	19	43.5	19	65	6	506	12
Italy	2.4	14	39.2	20	54	15	476	19
Germany	2.5	15	44.8	16	53	17	499	13
Japan	1.8	8	51.4	8	44	20	527	3
Switzerland	1.6	7	47.1	13	44	20	513	9
Catch-up states								
Ireland	1.3	4	48.4	11	70	5	508	11
Korea	2.0	10	44.9	15	61	10	538	2
Greece	2.9	19	31.3	22	60	11	466	21
Spain	3.1	21	47.4	12	60	11	484	18
Portugal	3.5	22	34.0	21	53	17	471	20
New Zealand	1.3	4	59.3	1	77	2	522	7
Average values for the measures of general conditions								
Growth stars	1.7	1	55.0	1	67	1	515	1
Laggards	2.2	2	45.6	2	55	3	506	2
Catch-up states	2.4	3	44.2	3	63	2	498	3

[1] Without labor market regulation. Range of the index: median of the individual measures = 50.
[2] Employees aged between 25 and 64.
[3] Results in literacy, mathematics and the natural sciences. OECD average = 500.
[4] Range of the index: from 0 (few regulations) to 6 (highly regulated).
Source: OECD (2004a, 2004d, 2004e), Matthes and Schröder (2004), Institut Der Deutschen Wirtschaft Köln.

The national economies with a high per capita real income level and low levels of economic growth (the laggards) are positioned at the bottom of the list. From this group, Belgium performs the best, followed then by Denmark, the Netherlands, Japan, Switzerland, France, Germany, Austria and, well behind the other countries, Italy. The group of catch-up states is led by Korea, New Zealand and Ireland. These three countries perform better than the majority of the laggards. The catch-up states

Table 8.4 The composite ranking of 22 OECD countries

Country	Average rank
Canada	5.7
USA	5.8
UK	6.2
Sweden	6.6
Finland	6.8
Australia	6.8
Korea	7.3
New Zealand	9.9
Belgium	10.5
Denmark	10.8
Netherlands	11.0
Japan	11.2
Switzerland	11.5
France	11.6
Norway	11.7
Ireland	12.1
Germany	12.9
Spain	13.8
Austria	14.3
Portugal	17.2
Italy	17.4
Greece	18.2

Source: Own calculation.

Spain, Portugal and Greece follow a long way behind, and are near the bottom of the list.

Finally, the core hypothesis of this evaluation should be assessed. Accordingly, the conditions for innovation should have a strong effect on the growth rates of those states close to the technological frontier. In other words, as explained in the basic theoretical approach above (Acemoglu et al., 2002), conditions for radical innovations are of significant importance for growth stars and laggards. Figure 8.2 supports this hypothesis. The average position for all of the measures of innovation used here (which mainly portray the conditions for innovation between 2000 and 2003) has a statistically significant effect on the growth rate in per capita GDP between 2000 and 2004.

The current study aims at selecting, according to theoretical considerations, indicators from a range of innovation benchmarks provided by the OECD; the selected measures can then be used in an analysis to determine whether a statistical relationship exists between the country's average rank on such indicators and its macroeconomic growth rate. If the measure of human capital alone were selected, the R^2 measure would take an even higher value. It can, therefore, be tentatively concluded that human capital plays a particularly important role in innovation and growth.

However, it can certainly be argued that those growth stars between 1994 and 2004 were also the fastest growing national economies between 2000 and 2004. The categorization of states according to their growth rates would therefore appear

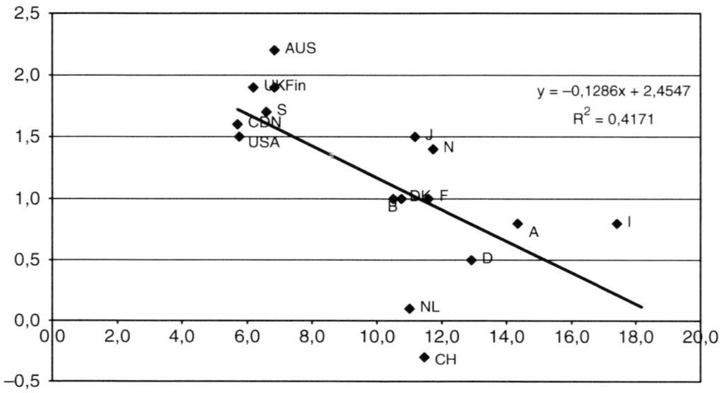

Fig. 8.2 Conditions for innovation and economic growth
Source: Own calculations based on EUROSTAT (2005); Growth in per capita GDP between 2000 and 2004 in percent and conditions for innovation of the countries at the technological frontier in year 2001.

not to have been affected by the ebb and flow of the new economy. Those countries that grew strongly over the whole time period considered here had good conditions for innovation at the start of this millennium and were also able to achieve a higher rate of growth between 2000 and 2004 than those states that had worse values on the innovation measures.

8.2.4 Do Growth Stars and Laggards Differ in Terms of Their Innovation Indicators?

In order to examine this question, a hierarchical cluster analysis will be conducted. After a Z standardisation of the 12 variables, the differences between the variables – in terms of squared Euclidean distances – were measured. The Ward method was used to form the clusters. This technique unites cases in a way that leads to the smallest increase in the sum of squared errors term. The purpose of this method is therefore to create groups that are as homogeneous as possible. The dendogram below provides the following result for this process (Fig. 8.3).

The growth stars and the laggards (plus Norway) form two separate clusters. Within the group of growth stars, the first two links formed are between, firstly, Australia and the UK, and secondly, Finland and Sweden. Australia and the UK are then joined by Canada, and then by the USA before the cluster is finally formed with the addition of Sweden and Finland. In terms of the specifications of the indicators, Germany has the greatest proximity to Switzerland. These two countries are next joined by the Netherlands, and then by Japan. The final cluster is formed when this group is joined by another cluster comprising the other growth laggards and Norway.

According to the cluster analysis, there are two ways to achieve the greater innovative capabilities of the growth stars. There would appear to be a Nordic way and an Anglo-Saxon path to success in this area. Both of these sub-groups within the

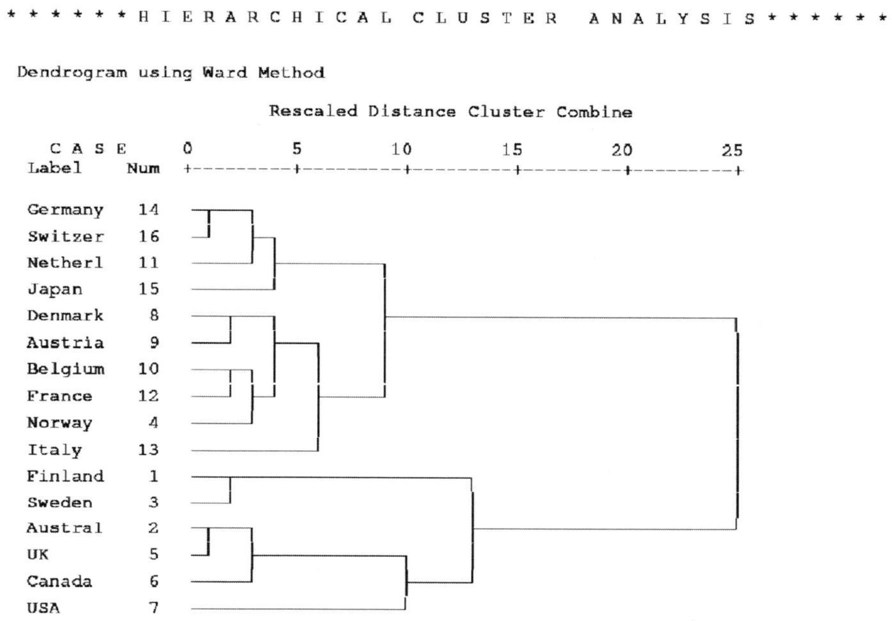

Fig. 8.3 Hierarchical cluster analysis
Source: Own calculations.

growth stars cluster have different strengths and weaknesses. Whereas Sweden and Finland, above all, perform exceptionally well in terms of investments in R&D, and in terms of Ph.D. graduates, the Anglo-Saxon countries are particularly successful in terms of labor-market regulations and the availability of venture capital.

Norway is the only country to fall out of the group of growth stars and join the laggards when the innovation indicators are considered. Norway's growth between 1994 and 2004 certainly profited from other factors, particularly the discovery and extraction of oil as well as successful policies to avoid potential negative side-effects that are not related to its innovative capabilities, as is demonstrated by different studies (Larsen, 2004; IMF, 2005).

8.3 Final Remarks

The 22 OECD countries in the current study grew, on average, by 2.4%. Leading positions were taken by Ireland and South Korea, which because of their relatively low starting positions in 1994 were, to a large extent, still able to profit from a process of catching up. This catching up was characterized, first and foremost, by investments in already existing and predominantly labor-intensive technologies. Such a strategy is, however, no option for countries at the technological frontier. In highly developed industrial countries, the ability to innovate and a fast diffusion of these innovations as well as adaptation to technical progress plays, in comparison to imitation, an increasingly important role.

If Germany or, indeed, any other highly developed country with a poor growth record wants to increase its growth potential, it needs to improve the conditions for innovations at the technological frontier and the fast diffusion of these innovations. Improvements in human capital and the de-regulation of product markets should be of the highest priority. In particular, government bureaucracy (including economically unsound subsidies) must be reduced considerably. In addition, more venture capital should be made available, and incentives to invest in ICT should be increased. A further level-headed de-regulation of the labor market and improved incentives in the welfare state which allow innovators to find workers matching their new technologies more quickly could make an additional contribution to an improvement in performance on growth and innovation.

References

Acemoglu, D., Aghion, P., & Zilibotti, F. (2002). Distance to frontier, selection and economic growth (NBER Working Paper No. 9066). Cambridge, MA: National Bureau of Economic Research.

Aghion, P., & Howitt, P. (1992). A model of growth through creative destruction. *Econometrica, 60*, 323–351.

Aghion, P., Meghier, C., & Vandenrusche, J. (2003). Growth, education and distance to the technological frontier. Mimeo, Harvard University: Boston.

Arnold, E., & Thuriaux, B. (2003). Introduction. In European Commission (Ed.), *Future directions of innovation policy in Europe* (pp. 1–10). Brussels: European Commission.

Bassanini, A., & Scarpetta, S. (2002). Growth, technological change, and ICT diffusion: Recent evidence from OECD countries. *Oxford Review of Economic Policy, 18*, 324–344.

BMBF – Bundesministerium für Bildung und Forschung (2004). *Technologie und Qualifikation für neue Märkte*. Bonn and Berlin: BMBF.

Boyer, R. (2004). New growth regimes, but still institutional diversity. *Socio-Economic Review, 2*, 1–32.

Brandt, N., Burniaux, J.-M., & Duval, R. (2005). Assessing the OECD jobs strategy: Past developments and reforms. (OECD Economics Department Working Paper No. 429). Paris: OECD.

EIS. (2005). *European Innovation Scoreboard.* http://trendchart.cordis.lu/scoreboards/scoreboard2005/pdf/EIS%202005.pdf [1-20-2006].

European Commission. (2004a). *Benchmarking enterprise policy.* Brussels: European Commission.

European Commission. (2004b). *The EU Economy: 2004 Review.* Brussels: European Commission.

EUROSTAT. (2005). *NewCronos.* http://europa.eu.int/comm/eurostat/newcronos/reference/display.do?screen=welcomeref&open=/&product=EU_MAIN_TREE&depth=1&language=de [Download: 1-20-2005].

Fagerberg, J. (2003). The innovative society. In European Commission (Ed.), *Future directions of innovation policy in Europe* (pp. 16–19). Brussels: European Commission.

Fagerberg, J. (2005). The dynamics of technology, growth and trade: A Schumpeterian perspective. In H. Hanusch & A. Pyka (Eds.), *Elgar Companion to Neo-Schumpeterian Economics.* Cheltenham: Edward Elgar, forthcoming.

Farhauer, O., & Henke, K.-D. (2002). Wachstumstheoretische Konzeptionen und wirtschaftspolitische Folgerungen. *Das Wirtschaftsstudium, 31*, 582–590.

Fehn, R. (2004). Strukturwandel und europäische Wirtschaftsverfassung: Gibt es einen Zielkonflikt zwischen Effizienz und Sicherheit? In W. Schäfer (Ed.), *Zukunftsprobleme der europäischen Wirtschaftsverfassung* (pp. 87–121). Berlin: Duncker & Humblot.

Gehrig, T. P. (2000). Zur Politischen Ökonomie des Technischen Fortschritts. In B. Külp & V. Vanberg (Eds.), *Freiheit und wettbewerbliche Ordnung. Gedenkband zur Erinnerung an Walter Eucken* (pp. 561–577). Freiburg, Berlin: Haufe Verlagsgruppe.

IMF – International Monetary Fund. (2005). *Norway – Staff report for the 2005 article IV consultation*. Washington, DC: IMF.

Jungmittag, A. (2004). Innovations, technological specialisation and economic growth in the EU. *International Economics and Economic Policy, 1*(2, 3), 247–273.

Larsen, E. R. (2004). *Escaping the resource curse and the Dutch disease? When and why Norway caught up with and forged ahead of its neighbours, statistics Norway* (Discussion Paper No. 377). Research Department.

Levine, R. (1997). Financial development and economic growth: Views and agenda. *Journal of Economic Literature, 35*, 688–726.

Matthes, J., & Schröder, C. (2005). Business regulation in international comparison – Aggregating world bank. "Doing business" data. *CESifo Forum, 6*(1), 42–49.

Murphy, K. M., Shleifer, A., & Vishny, R. W. (1991). The allocation of talent: Implications for growth, *Quarterly Journal of Economics, 106*, 503–530.

Nicoletti, G., & Scarpetta, S. (2003). *Regulation, productivity and growth: OECD evidence* (OECD Economics Department Working Papers No. 347). Paris: OECD.

OECD. (2004a). *Employment outlook*. Paris: OECD.

OECD. (2004b). *OECD science, technology and industry scoreboard 2003*. Paris: OECD. http://www1.oecd.org/publications/e-book/92-2003-04-1-7294 [Download: 11-22-2004].

OECD. (2004c). *Wirtschaftsbericht Deutschland*. Paris: OECD.

OECD. (2004d). *Lernen für die Welt von morgen*. Erste Ergebnisse von PISA 2003, Paris: OECD.

OECD. (2004e). *Education at a glance*. Paris: OECD.

OECD. (2004f). *OECD-education database*. http://www1.oecd.org/scripts/cde/members/EDU_UOEAuthenticate.asp [Download: 1-20-2005].

Romer, P. M. (1990). Endogenous technological change. *Journal of Political Economy, 98*, 71–102.

Romer, D. (1996). *Advanced macroeconomics*. New York: McGraw Hill.

Sapir, A., Aghion, P., Bertola, G., Hellwig, M., Pisani-Ferry, J., Rosati, D., et al. (2004). *An agenda for a growing Europe*. The Sapir Report. Oxford: Oxford University Press.

Scarpetta, S., & Tressel, T. (2004). Boosting productivity via innovation and adoption of new technologies: Any role for labor market institutions? (Policy Research Working Paper Series No. 3273). Washington, DC: The World Bank.

Van Ark, B. (2003). *The productivity problem of the Dutch economy: Implications for economic and social policies and business strategy*. Groningen: Research Memorandum GD-66, Groningen Growth and Development Centre.

World Bank (2004). *Doing business in 2005 – Removing obstacles to growth*. Washington, DC: World Bank.

Chapter 9
The Innovation Imperative and the Role of SBIR Innovation Awards

Charles W. Wessner

Policymakers in the United States and Germany share the recognition that innovation remains the key to international competitiveness in the 21st century. Moreover, policy makers in both countries increasingly recognize that equity-financed small firms are an effective means of capitalizing on new ideas and bringing them to the market.

Small firms, however, face a variety of obstacles as they seek to bring new products and processes to the market. In this context, public policies that reduce the structural and financial hurdles facing such innovative small firms can play a useful role in enhancing a nation's innovative capacity.

In the United States, the Small Business Innovation Research (SBIR) program has proven effective in helping small innovative firms overcome these hurdles while also enhancing networking among US universities, large firms, and small innovative companies. Innovation award programs similar in concept and scale to SBIR might also help Germany realize higher returns to its substantial and growing investments in research and development. As Germany increases its investments in R&D as a part of the Lisbon process, it becomes more important to be able to show some returns in terms of company formation and growth.

9.1 The New Innovation Imperative

This imperative to innovate more rapidly comes as new entrants from China and India expand their presence in the global economy. While this expansion provides opportunities for businesses around the world to lower costs, develop new ideas and business processes, and develop new markets, it also poses new challenges for

Charles Wessner directs the National Academies' Program on Technology, Innovation, and Entrepreneurship that includes the Board on Science, Technology, and Economic Policy's study on Comparative Innovation Policies, Best Practice for the 21st century. He also directs the Academies' major assessment of the Small Business Innovation Research Program. The author would like to recognize Dr Sujai Shivakumar many contributions to the preparation of this article.

C.W. Wessner (✉)
National Academies, Washington, DC, USA

P.J.J. Welfens, J.T. Addison (eds.), *Innovation, Employment and Growth Policy Issues in the EU and the US*, DOI 10.1007/978-3-642-00631-9_9,

countries like Germany and the United States to maintain their competitiveness and preserve their standard of living by accelerating their innovative potential.

China brings not only scale advantages, but also a remarkable high-level focus to the challenge of competitiveness. The clear goal of China's leaders is the acquisition of technological capabilities and control of national market as a means of maintaining national autonomy and generating political and military strength. As President Jiang Zemin (August 23, 1999) stated, "In today's world, the core of each country's competitive strength is intellectual innovation, technological innovation and high-tech industrialization." This high level commitment is evident in the rapid rise in Chinese R&D expenditure. In 1999, China's R&D spending accounted for 6% of the total world expenditures in R&D ($618 billion). By 2005, China accounted for 13% of the world total of $836 billion spent on R&D.[1]

India is also an increasingly important locus of innovation.[2] Gaining momentum from a decade of economic liberalization, India today is changing rapidly from a locus for business process outsourcing to a global center for advanced research and development.[3] Even as US and other multinational companies are increasingly locating their advanced research and development operations in India, Indian companies are themselves – drawing on their nation's vibrant entrepreneurial class and a critical mass of capable, highly trained scientists and engineers – seeking to become globally competitive, with active international partnering and acquisition strategies now underway.[4]

The rise of new competitors and new markets alerts us to the need to invest in our nations' innovation potential. Growing our capacity for rapid and productive innovation is essential for our future security and economic well-being.[5] A key element in enhancing a nation's innovation capacity is its small firms. They play a catalytic role in capitalizing on existing public investments in research to bring new ideas to the market.[6]

---

[1]OECD, Main Science and Technology Indicators, 2006.

[2]For an analysis of India's economic potential compared to China, see Yasheng Huang and Tarun Khanna, "Can India Overtake China?" *Foreign Policy*, July–August, 2003. The authors argue that India's development strategy, while initiated later than China's and thus lagging China, is more sustainable because it is more strongly based on fostering bottom-up entrepreneurial capacity.

[3]National Research Council, *India's Changing Innovation System*, C. Wessner and S. Shivakumar, eds., Washington, DC: National Academies Press, 2007.

[4]Tata's takeover of Corus Steel is among a series of recent global acquisitions by Indian firms. Other Indian firms recently acquiring assets overseas include Bharat Forge, Ranbaxy, Wipro, and Nicholas Piramal. According to the *Economist*, Indian companies announced 115 foreign acquisitions, with a total value of $7.4 billion in the first 3 quarters of 2006. See *Economist*, "India's acquisition spree," October 12, 2006.

[5]France, for example, has allocated 1 billion euros towards a new Industrial Innovation Agency, following the release of the Beffa Report. Similarly, Japan is restructuring and making significant investments its innovation system. For a review of innovation policies around the world, see National Research Council, *Innovation Policies for the 21st Century*, C. Wessner, ed., Washington, DC: National Academies Press, 2007.

[6]Zoltan J. Acs and David B. Audretsch, *Innovation and Small Firms,* Cambridge, MA: MIT Press, 1990.

9.2 The Importance of Small Business Innovation

It is now widely recognized that, small firms are a leading source of employment growth in the United States, generating 60–80% of net new jobs annually over the past decade. What is less widely recognized is that these small businesses also employ nearly 40% of the United States' science and engineering workforce.[7] What is more, scientists and engineers working in small businesses, produce 14 times more patents than their counterparts in large patenting firms in the United States – and these patents tend to be of higher quality and are twice as likely to be cited.[8]

In the United States, firms like Microsoft, Intel, AMD, FedEx, Qualcomm, Adobe, all of which grew rapidly in scale from small beginnings, continue to transform how people everywhere work, transact, and communicate. The resulting economic growth and social benefits underscore the need to encourage new equity-based high-technology firms in both the United States and Germany in the hope that some may develop in to larger, more successful firms that create the technological base for future competitiveness.

9.3 US Strengths in Innovation

This capacity to renew the economy by developing numerous large firms from small beginnings is a significant comparative advantage for the United States. It draws on the nation's large and integrated domestic market, and an economic and institutional infrastructure that is able to re-deploy resources rapidly to maximize their efficient use. A strong, highly-developed, and highly diverse educational infrastructure involving both important public and private entities is another key advantage. Deep and flexible capital and labor markets permit the rapid reallocation of resources mentioned above. These competitive strengths are buttressed by highly-distributed and highly-developed S&T institutions that are endowed with significant resources and charged with applications-oriented missions ranging fro space exploration to health to national security. Flexible managerial and organizational structures and a willingness to adopt innovative management practices and products are further distinguishing features of the US economy.

The United States also benefits from an entrepreneurial culture that accepts failure as a byproduct of new entrepreneurial initiatives and a willingness of investors to provide second opportunities to experienced, if initially unsuccessful, managers. Bankruptcy laws that limit the liability that an entrepreneur can incur from an initial bankruptcy further support this "second-chance" cultural and business perspective on entrepreneurial failure and success.

These American characteristics – the diversity, decentralization, and willingness to take initiative and tolerate failure – are not new. Writing in the 19th century,

[7] Small Business Administration, Office of Advocacy, "Small Business by the Numbers," June 2004.

[8] Ibid.

Alexis de Tocqueville noted in *Democracy in America* that "In America, the social force behind the state is much less well regulated, less enlightened, and less wise, but it is a 100 times more powerful than in Europe. Without doubt, there is no other country on earth where people take such great efforts to achieve social prosperity. So it is no good looking in the United States for perfection of administrative procedures; what one does find is a picture of power, somewhat wild perhaps, but robust, and a life liable to mishaps but full of striving and animation."[9]

Yet, despite these many national strengths, many regions in the United States, as in Germany, do not participate as effectively as they might in the national innovation system. Often, this divergence is described in terms of the performance of individual states. Yet, the reality is that there is great diversity in the innovative capacity of the various cities and regions of the United States. Much of the innovative activity in the US is geographically concentrated in only parts of states like California and Massachusetts, reflecting in part the polycentric nature of the American federal system of government and the local concentrations that characterize advanced innovation systems (see Fig. 9.1 below). As the figure below highlights, patenting activity varies dramatically by city.

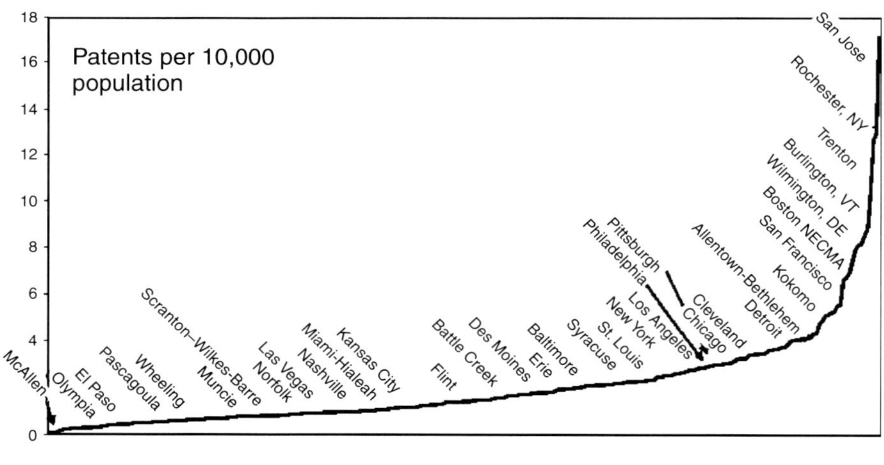

Fig. 9.1 Urban density and the rate of innovation[10]

[9]Alexis de Tocqueville, *Democracy in America,* Translated by Henry Reeve, Esq. With an Original Preface and Notes by John C. Spencer. New York: Adlard and Saunders, 1838.

[10]Source: Gerald, Carlino, Satyajit Chatterjee, and Robert Hunt, Ferrara OECD Conference on IAVC, October 2005.

9.4 Growing Concerns About US Competitiveness

These underlying strengths in the US system suggest that the overall economic prospects in the United States today will remain healthy. Nonetheless, there are clouds on the horizon. Many business leaders, senior academics, and experienced policymakers in the United States believe that the country is beginning to face major challenges to its technological leadership. They point, for example, to inadequacies in the education system, especially at the secondary level where US students score below their peers abroad in science and mathematics. These concerns have spawned recent studies that highlight troubling trends in publications, foreign student retention, high-technology exports and the production of information technology products.[11] It is also true that fewer American students are pursuing science careers, and that the United States may be losing some of its attraction as a destination for the best students from around the world.[12]

Responding to this and other concerns about the nation's innovation capacity, the United States Congress recently requested the National Academies to assess the nation's competitive situation and identify concrete steps to ensure US economic leadership. The resulting National Academies report, "Rising Above the Gathering Storm" notes that weakening federal commitments to science and technology places the future growth and prosperity of the United States in jeopardy:

> Although many people assume that the United States will always be a world leader in science and technology, this may not continue to be the case, inasmuch as great minds exist throughout the world. We fear the abruptness with which a lead in science and technology can be lost – and the difficulty of recovering a lead one lost, if indeed it can be regained at all.[13]

To overcome this growing vulnerability, the report calls for (among other measures) increasing America's talent pool by providing greater incentives for science and mathematics teachers. The report also calls for increasing federal investments in long-term basic research by 10% per annum over the next 7 years. In addition, it recommends a variety of steps to make the United States a more attractive place to study and perform research for foreign students, including actions to increase the

---

[11] See for example, recent reports by the President's Council of Advisors on Science and Technology, "Sustaining the Nation's Innovation Ecosystems," January 2004; the Council on Competitiveness, *Innovate America: Thriving in a World of Challenge and Change*, Washington, DC, 2005; and the National Academy of Sciences, *Rising Above the Gathering Storm, Energizing and Employing America for a Brighter Economic Future,* Washington, DC: National Academies Press, 2005.

[12] The drop off in foreign students was largely self-inflicted. Following the 9/11 attacks, the US government imposed much tighter controls on foreign students without having the necessary procedures and adequate staff in place to implement them. The result was long delays, lengthy travel, and often arbitrary rulings. In the last few years, procedures have become relatively more stabilized and the US share of foreign students is again rising.

[13] National Research Council, *Rising Above the Gathering Strom*, 2005, op. cit.

number of visas that permit US trained foreign students to remain and work in the United States after their studies are completed.[14]

The Academies report has helped to advance the policy debate. In his January 2006 State of the Union Speech, President Bush called for a competitiveness initiative that would, inter alia, double the federal commitment to basic research programs in physics and engineering over 10 years, improve K-12 education in math and science, reform and expand workforce training programs, and support immigration reform to compete for the world's best and brightest high-skilled workers.[15] More recently, the US Senate passed "America Competes Act" on April 26, 2007 by a significant margin. This legislation, which also draws specifically on the Academies' Gathering Storm report, increases the federal government's research investments, and calls for steps to strengthen educational opportunities in science, technology and engineering.[16]

9.5 Policy Myths and Innovation Realities

While these new measures are welcome, commonly held myths about the innovation process remain an obstacle to developing and maintaining policies that encourage small business innovation. Many American policymakers have ideological convictions about the primacy of the market and a corresponding reluctance to recognize the limitations of markets despite ample evidence concerning the close interactions between markets and public policy initiatives to encourage innovation.

In the case of early-stage finance, a common myth, at least among Washington policymakers, is that "if it's a good idea, the market will fund it." In reality, markets are not perfect. Unlike the abstract models found in introductory economics texts, markets in the real world always operate within specific rules and conventions that lend unique characteristics to particular markets. What is more, market participants necessarily operate with less than perfect information.

Indeed, the problem of imperfect capital markets is particularly challenging for fledgling entrepreneurs. The knowledge that an entrepreneur has about his or her product is normally not fully appreciated by potential customers – a phenomenon that economists call *asymmetric information*. This asymmetry can make it hard for small firms to obtain funding for new ideas; as Michael Spence, a recent winner of the Nobel Prize in economics points out, market noise often obscures the significance of promising new ideas.[17]

Market entry is thus a challenge for new entrepreneurs, especially those with new ideas for a potentially disruptive product. Access to government procurement

[14]Ibid, p. ES-2.

[15]Access at http://www.whitehouse.gov/stateoftheunion/2006/aci/#section4

[16]Access at http://thomas.loc.gov/cgi-bin/query/D?c110:1:./temp/~c110FrdsGL

[17]The Nobel Committee cited Spence's contribution in highlighting the importance of market signals in the presence of information asymmetries. For his seminal paper on this.

markets can be particularly difficult for new, small entrepreneurial firms. These entrepreneurs tend to be unfamiliar with arcane government regulations and complex procurement procedures that are often referred to as a "procurement thicket." In addition, academic researchers and novice entrepreneurs may be unacquainted with government commercial accounting and business practices, a more prosaic but important obstacle.[18] Many small firms are, therefore, at a disadvantage – especially compared with more established suppliers in the defense procurement process – in dealing both with the regulations encountered in government contracting and scale requirements. Such firms face especially high challenges with regard to market access and finance.[19]

Another hurdle for entrepreneurs is *the leakage of new knowledge* that escapes the boundaries of firms and intellectual property protection. The creator of new knowledge can seldom fully capture the economic value of that knowledge for his or her own firm. This spillover can inhibit investment in promising technologies for large and small firms – though it is especially important for small firms focused on a particularly promising product or process.[20]

The challenge of incomplete and insufficient information for investors and the problem for entrepreneurs of moving quickly enough to capture a sufficient return on "leaky" investments pose substantial obstacles for new firms seeking private capital. The difficulty of attracting investors to support an imperfectly understood, as yet-to-be-developed innovation is especially daunting. Indeed, the term, *Valley of Death* has come to describe the period of transition when a developing technology is deemed promising, but too new to validate its commercial potential and thereby attract the capital necessary for its development[21] (see Fig. 9.2).

This simple image of the Valley of Death captures two important points. The first is that while there are substantial national R&D investments in the US, Germany, and elsewhere, transitioning these investments in research to create valuable products is not self-evident, given the informational and financial constraints noted above. A second, related point is that technological potential does not lead inevitably

[18]See National Research Council, *SBIR and the Phase III Challenge of Commercialization*, C. Wessner, ed., Washington, DC: National Academies Press, 2007.

[19]With regard to the challenges small firms face in obtaining funding, see Branscomb and Auerswald, *Taking Technical Risks*, op. cit. See also Josh Lerner, "Public Venture Capital," in National Research Council, *The Small Business Innovation Program: Challenges and Opportunities,* C. Wessner, ed. Washington, DC: National Academy Press, 1999.

[20]Edwin Mansfield, "How Fast Does New Industrial Technology Leak Out?" *Journal of Industrial Economics,* Vol. 34, No. 2, pp. 217–224.

[21]The analyses carried out at the Academies on early stage finance programs have highlighted the importance of these challenges and helped policymakers appreciate the Valley of Death phenomenon. For an overview of the Academies' analyses, see National Research Council, *Government-Industry Partnerships for the Development of New Technologies*, C. Wessner, ed., Washington, DC: National Academies Press, 2003. For an early articulation of this phenomenon, see Vernon J. Ehlers, *Unlocking Our Future: Toward a New National Science Policy, A Report to Congress by the House Committee on Science* (Washington, DC: GPO, 1998). Accessed at http://www.access.gpo.gov/congress/house/science/cp105-b/science105b.pdf

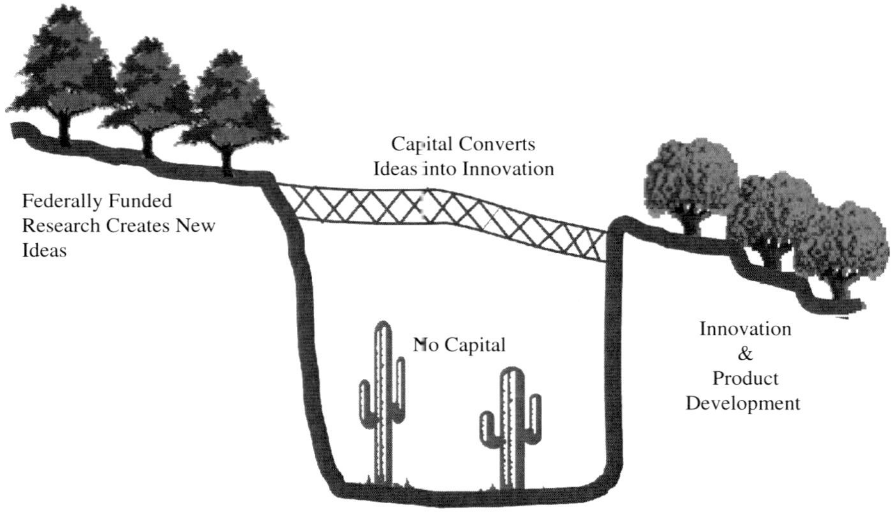

Fig. 9.2 The valley of death

to commercial reality. Many good ideas perish on the way to the market. The challenge for policymakers is to help firms create additional, market relevant information by supporting the development of promising ideas through this difficult early phase.

9.6 Does Venture Capital Provide the Bridge?

Notwithstanding the reality of these early-stage financing hurdles, many policy makers in the United States believe that the US venture capital markets are so broad and deep that entrepreneurs can readily access the capital needed to cross the Valley of Death. In fact, venture capitalists not only have limited information on new firms but are also prone to herding tendencies, as witnessed in the dot.com boom and bust.[22] Venture capitalists are also, quite naturally, risk averse. Their primary goal, after all, is not to develop the nation's economy but to earn significant returns for their investors.[23] Most venture funds tend to focus on later stages of technology development because there is more information at this stage in the process about the commercial prospects of the innovation (and hence less risk to their investment.) As the figures below show, the result is that the US venture capital market, although

---

[22]See Tom Jacobs, "Biotech follows dot.com boom and bust," *Nature*, Vol. 20, No. 10, p. 973, October 2002.

[23]"The goal of venture capitalists is to make a return for our investors – it is not to develop the economy." Personal communication with David Morgenthaler, founder Morgenthaler Ventures and past President of the National Venture Capital Association.

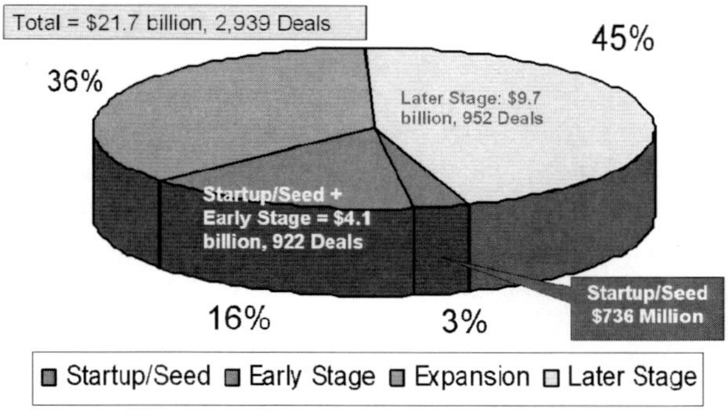

Fig. 9.3 Breakdown of US venture capital by stage of development (2005)

large, is not focused on early-stage firms: In 2005, startups in the United States received only $736 million or about 3% of the $21.7 billion of available venture capital (see Figs. 9.3 and 9.4).

VCs (PWC MoneyTreeTM data)	**~$0.3 billion (<200 companies)**
State Funds (Estimate by S. Weiss)	**<$0.5 billion**
Angel investors (Center for Venture Research)	**$20 billion**
SBIR/STTR Funds	**$2.2 billion**

Fig. 9.4 Multiple sources of early-stage finances

9.7 Bridging the Funding Gap

The limitations of the market for venture capital mean that small innovative firms seek funding from a variety of sources.[24] In addition to business angels and venture capital firms, early stage technology firms also seek development funding from industry, federal and state governments, and universities. Indeed the diversity of these sources for early-stage funding represents one of the strengths of the US system.

As the figure above shows, venture funding is a small proportion of early-stage finance. State funds play a larger role than many realize and this source of investment is growing rapidly. By far the most important contributor to early stage funding

[24]Branscomb and Auerswald, *Between Invention and Innovation: An Analysis of Funding for Early-Stage Technology Development*, NIST, 2002

are angel investors, at more than 20 times the contribution of state and venture funding combined.

Interestingly enough, among these funding sources, the role of the federal government is significant both in size and importance. Research by Branscomb and Auerswald estimates that the federal government provides between 20 and 25% of all funds for early stage technology development – a substantial role by any measure and one often surprising to Americans in its dimensions[25] (see Fig. 9.5). This federal contribution is rendered more significant in that competitive government awards address segments of the innovation cycle that private institutional investors often (quite rightly) find too risky for investment.

The availability of early stage financing and its interaction with other elements of the US innovation process are the focus of growing analytical efforts.[26] We examine below, the case of the Small Business Innovation Research Program.

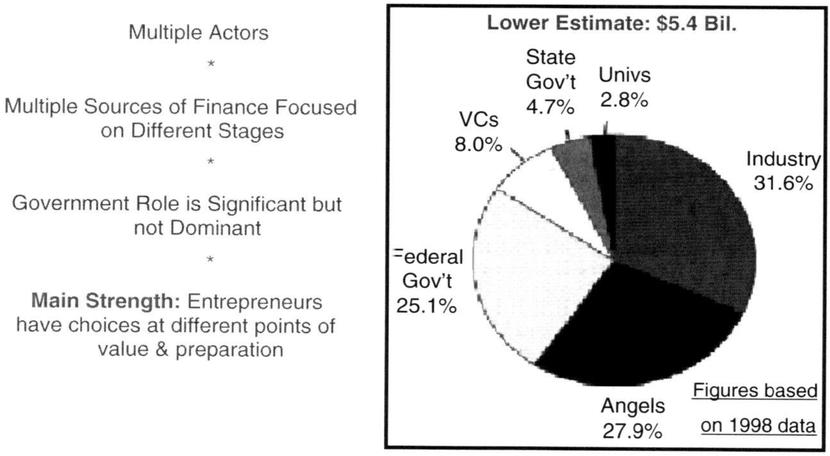

Fig. 9.5 Estimated distribution of funding sources for early-stage technology development

[25]It is important to remember that these are estimates. The authors stress the "limitations inherent in the data and the magnitude of the extrapolations. . ." and urge that the findings be interpreted with caution. They note further that while the funding range presented for each category is large, these approximate estimates, nonetheless, provide "valuable insight into the overall scale and composition of early-stage technology development funding patterns and allow at least a preliminary comparison of the relative level of federal, state, and private investments." For further discussion of the approach and its limitations, see Lewis M. Branscomb and Philip E. Auerswald, *Between Invention and Innovation, An Analysis of Funding for Early-Stage Technology Development,* Gaithersburg, MD: NIST GCR 02–841, pp. 20–24, November 2002.

[26]The growth and subsequent contribution of venture capital have begun to attract the serious study needed to illuminate the dynamics of high-technology firm evolution. See for example, the work of Jeffrey Sohl and colleagues and the University of New Hampshire's Center for Venture Research, described at http://www.unh.edu/cvr

9.8 The Small Business Innovation Research Program (SBIR)

Created in 1982 through the Small Business Innovation Development Act, SBIR leverages small business innovation to address government and societal needs in such areas as health, security, environment, and energy. The strength of the SBIR concept lies in aligning the interests of each of the participants in the program with the goals of the program. SBIR proposals are industry-initiated based on broad solicitations posted by federal agencies. This bottom-up design promotes a positive interest by small businesses in the outcome of their research. Similarly, the federal agencies can each use the program to advance their own missions; ownership rests with the many agencies, not a single "tech agency."

Since its establishment in 1982, the SBIR program budget has grown to some $2 billion per year. Eleven federal agencies are currently required to set aside 2.5% of their extramural research and development budget exclusively for SBIR contracts for small companies.[27,28] Each year these agencies identify various R&D topics for pursuit by small businesses under the SBIR program, representing scientific and technical problems requiring innovative solutions.

Features that make SBIR grants attractive from the perspective of the entrepreneur, aside from the funding itself, include the fact that there is no dilution of ownership or repayment required. Importantly, grant recipients retain rights to intellectual property developed using the SBIR award, with no royalties owed to the government. The government retains royalty free use for a period, but this is very rarely exercised. Selection to receive SBIR grants also tend to confer a certification effect – a signal to private investors of the technical and commercial promise of the technology.[29] The awards usually provide a faster route to new technology than traditional procurement processes allow. Importantly, firms with successful SBIR awards add to the diversity and competitiveness of the supplier base for US companies.

From the perspective of the government, the SBIR program helps officials draw on private sector ingenuity to achieve their respective agency missions.[30]

[27]A small business is defined in the US as having less than 500 employees, significantly larger than the 250 employee cap applied in Europe.

[28]These include the Department of Defense, the Department of Health and Human Services, the National Aeronautics and Space Administration, the Department of Energy, the National Science Foundation, the Department of Agriculture, the Department of Commerce, the Department of Education, the Department of Transportation, the Environmental Protection Agency, and, most recently, the Department of Homeland Security.

[29]This certification effect was initially identified by Josh Lerner, "Public Venture Capital," in National Research Council, The Small Business Innovation Program: Challenges and Opportunities, C. Wessner, ed. Washington, DC: National Academy Press, 1999. For a similar concept from the Advanced Technology Program, see Maryann Feldman and Maryellen Kelly, "Leveraging Research and Development: The Impact of the Advanced Technology Program," in National Research Council, *The Advanced Technology Program, Assessing Outcomes*, C. Wessner, ed., Washington, DC: National Academy Press, 2001.

[30]See National Research Council, *The Small Business Innovation Research Program, Program Diversity and Assessment Challenges,* Washington, DC: National Academies Press, 2004.

By providing a bridge between small companies and the federal agencies, especially for procurement, SBIR serves as a catalyst for the development of new ideas and new technologies to meet federal missions in health, transport, the environment, and defense.

9.8.1 The National Academies Assessment of SBIR

As the Small Business Innovation Research (SBIR) program approached its 20th year of operation, the US Congress asked the National Research Council (NRC) to carry out a "comprehensive study of how the SBIR program has stimulated technological innovation and used small businesses to meet Federal research and development needs" and make recommendations on improvements to the program.[31] The resulting $5 million, multi-year effort has yielded new, in-depth information about the advantages and limitations of the SBIR concept, as well as identified key factors necessary (but not sufficient) for the program's successful implementation.[32] Drawing from this study, the program's key conceptual attributes are summarized below.

9.8.2 Some Advantages of the SBIR Concept

- *SBIR plays a catalytic role at an early-stage in the technology development cycle.* The awards have the virtue that they are not repayable and they do not dilute ownership or control of a firm's management. The awards enable the firm to explore new technological options and in Phase II can often demonstrate a technology's potential. The intellectual property rights remain with the firm, creating an opportunity for downstream contracts. Perhaps most importantly, the awards provide a signaling effect affirming both the quality and a potential market for the technology. These advantages are significant and not matched by private finance or other public-sector mechanisms.
- *Few private sector substitutes exist to the SBIR program.* The SBIR contributions are also quite distinct from both bank lending and private equity financing of small technology firms. Commercial lending places a financial burden on small businesses with a long product-development cycle. Private equity funding does not require the small business to keep up with interest payments, but it does require the small business to give up a share of ownership and an associated measure of control. Firms that seek equity funding early in their development may be

[31] See Public Law 106-554, Appendix I – H.R. 5667, Section 108.

[32] National Research Council, *An Assessment of the Small Business Innovation Research Program*, C. Wessner, ed., Washington, DC: National Academies Press, forthcoming.

compelled to accept lower valuations that result in a greater loss of control for the same amount of funding than would be the case at a later stage of development.[33]

- *SBIR does not compete with the private financial markets.* On the contrary, often it facilitates the functioning of financial markets by signaling quality – reducing information asymmetries that complicate contracting. Such quality signaling partially reduces the impediments typically faced by technology entrepreneurs in seeking financing from both private and public sources.
- *Few public-sector substitutes exist to the SBIR program.* SBIR continues to offer small firms an opportunity to perform Federal R&D, while minimizing the burdens of government contracting regulations. No other government-wide program offers technology-development awards to small business in amounts, or with objectives, comparable to the SBIR program.[34]
- *SBIR provides a bridge between universities and the marketplace.* Surveys and firm case studies commissioned over the course of the National Academies' assessment reveal the existence of extensive networking between SBIR-funded projects and universities. University staff and faculty often assist with proposal preparation, provide facilities and equipment, and make ongoing contributions to the intellectual capital underpinning company innovations. Faculty also often served as proposal reviewers. In addition, university faculty and students use SBIR to establish businesses, start projects, and work on projects. SBIR has facilitated technology transfer out of universities.

9.8.3 Current Limitations of the SBIR Concept

- *Overhead costs.* The task of providing many small awards to small businesses involves high overhead costs for the administering agency. Agency managers often require more time and energy to solicit, evaluate, and monitor multiple small awards than for larger contracts.

---

[33]Paul A. Gompers and Josh Lerner, *The Venture Capital Cycle*. MIT Press, 1999. Lewis M. Branscomb and Philip E. Auerswald, *Between Invention and Innovation: An Analysis of Funding for Early Stage Technology Development*. Advanced Technology Program, National Institute for Standards and Technology (NIST), US Department of Commerce report, NIST GCR 02–841, November 2002; Bronwyn H. Hall, *The Financing of Research and Development*, Working Paper 8773, National Bureau of Economic Research (NBER), 2002. Branscomb and Auerswald (2002) report that overall, of $266 billion that was spent on national R&D by various sources in the US in 1998, substantially less than 14% flowed into early stage technology development activities. The exact figure is elusive, because public financial reporting is not required for these investments.

[34]A program that is comparable along a number of dimensions – although with important differences in objectives, award size, and its emphasis on partnering – is the Advanced Technology Program (ATP), administered by the National Institute of Standards and Technology at the Department of Commerce. While the ATP program similarly stresses the development of novel technologies, that program does not focus exclusively on small firms. Furthermore, award amounts in the ATP program are typically at a factor of 10 larger than the standard SBIR phase I award; and awards are not directly related to agency mission.

- *Uncertainty.* New ideas and new firms involve higher levels of uncertainty than with more established (if less innovative) suppliers. As a result, many awards, especially those associated with higher-risk, less proven technologies, will not be successful.
- *Processing time.* The time agencies take to solicit, assess, select, and make awards can be a challenge for small firms, especially those newly established. Making the awards in a timely fashion that meets the needs of the small companies, in terms of technology development and potential markets of their product or service, remains a challenge for the program.
- *The commercialization challenge.* Transitioning products and processes developed under the SBIR program into the procurement processes of agencies or into the private markets remain a challenge for the program. The need for appropriate timing for inclusion of a new product in government procurement activities, sufficient demonstration of project potential and firm capability, and sometimes the inability to communicate a project's potential to acquisition officials are all related challenges.

SBIR – Key Factors for Success

- *Competitively awarded grants.* SBIR is intensely competitive, with multiple-stage reviews and a limited number of successful applicants. In this regard, it may be compared aptly to leading scholarship programs for outstanding students, not only in terms of the success rate but more profoundly in terms of the social investment in private individuals based on the rationale of long-term public gain.
- *Limited in time and amount.* It is important that innovation award programs remain open to new entrants and stay competitive for each round of funding. This does not mean that companies cannot re-apply for additional work. It does mean, however, that there are no "politically favored firms" able to draw regularly on government support.
- *Industry ownership through shared costs and risks.* SBIR awards are sufficiently small to mitigate some of the risk and preserve the benefits of ownership. In this way, the appropriate incentives are in place to encourage successful entrepreneurship.
- *No direct recoupment.* The federal government does not seek to recoup the funds that they grant to the companies that are successful in the SBIR competition. This is sound practice because it is often difficult to distinguish the relative contribution of an award for a particular project. (Paradoxically, even unsuccessful projects can prove valuable in that they illuminate technological dead ends while imparting knowledge that can be useful on a related technological path.) The cost of determining the contribution of an award can be both a poor use of public funds and a deterrent to would-be

participants. The US system relies, very simply, on the tax system to obtain funds from salaries paid to workers and managers and on firm earnings.

- *Learning through experimentation and regular evaluation.* SBIR's culture of evaluation is developing. The agencies responsible for the SBIR program were required, at the last program reauthorization, to undertake a major evaluation by the National Academies. This review the program's achievements, operations, and challenges have contributed to the development of a "culture of evaluation" among the agencies responsible for managing the SBIR program. Additional studies have been commissioned by the agencies and greater program experimentation has occurred.

9.9 Our Common Innovation Challenges

While national innovation systems differ in scale and flexibility, both Germany and the United States face similar challenges in innovation. We have to address the new competition from low-wage, high-skill countries by becoming more innovative and productive and we have to justify R&D expenditures by creating new jobs and new wealth.

Although already one of the world's most innovative countries, German policymakers recognize the need to spur innovation to sustain their nation's economic growth and standard of living. In this regard, innovation award programs, similar to the SBIR concept, may provide Germany higher returns on its significant investments in R&D.

In the United States, SBIR serves as a catalyst for innovation by motivating entrepreneurs and by generating new information for the capital markets about the commercial potential of new ideas. They encourage networking by bringing together university researchers and the small and big business communities. They also introduce better solutions for government mission – thereby keeping down the cost of government services. Lastly, by introducing new technologies to the marketplace, and by creating more competition with beneficial effects on price and quality, SBIR enhances the nation's economic growth and competitiveness.

The SBIR program illustrates that effective partnerships require that entrepreneurs, firms, government agencies, and other organizations be able to each work towards a common goal. For this cooperation to occur successfully, effective industry-led leadership, shared costs and shared stakes in a positive outcome, as well as regular evaluation and learning are necessary.

Chapter 10
Trips, Pharmaceutical Innovation and Access to Medicines

Mario Villarreal and Elizabeth DuPre

10.1 Introduction

In the late 1970s, creative minds began harnessing the power of genetics and molecular biology to fuel an emerging revolution in drug development. Ever since, pharmaceutical innovations have been prolonging human life and alleviating human suffering, by treating what were once intractable diseases and medical conditions. In the years ahead, patients will likely benefit further from a growing arsenal of precision-guided weapons that more effectively target the lifelines of disease while minimizing the collateral damage of side effects.

While patients in developed countries will soon have access to a plethora of next-generation pharmaceuticals, ailing citizens of poor and developing nations will still be struggling to procure drugs of last generation. Already, in industrialized countries, antiretroviral therapies are helping the majority of HIV-positive individuals live longer, often symptom free, lives; yet, in sub-Saharan Africa only 11% of the HIV-infected population can even acquire these life-saving treatments.[1] Amidst the multitude of barriers that impede access, new international political considerations may be making prospects worse for those least able to help themselves.

One such consideration centers upon the issue of patents for pharmaceutical products. International disputes over access to medicines and property rights have already led to the establishment of the WTO's Agreement on Trade-Related Aspects of Intellectual Property Rights (TRIPs), which sets minimum standard intellectual property rights (IPRs) that member nations must respect and enforce.

Further public health considerations regarding developing and least-developed countries were taken into account in the Doha Declaration, a WTO ministerial

[1] WHO/UNAIDS (2005).

M. Villarreal (✉)
American Enterprise Institute, Washington, DC, USA

P.J.J. Welfens, J.T. Addison (eds.), *Innovation, Employment and Growth Policy Issues in the EU and the US*, DOI 10.1007/978-3-642-00631-9_10,

declaration that provides a framework for negotiations and implementation of the guidelines outlined in the original TRIPS agreement.[2]

Public health issues can be controversial. After all, directly or indirectly, we are talking about the quality of human life. It is no surprise that the discussion about patents and the pharmaceutical industry has generated a rich and wide range of academic work that is often based on moral arguments.

The debate stems from two apparently contradicting goals: promoting pharmaceutical innovation and making the resulting therapeutics available to those that need them, despite their ability to pay. Determining what approach will effectively achieve both objectives always elicits intense debate involving not only academics, but international governments and pharmaceutical companies as well, with each party having its own opinion on what the "right" or "humane" course of action should be.

In our view, these heated debates spring largely from a time-inconsistency problem, where a number of incentives are misaligned and moving in opposite directions. Economic incentives (in the form of patents and IPR protection) favor innovation in the long-run, while political incentives favor action (in the form of compulsory licensing or parallel imports) in the short-run. In order to save 1 million lives today, governments may opt to engage in actions that ultimately deny the world of innovations that could save 3 million lives tomorrow. A clear understanding of this situation would aid policymakers in predicting the overall benefits and costs of any public health policy reform at both the domestic and international level.

Our paper explores this time inconsistency problem and discusses its importance in determining how to bridge divergent political and economic motives. In doing so, we explore the role of economic and political incentives in dealing with patents and access to medicines? Within this context we address the patent system, the unique dynamics of the pharmaceutical industry and its dependence on intellectual property, and the role of the TRIPS Agreement.

10.2 The Economics of Patents

> The patent system added the fuel of interest to the fire of genius.
>
> Abraham Lincoln

A great deal of the value of a unique idea lies in the possibility of realizing economic gains. The process involved in the development of a novel invention requires a significant amount of intellectual and financial resources. In order for an inventor to reap the benefits of these investments, a property right system must be in place.

Intellectual property rights reflect the real value of a new invention and compensate the inventor for the amount of resources that were expended to make that new idea or product useful and suitable for human consumption. A patent is an exclusive

[2]The Doha Declaration includes considerations regarding not only pharmaceutical products, but 20 other subjects, including agriculture, trade facilitation, and the environment.

right granted by a country to an inventor that allows him or her to exclude others from making or selling the invention in that country during the life of the patent. Once a patent for an invention is granted, the holder may directly market the product or sell the product's rights to a third party in order to recoup development costs and obtain profit from investments.

There are several potential benefits associated with intellectual property rights (IPRs) and patents, but before discussing them, an important economic point needs to be made. Patents grant temporary monopoly power and lead to some economic inefficiency, since a monopoly reduces output and increases prices above marginal cost for the sake of maximizing profit. Because of this waste, it makes sense to limit the term of a patent. But, how long a patent term should be is a question that can only be arbitrarily answered.

Theoretically, the term of a patent should optimally end when the marginal benefits are equal to the marginal costs. The term should be long enough to make the investment profitable. If the term is too short, the incentive to innovate will diminish or disappear, and society will advance at a much slower pace. Conversely, if the term is too long, producers will enjoy excessive, and perhaps unwarranted, profits at the expense of consumers, via prices above marginal cost.

In practice, however, it is nearly impossible to figure out what that optimal term should be. The cost of producing a new invention is often unpredictable, and differs from one product to another. And even if this cost is known, a single patent term may be too long for one invention, but too short for another. In addition, it is unfeasible to justly and accurately predict how much profit a company is entitled to for a particular product, and thus to specify absolute patent terms.

Nevertheless, a patent system is a necessary, albeit imperfect, solution to the incentive problem. Despite the economic inefficiencies that may stem from patents, consumers are better off than they would have been had the invention not occurred. Adequate patent protection has other positive effects as well.

Patents promote innovation. As we discussed earlier, they provide an incentive mechanism that favors the development of new ideas and products. The development of new technologies and productive discoveries is a common characteristic of societies with strong IPRs (Shapiro & Hassett, 2005). While patents directly incentivize innovation, they also indirectly promote the transfer of knowledge to society at large. In exchange for exclusive commercial rights for a limited period of time, patent holders have to disclose the intricacies of how the invention was developed. This encourages further novel developments from both the patent holder and those using the product or process, since additional improvements can also be subject to patent protection.

Are patents important just for developed economies? After all, developing and poor countries are primarily "technology consumers" and desire to reduce the costs involved with the use of new products or technologies, even at the expense of other's property rights. Moreover, since these countries consist of relatively small economies, the overall amount of innovation coming from less developed nations may be negligible. The short answer to this question is no. Patents are really important for developing countries. The long-term economic benefits of respecting IPRs greatly offset the potential short-term costs.

However, this is not to say that short-run economic and political incentives are irrelevant, and therefore unworthy of careful consideration. On the contrary, it is our view that this tension between short-run objectives and long-run incentives is a major component of the present international controversy regarding patents and the pharmaceutical industry.[3] We elaborate on this point in Sect. 10.4. What is important to establish in this section is that strong IPR protection in developing countries benefits not only the patent holder, but these countries as well.

Patents encourage trade and foreign direct investment (FDI). Fink and Primo Braga (2005) suggest that stronger IPRs have a positive effect on total trade, although they have ambiguous effects on high-technology industries (Maskus & Penubarti, 1995). Under an adequate IPR protection structure, trade is just one of the options a patent holder can utilize to serve a given market. Instead of exporting their products, they may choose to produce them abroad. If the risk of loosing their property rights is low, companies may choose to use FDI in the country, reducing trade flows but transferring know-how.

Conversely, FDI is unlikely to occur in countries with weak IPRs, and if it does, its level of technological sophistication is lower than under adequate IPR protection (Javorcik, 2005). Stronger IPR protection makes investments in local manufacturing more likely. Under weaker IPR regimes, companies tend to avoid more substantial long-term production commitments, focusing on investments in distribution facilities.

All these benefits have a positive impact on economic growth and productivity. Gould and Gruben (1996) determined that strong property right protection promotes growth in countries with low per capita incomes, mainly by encouraging trade and foreign direct investment. Moreover, the relative gains from these benefits are much larger for poorer countries than countries with high per capita incomes.

In addition, recent research on economic growth determinants has stressed the importance of institutional arrangements that favor productive investments. Innovation flourishes in developed economies characterized by institutions that reduce the transaction costs associated with engaging in productive endeavors. Such institutions are related to high levels of political and economic stability, and include a strong rule of law and protection of intellectual property rights. Therefore, an environment that favors economic growth and progress is one that enables individuals and companies to invest and profit from their investment.

Can the rights granted by patents be abused? Yes. Aside from excessive monopoly power, patents can, in some situations, block innovation. If a patent holder makes it difficult (by, for example, charging prohibitive prices) for others to have access to knowledge, improvements on the product or process may not be achiev-

[3]This tension has also affected the intellectual debate and policy making process in the United States. See Comanor (1986).

able. Some evidence suggests that innovators use patents to protect monopolies and prevent competitors from entering the market.

Nevertheless, these potential harmful effects are outweighed by the positive incentives for innovation and the consequences they have on economic growth and productivity. This is particularly true for the pharmaceutical industry.

10.3 Patents and the Pharmaceutical Industry

Of all the global markets that are defining the 21st century, none are more in the political fray than the pharmaceutical industry. In the international arena, Big Pharma and the up and coming Bio are slugging it out with world governments to defend their most prized Title – the patent. With such bad odds of emerging unscathed, why is this industry continuing to fight in the name of intellectual property rights?

The answer lies in part in the vast financial and regulatory risks that compound the already unpredictable nature of science and pharmaceutical research and development (R&D). Given this uncertain climate, it is no surprise that the cost of drug development has increased by 150% since the late 1980s:

While contentious, a widely cited analysis of proprietary industry data concluded that the average cost of bringing one new drug to market during the 1990s was 802 million in 2000 US dollars (Dimasi et al., 2003). This figure includes the opportunity

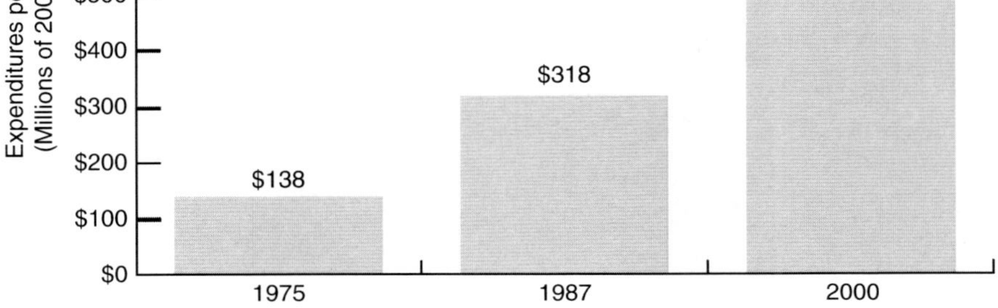

Fig. 10.1 Cost of drug development
Source: Dimasi, Hansen, and Grabowski (2003).

costs of capital and more importantly the losses from numerous prospects that fail to survive the developmental process.

Despite a growing understanding of genomics and disease progression, only one out of 10,000 promising compounds successfully make the journey from benchtop to bedside.

Those that succeed do so only after a series of grueling pre-clinical and clinical evaluations that can take more than a decade to complete[4] (Fig. 10.2).

Government agencies that assure the safety and efficacy of drug products have recently heightened regulatory hurdles by increasing the requisite length and size of clinical trials. Today, a new drug application submitted to the US Food and Drug Administration typically entails, among other stringent requirements, a study of more than 4300 patients, up from 1300 during the mid-1980s (Gottlieb, 2005). While the overall benefit to patients is unclear, regulatory scrutiny has certainly made drug studies more difficult and costly to perform. No pharmaceutical company

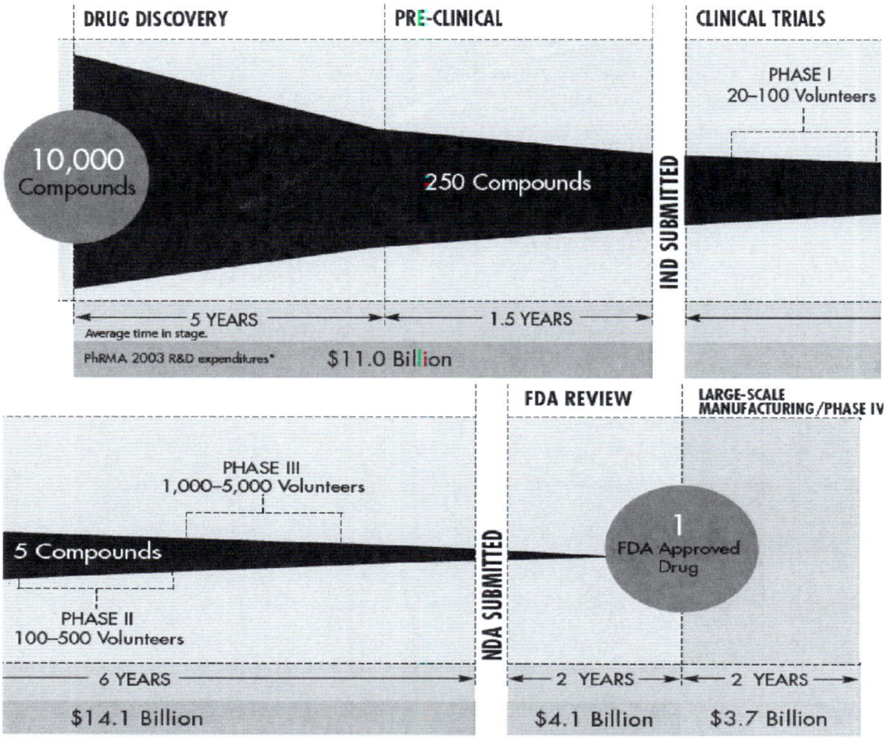

Fig. 10.2 The process of developing a new drug

---

[4]Pharmaceutical Research and Manufacturers of America (2005).

could endure such a precarious and challenging environment without the protection that patents provide.

This protection allows companies to exclusively, but temporarily, market their innovation and garner a profit that rewards them for undertaking risky and expensive endeavors if indeed they create a drug of value to society. Absent patent protection, competitors could and would quickly copy and sell a breakthrough drug at a lower price and obtain essentially a free ride on the originator's enormous investment toward the drug's development.

Unlike other research-based industries that generally have complex and lengthy manufacturing processes, the pharmaceutical industry produces a good that can be easily reverse engineered by an imitator company with minimal capital or technological skill. Incidentally, a number of cross-industry surveys have found that pharmaceutical executives depend more on the patent system as a means of recouping R&D costs than any other industrial manager (Levin, Klevorick, Nelson, & Winter, 1987; Cohen, Nelson, & Walsh, 1997). Mansfield (1986) estimated that, without adequate patent protection, 60% of pharmaceutical drugs that were introduced between 1981 and 1983 would never have been developed, compared to 14% across all other industries.

The integral role that intellectual property rights play in attracting pharmaceutical investment is also demonstrated in countries that have recently adopted effective patent systems. In 1976, Japan modified its patent laws to include protection of products and not just of the developmental process as was priorly the case (Neary, 1995). By taking advantage of these strengthened standards, Japan has evolved from a domestic copier of foreign pharmaceuticals to one of the most cutting-edge drug developers on the global stage.

In summary, developing a new drug involves a serious financial commitment that can be seriously undercut by copiers (developing biotech medicines may be an exception). Without patent protection, innovators would be unable to profit from their inventions or even recover the research and development expenses of not only the drugs that successfully make it to market, but of those that fail along the way. Without patent protection, profit incentives for pharmaceutical innovation would be severely drained and life-saving drugs and treatments would not be developed.

Furthermore, the granting of patents in the pharmaceutical industry and the profit-based incentive mechanism for developing new drugs are related to two issues of international importance: access to essential medicines and the development of orphan drugs for relatively rare diseases.

As we reviewed in Sect. 10.2, patents grant a certain degree of monopoly power. Therefore, the sale price of a new drug is typically many times greater than the drug's cost of production. This differential may prevent patients from accessing the drug, who are willing and able to pay more than the cost of production but not the monopoly price. Such a dynamic makes it difficult for impoverished patients to access life-saving drugs.

Additionally, pharmaceutical companies have an incentive to develop medicines and treatments that will command high enough prices for the company to recover research costs and make a profit. Conversely, they have little or no incentive to develop a product with low profit prospects, even if patient need is great. As a result, many diseases that affect the poor, a high-demand population with limited ability to pay, remain unaddressed. In fact, out of all the therapeutic interventions licensed from 1975 to 1997, only 1% were specific for tropical diseases that afflict the poor (Pecoul, Chirac, Trouiller, & Pinel, 1999).

Both situations give way to an important human rights issue with moral connotations that go beyond the scope of this paper. Suffice it to say that making more pharmaceuticals available to poor people that are tailored to their needs is a matter that should not be overlooked and that deserves careful consideration. The suffering of millions of human beings from preventable or curable medical conditions has not only ethical ramifications, but also economic ones. These serious public health problems tend to create a vicious cycle of misery in poor countries where sick individuals cannot cure their illnesses and be productive, thus perpetuating their poverty.

A similar situation exists in developing countries, but with an additional caveat: the political process seems to have a bigger effect on how these countries choose to deal with public health challenges and the related pharmaceutical patent rights.

In dealing with the access challenge, it is our view that profit-based incentive mechanisms are not part of the problem, but rather part of the solution. Certainly, there is a natural tension between the patent mechanism and the health needs of those who are sick and unable to pay for critical medicines. However, the important questions to ask are whether we can do better than the present patent system and how we can achieve any gains in performance. Moreover, in assessing the trade-offs involved in potential reforms, the political economy of pharmaceuticals should be a pivotal part of the debate surrounding trade related aspects of intellectual property rights.

10.4 The Political Economy of Trips and Pharmaceuticals

10.4.1 The WTO and Trade Related Aspects of IPRs

Not surprisingly, political controversies derived from the situations described in the previous sections are common. Public debates over the pharmaceutical industry's interests and its social impact have taken place even within the United States, a country renowned for its strong IPR protection.

In the late 1950s, the Kefauver report attacked the US pharmaceutical industry on the grounds of excessive prices and profits. The companies defended their level of profits based on their contribution to the development of new treatments, highlighting the importance of patents in spurring this process. Important regulatory changes came out of this dispute, giving the Food and Drug Administration a larger role in the drug assessment process.[5] With the emergence of a more global economy and

[5] Before the Kefauver investigations, new drugs were only tested for toxicity, while after the debate companies were also required to show efficacy. Comanor (1986) offers an excellent account of the political economy of this debate.

higher levels of international trade, the IPR debate became an issue of international relevance. Ideas and knowledge became an increasingly important part of trade.

However, before the 1986–1994 Uruguay Round negotiations, there was no specific agreement on IPRs within the structure of the GATT multilateral trading system. At the same time, the degree of protection and enforcement of these rights varied widely around the world. The need for an international agreement that established certain IPR standards and dispute settlement mechanisms became evident. The Uruguay Round addressed this need with the creation of the Trade Related Aspects of Intellectual Property Rights (TRIPS) agreement.

Broadly speaking, the agreement covers five issues:

1. How basic principles of the trading system and other international intellectual property agreements should be applied
2. How to give adequate protection to intellectual property rights
3. How countries should enforce those rights in their own territories
4. How to settle disputes over intellectual property between members of the WTO
5. What special arrangements would be needed during the transitional period when the new system is being introduced.

Essentially, by establishing minimum levels of protection, the TRIPS Agreement gives the WTO authority to enforce intellectual property rights. The agreement is an attempt to bring a certain level of homogeneity in the way and to what extent IPRs are protected around the world.

The TRIPS agreement took effect on January 1, 1995 and established a timeline for WTO – member countries to comply with the minimum levels of protection. Developed and developing countries were given 1 and 5 year periods, respectively, while least-developed countries were granted an 11 year periods, until 2006, although the transition period has now been extended to 2016 for pharmaceutical patents.

Ultimately, the agreement seeks to provide a balance between the long-term benefits of protecting IPRs and the particular needs of developing and least-developed nations. In doing so, the agreement allows for some flexibility in the application of the terms of the agreement, particularly when dealing with public health issues. These flexibilities allow for the use of compulsory licensing and parallel importing.

According to WTO definitions, compulsory licensing "is when a government allows someone else to produce the patented product or process without the consent of the patent owner," and parallel importing is the importation of "products marketed by the patent owner from one country into another, without the approval of the patent owner." It is precisely within the realm of these flexibilities where major controversies over pharmaceutical patents have emerged.

Prior to the 2001 WTO Doha Ministerial Conference, some countries were unsure of how to interpret the language of these flexibilities. The African Group took the lead in demanding a clarification of these issues and a more detailed roadmap on how and when exceptions to the TRIPS agreement would be allowed. The final result was the inclusion of a special declaration on TRIPS and Public Health, along with the infamous Doha Declaration.

In this declaration, the WTO stressed the importance of interpreting the TRIPS agreement in a way that would support public health, by promoting both access to existing medicines and the creation of new ones. Basically, countries agreed that the TRIPS Agreement "does not and should not prevent members from taking measures to protect public health." While the declaration tried to delineate the scope of the TRIPS agreement with regard to pharmaceutical products and patents, it just added more elements to the political debate.

In particular, paragraph 6 of the declaration emphasized the need to help poorer countries which are unable to make medicines domestically, by granting them access to drugs made under compulsory licensing. On August 2003, WTO agreed to waive the requirement that "production under compulsory licensing must be predominantly for the domestic market" (Article 31f of the TRIPS agreement). This was intended to make it "easier for poorer countries to import cheaper generics made under compulsory licensing if they are unable to manufacture the medicines themselves." After a series of negotiations, on December 6, 2005, WTO members approved these changes to the TRIPS agreement and made the 2003 waiver a permanent amendment.

The Doha Declaration and the permanent amendment on medicine parallel importation rules have considerably influenced the behavior of both governments and pharmaceutical companies. These modifications have opened the door for an even wider interpretation of the TRIPS language concerning compulsory licensing and parallel importation.

10.4.2 The Political Economy of TRIPS and Pharmaceuticals

Kydland and Prescott (1977) won the Nobel Prize in Economics for developing a simple, yet powerful idea, the so-called "time inconsistency" problem. In a nutshell, politicians have the incentive to heavily disregard the future and instead focus on short-term gains, sometimes at the expense of damaging consequences in the long-term. This "time inconsistency" creates an incentive to adopt public policy with short-term gains, even if it would have long-term costs.

A classic example is the adoption of monetary policy that boosts short-term economic growth, but that disregards long-term inflationary pressures.

We can see a similar pattern in the international debate over access to medicines and incentives for new drug development. On the one hand, there are clear short-term political gains in acting on behalf of those who are ill and who do not have the means to pay for treatment. However, on the other hand, some of the interventions for dealing with the access problem may hinder long-term incentives that favor innovation. Within this struggle, it may be the case that the short-term political gains are attractive enough to make long-term considerations a secondary argument in deciding what policies to implement.

This is not to say that those who care about solving public health problems have strictly political motivations or that we wish to downplay the ethical issues that arise from the situation. What we want to highlight is that these short-term political incentives deserve careful consideration, especially when discussing the type of solutions that will help align diverging incentive mechanisms. Changes in the TRIPS agreement's structure after the Doha Declaration and the paragraph 6 amendments fall short in dealing with this "time inconsistency" problem, and may even encourage the short-term/long-term incentive biases.

10.4.2.1 Compulsory Licensing

The Doha Declaration clearly says that it is a country's right to determine under what grounds a compulsory license can be issued. The TRIPS agreement intentionally does not list specific circumstances that would merit the use of compulsory licensing. It only mentions that national emergencies, other circumstances of extreme urgency, and the existence of anti-competitive practices justify the use of a patent without the authorization of the right holder. It has yet to be defined as to what constitutes a national emergency, extreme urgency, or monopoly power abuse. Leaving such a wide window for interpretation only fosters the creation of disputes, instead of solving them.

The TRIPS agreement also grants governments the right to, subject to certain conditions, "prevent patent owners and other holders of intellectual property rights from abusing intellectual property rights, 'unreasonably' restraining trade, or hampering the international transfer of technology." The definition of "unreasonably" is theirs to decide.

It is also true that the agreement attempts to protect long-term incentives for innovation. For example, it states that "normally, the person or company applying for a license must have first attempted, unsuccessfully, to obtain a voluntary license from the right holder on reasonable commercial terms" (Article 31b) and that "if a compulsory license is issued, adequate remuneration must still be paid to the patent holder" by "taking into account the economic value of the authorization" (Article 31 h.)

Despite the noble intentions of some of these provisions, they are problematic in that their language is extremely vague. What represents a "reasonable commercial term" or an "adequate remuneration" invariably depends on which party carries the most weight in the negotiation.

The flexibilities put forth in the TRIPS exemptions were designed to serve the interests of poor countries. However, by allowing for a liberal interpretation of the compulsory licensing clauses, the Doha Agreement now serves more as a platform from which government officials of middle income countries can launch their political agenda. In the case of Brazil, which provides free treatment for its HIV-infected citizens, political fervor fueled by left-wing interest groups has compelled not only pharmaceutical companies, but the US government as well to withdraw from patent disputes.

In 1996, well before the TRIPS deadline of January 2000, Brazil – a country known for its thriving generic drug industry – surprisingly decreed a 20-year window of patent exclusivity for originator products. The catch of course was that a patent would only be granted for products manufactured within the borders of Brazil.

If a foreign manufacturer could not supply the entire Brazilian market through local facilities within 3 years of a patent filing, the Brazilian government would be allowed by law to issue a compulsory license for domestic production of the drug at stake.[6]

According to the Office of the US Trade Representative, such a provision clearly conflicted with a founding principle of the TRIPS Agreement – that patents shall be awarded on a nondiscriminatory basis (Yerkey & Pruzin, 2001). Therefore, in 2001, the United States issued a complaint against Brazil through the WTO, declaring that Brazilian patent law was "inimical to the principles of free trade and inconsistent with various WTO rules, including the TRIPS Agreement."[7]

This pronouncement was immediately followed by a wave of trenchant criticism from NGOs to patient advocacy groups, and was further amplified in intensity by politicians and media personnel. At risk of being portrayed as a puppet for the pharmaceutical industry, the US retracted the complaint, citing that Brazil had implicitly agreed to consult with US officials before taking action under the 1996 law.[8]

Since then, Brazil has twice evoked this law as leverage in price negotiations of US AIDS drugs, without once soliciting the counsel of US authorities.[9] Two companies, Gilead and Abbott, have already given in to Brazil's threats and reduced the price of their respective antiretroviral drugs, Viread and Kaletra, by half.[10] The Brazilian Health Minister has also threatened to abrogate the patent rights of Swiss-based Roche, as a means of exacting price cuts on its antiretroviral therapies.[11]

Faced with an increasingly stretched budget, the Brazilian government continues to demand that pharmaceutical companies either reduce already discounted prices to the levels charged in least developed nations, or succumb to compulsory licensing.

Such a demand undoubtedly violates the TRIPS Agreement, in that patent holders "shall be paid adequate remuneration . . . taking into account the economic value of the authorization."[12] The "economic value" of a patent in Brazil – a country with one of the world's ten largest economies and an HIV prevalence rate of only 0.7%[13] – is clearly not at par with that of a sub-Saharan African country where 20%

[6] Brazilian Law No. 9.279/96 was passed on May 14, 1996 [available at http://www.sice.oas.org/int_prop/nat_leg/Brazil/ENG/L9279eI.asp#nonpatb]

[7] Ibid quotation by Zoellick, R.

[8] Ibid.

[9] *New York Times*, "Brazil May Defy US and Make More AIDS Drugs," February 14, 2001; M. White, "Drug Patents are Good for Our Health," *The Financial Times*, November 30, 2005.

[10] Kaiser Daily HIV/AIDS Report, "Brazilian Health Ministry, Gilead Reach Price Reduction Agreement for Antiretroviral Tenofovir," May 11, 2006.

[11] Reuters, "Brazil to Violate Roche Patent on AIDS Drug," August 22, 2001.

[12] Article 31(h) of the TRIPS Agreement.

[13] Estimate ending 2003 from the 2004 Report on the Global AIDS Epidemic, UNAIDS, June 2004.

of the population may be afflicted by AIDS.[14] Nonetheless, few companies are willing to battle a media and political firestorm for the sake of raising these points in an international tribunal.

If the TRIPS language continues to be unilaterally defined by government officials, the use of compulsory licensing may spill over to patented drugs for other illnesses that are neither prevalent among the poor, nor what most would recognize as national health emergencies.

10.4.2.2 Parallel Importing

Closely related to compulsory licensing, the now permanent paragraph 6 amendment also addresses parallel importation. While Article 31(f) clearly establishes that compulsory licensing must be "predominantly for the supply of the domestic market", the new amendment waives this requirement. Now, any member country can export generic drugs manufactured under compulsory license to other countries that are unable to manufacture needed drugs on their own.

Provisions to prevent the abuse of the waiver were included in the amendment, mainly with the objective of protecting the patent system. These measures require that members inform the WTO when they choose to use parallel importation and take actions that prevent medicines from being diverted to the "wrong markets".

However, the Agreement also states that countries cannot bring legal disputes to the WTO on this issue. Adding to the controversy, the Doha Declaration stated that countries are free to set up their own rules and procedures for dealing with parallel imports. In fact, phrases such as "reasonable measures within their means" and "proportionate to their administrative capacities" regarding importing countries were deliberately included in the agreement to facilitate the use of parallel importation.

Recently, a number of pharmaceutical companies came under fire for challenging, and ultimately suing, the South African government on its interpretation of these clauses. In 1997, then President Nelson Mandela amended the South African Medicines and Related Substances Control Act to allow his country to import cheaper generic equivalents of patented medicines from Brazil and India without permission of the patent owner.[15] The pharmaceutical industry argued that the Act was incompatible with the country's TRIPS obligations in that it gave the Minister of Health unchecked power to arbitrarily define when and what to parallel import, and failed to set guidelines for compensating the patent holder.[16]

---

[14]UNAIDS. AIDS epidemic update, December 2005 [available at http://www.unaids.org/epi/2005/doc/EPIupdate2005_pdf_en/epi-update2005_en.pdf].

[15]Department of Health, "Amendments to the South African Medicines and Related Substances Control Act No. 101 of 1965, Section 15C," 1997 [available at http://www.doh.gov.za/docs/legislation/acts/1997/act90.pdf].

[16]The International Review, NYLS Center for International Law [available at http://www.nyls.edu/miscdownloads/vF01news.pdf].

AIDS activists, government officials, and the media at large viewed the Act as the only mechanism to indirectly secure lower prices of much needed AIDS drugs from the "profit mongering" drug industry. With South Africa having one of the world's largest HIV infected populations, the dispute turned into a public relations debacle for the pharmaceutical companies involved. After the companies withdrew from the case and the media hysteria died down, it became clear that the industry's primary reason for pursuing litigation was not to maintain monopoly prices, but to maintain control over trans-shipment of drugs from low-price to high-price countries.[17]

The challenge of enforcing the appropriate use of parallel importation is enormous. How can poor countries possibly prevent lucrative drugs from reaching unintended markets? Goods triangulation in international trade is an ancient problem. The economic gains of engaging in this activity are large enough to make participation extremely attractive, despite the risks associated with being caught by law enforcement authorities. Trafficking of AIDS drugs on the world-wide gray market is not only real, but is bound to intensify as the use of TRIPS exemptions becomes more commonplace.

In 2002, deeply discounted AIDS drugs intended for impoverished patients in five African countries wound up on the shores of Europe, at what amounted to huge losses ($15.75 million) for one company GlaxoSmithKline, but huge profits for the middlemen smugglers.[18] The sheer volume of diverted medicine points to the existence of an intricate web of African sellers and European buyers who are far from amateurs.

Furthermore, institutional and administrative capacity in poor importing countries tends to be seriously limited and in some cases rampant with corruption. Therefore, the proper use and enforcement of the parallel importation provision are extremely difficult goals to achieve for those that most need to use the TRIPS exemptions.

It is a fact that some countries have chosen not to provide effective protection of IPRs. A mix of economic and political incentives has influenced their decision, and the Post-Doha TRIPS Agreement has added incentives that may favor this behavior. These countries' short-term outlook and actions may seriously hinder innovation in the pharmaceutical industry, putting the health of their own citizens at risk.

It is precisely because of the seriousness of public health challenges that it is imperative to maintain adequate protection of IPRs. For the majority of drugs that treat third world diseases, profits are primarily to be made in middle income countries such as Brazil and South Africa. If developing countries continue to forswear their share of R&D costs by abridging IPRs, impoverished nations will suffer all the more as biotechnology and pharmaceutical companies flee from ventures of no return. Even though a profitable market for AIDS drugs still exists in developed countries, there is reason to suspect that mounting political pressure – particularly on patent rights in developing countries – has propelled 27% of companies involved in HIV research to abandon the field from 1997 to 2003 (Bate, 2003).

---

[17] PhRMA, "Drug Companies Suspend Action Against South African Law," press release, April 18, 2001.

[18] *New York Times*, "Europeans Investigate Resale of AIDS Drugs," October 29, 2002.

In markets where the potential profit is not large enough to encourage the development of urgently needed treatments for orphan diseases, abusing the provisions of the TRIPS agreement only further reduces the attractiveness of investing in these markets. In addition to significant financial risks, pharmaceutical companies must now factor political risk assessments into their investment decisions.

To some extent, pharmaceutical companies are victims of their own success. Because overcoming illness is a matter of life and death, the more efficacious medicines become, the more people will demand no profit prices that ultimately cause innovation to stagnate. The debate over access versus innovation is highly contentious, as exemplified by the current discussion over IPRs and TRIPS.

Will a balance be struck between rewarding innovation and maximizing access? While new incentives schemes need to be developed in order to tackle this challenge, strong IPR protection and sound economic incentives must be core components of the mechanisms used to address these important public health issues.

10.5 Concluding Remarks

> Consumption is the sole end and purpose of all production; and the interest of the producer ought to be attended to, only so far as it may be necessary for promoting that of the consumer.
>
> Adam Smith, The Wealth of Nations, Book IV, Chap. VIII

The main policy implication of the analysis presented in this paper is that political motivations seem to figure into the debate over access and innovation just as much as economic and patient interests. A thorough understanding of these incentives is needed before effective policy reforms regarding TRIPS and the pharmaceutical industry can be designed.

Access and innovation are two sides of the same coin. Promoting innovation is in the interest of not only "greedy" pharmaceutical companies, but consumers as well. In the same way, heightening access to new and existing medicines is not just in the interest of the consumers who benefit from them, but is also in the interest of the companies that profit from developing them. What Adam Smith wrote 200 years ago still applies today. The challenge resides in how to bridge the two sides of the debate.

The standard solution for the time-inconsistency problem is: (1) to create long-term political incentives and (2) to tone down the political component by granting independence to key decision makers. Central Bank independence and monetary policy provides a great example of this type of solution. However, the nature of the problem surrounding pharmaceuticals is very different. The international component of the subject changes the type of available and feasible strategies.

Academic debate has generated a significant amount of work and several proposals on how to modify the incentives structure. Clearly, this short piece can not possibly give due justice to the richness and quality of this work, but it is worthwhile to highlight some of the main mechanisms that have been discussed as potential solutions for the access/innovation dichotomy.

Table 10.1 Policy alternatives

Policy alternatives	Potential drawbacks
Use economic and trade sanctions against those that abuse the TRIPS agreement	Risk of abusing the use of these sanctions. Who decides?
Shorten patent terms	What is the "right" term for a patent?
Use an auction to estimate the private value of patents and then offer to buy out patents plus a fixed markup	Mechanism for revealing true value is not perfect. Does not deal with the orphan disease problem
Grant economic rewards for an invention in proportion to its global health impact	How can one estimate global health impact?
Encourage public/private partnerships in drug development	Different public and private incentives. Coordination problems
Prize incentives for the development of treatments that address orphan diseases	How does one determine the adequate size of the prize and the terms for winning it? Where do the financial resources come from?
Advance purchase commitments	How to forecast adequate amounts and establish credible commitments
Loosening regulatory requirements to cut down drug development costs	What is the appropriate balance between safety and cost-effectiveness?
Open-source projects that create a medical patent pool allowing different actors to make free use of them, in return for a royalty based on sales	Potential for free riding

Source: Authors' compilation.

All of these proposals have their own merits and deserve an extensive evaluation of probable outcomes. Engaging in this analysis in light of both economic and political forces is necessary in order to fully understand the strengths and weaknesses of each of these plans, and more importantly, the kinds of trade-offs that each reform entails.

Yet, the debate surrounding TRIPS and the Doha Declaration heavily lends itself toward one argument: those who care about innovation and profit based incentives invariably don't care about the poor who lack access to medicines. This fallacious and misguided reasoning is dangerous and ignores a basic fact: almost all of the time, domestic and international public policy reforms involve considerable trade-offs.

Additionally, some public health analysts have voiced concern that the debate over TRIPS diverts attention away from more pertinent problems that plague the health care system in developing and poor countries.

The debate over the Doha Declaration and its consequences is part of a much wider problem and the solution requires a mix of policy initiatives that foster health education, prevention programs, and improved sanitation within these countries.

Institutional factors will only complement the framing of this solution. Recent studies in New Institutional Economics have highlighted the importance of institutions such as private property and the rule of law in promoting economic development. The institution of intellectual property, and more specifically the patent system, plays a crucial role in the development of any country.

At the end of this paper, one may have more questions than answers. But hopefully we conveyed that, in the analysis of the access/innovation dichotomy of the global pharmaceutical market, politics and economics go hand in hand. Certainly, this is a global challenge that demands a global solution. It is the world's responsibility to ensure that more and better medical treatments are developed and that more people, especially the poor, have access to them.

References

Bate, R. (2003). *Saving lives today and tomorrow* (Working Paper). South Africa: Africa Fighting Malaria.

Cohen, W. M., Nelson, R. R., & Walsh, J. (1997). *Appropriability conditions and why firms patent and why they do not in the American manufacturing sector* (Working Paper). Pittsburgh, PA: Carnegie-Mellon University.

Comanor, W. (1986). The political economy of the pharmaceutical industry. *Journal of Economic Literature, 24*, 1178–1217.

Dimasi, J. A., Hansen, R. W., & Grabowski, H. G. (2003). The price of innovation: New estimates of drug development costs, *Journal of Health Economics, 22*(2), 151–185.

Fink, C., & Primo Braga, C. A. (2005). How stronger protection of intellectual property rights affects international trade flows. In C. Fink & K. E. Maskus (Eds.), *Intellectual property and development, lessons from recent economic research*. New York: World Bank and Oxford University Press.

Gottlieb, S. (2005). The great shifty to specialty drugs, *On the Issues*, American Enterprise Institute.

Gould, D. M., & Gruben, W. C. (1996). The role of intellectual property rights in economic growth, *Journal of Development Economics, 48*(2), 323–350.

Kydland, F. E., & Prescott, E. C. (1977). Rules rather than discretion: The inconsistency of optimal plans, *Journal of Political Economy, 85*, 473–490.

Javorcik, B. (2005). The composition of foreign direct investment and protection of intellectual property rights: Evidence from transition economies. In C. Fink & K. E. Maskus (Eds.), *Intellectual property and development, lessons from recent economic research*. New York: World Bank and Oxford University Press.

Levin, R. C., Klevorick, A. K., Nelson, R. R., & Winter, S. G. (1987). Appropriating the returns from industrial research and development, *Brookings Papers on Economic Activity, 3*, 783–820.

Mansfield, E. (1986). Patents and innovation: An empirical study. *Management Science, 32*(1).

Maskus, K. E., & Penubarti, M. (1995). How trade-related are intellectual property rights? *Journal of International Economics, 39*, 227–248.

Neary, I. (1995). Japanese industrial policy and the pharmaceutical industry.In A. Towse (Ed.), *Industrial policy and the pharmaceutical industry* (pp. 12–29). London: Office of Health Economics.

Pecoul, B., Chirac, P., Trouiller, P., & Pinel, J. (1999). Access to essential drugs in developing countries: A lost battle? *Journal of the American Medical Association, 281*, 361–367.

Pharmaceutical Research and Manufacturers of America. (2005, March). *Pharmaceutical industry profile 2005*. Washington, DC: PhRMA.

Shapiro, R. J., & Hassett, K. (2005). *The economic value of intellectual property.* USA for Innovation.

Yerkey, G., & Pruzin, D. (2001, June 26). *United States drops WTO case against Brazil over HIV/AIDS patent law*, WTO Reporter.

WHO/UNAIDS. (2005, June). *Progress on global access to HIV antiretroviral therapy: An update on "3 by 5"*. Geneva: WHO/UNAIDS.

Chapter 11
The Knowledge Society

Ernst Helmstädter

11.1 The Knowledge Society as a Political Subject

It is useful to give a short overview about the *Knowledge Society* as a subject of current political considerations. In any case it is a favorite theme of political authorities, in Europe and all over the world, apart from USA. The general idea is that scientific knowledge helps the economy grow, depending on the amounts to be spent for research and development. Those ideas are not completely wrong, but a little too naïve. It has been often said that the economics discipline has as a duty to deliver useable analytical instruments for reasonable political measures. But this new field of political action has not yet found the full interest of the economics discipline. Other social sciences have taken the challenge.

The paper outlines in part I the political discussions on the *Knowledge Society*. That can only be done very restrictively. The aim is to show the political importance of the subject and to give an impression of the EU strategy to establish the "most competitive and dynamic knowledge-based economy of the world" by 2010.

The larger Part II aims at analyzing the role of knowledge that we owe to single authors in the long history of economic thought. Indeed they have not been able to establish a continuing discussion on this subject; it always got lost again and again. However, we will try to draw a connecting line through all these attempts and then take it over into the institutional considerations of part III. Our last part IV provides conclusions and a summary for political application.

At the beginning of the 21st century, two international organizations stressed in detailed declarations the importance of science and knowledge for the economic development in the dawning new age. Since then, many further official documents took the same line. One must say that the general tenor of political strategies considered the conditions for a "knowledge society" to be a promising cause.

This may very well be the first time that science and knowledge remained in the center of political strategies. Over the last several centuries, there was nevertheless

E. Helmstädter (✉)
Institut für Arbeit und Technik (IAT), Gelsenkirchen, Germany

P.J.J. Welfens, J.T. Addison (eds.), *Innovation, Employment and Growth Policy Issues in the EU and the US*, DOI 10.1007/978-3-642-00631-9_11,

no doubt about the role of science and knowledge as an important source for economic progress. So, what is new in the current discussion compared to former times? We will answer this question in connection with a short outline on the issue of *globalization* later.

11.1.1 UNESCO: Declaration on Science and the Use of Scientific Knowledge

The **World Conference on Science for the Twenty-first Century: a New Commitment** (26 June – 1 July 1999, Budapest, Hungary) convened by the United Nations Educational, Scientific and Cultural Organization (UNESCO) and the International Council for Science (ICSU). The results of the negotiations may be characterized by the following statement: "The Conference established what efforts should be invested to make science advance in response to both social expectations and the challenges posed by human and social development. In other words, it negotiated a new 'social contract' for science as we enter the 21st century" [UNESCO 2000, 5]. This appeal is directed to all "stakeholders" (governments, civil society, productive sector, scientists) for a stronger "commitment to science". It calls for a "new relationship between science and society ... to cope with such pressing global problems as poverty, environmental degradation, inadequate public health, and food and water security, in particular those associated with population growth" (number 27). The "sharing of knowledge" should be better balanced, and the gap between developing and developed countries should de reduced. All in all, the declaration is an enthusiastic pamphlet for more science in a better world. However, it also mentions a "scientific endeavor" to carry out "a comprehensive and thorough inquiry into nature and society, leading to new knowledge" (number 19). Naturally, analytical points of view are not discussed.

11.1.2 Lisbon European Council (23 and 24 March 2000), Preparing the Transition to a Competitive, Dynamic and Knowledge-Based Economy

In contrast to the above-mentioned declaration by UNESCO, the *Lisbon Strategy* of the European Council is a concrete EU program "for the next decade: to become the most competitive and dynamic knowledge-based economy in the world, capable of sustainable economic growth with more and better jobs and greater social cohesion. Achieving this goal requires an overall strategy aimed at:

- Preparing the transition to a knowledge based economy and society by better policies for the information society and Research and Development (R&D), as well as by stepping up the process of structural reform for competitiveness and innovation and by completing the internal market;

- Modernising the European social model, investing in people and combating social exclusion;
- Sustaining the healthy economic outlook and favorable growth prospects by applying an appropriate macro-economic policy mix."

The goal of this very ambiguous strategy can presumably not be reached in only one decade, if at all. Surely it will need more time. Nevertheless, the goal itself seems quite admirable, but while the strategic means are presumably adequate, the instruments are not united in the hands of the EU Commission. An adequate collaboration of the member states is necessary. And, it is clear that the fundamental movement must come from the efforts and attitudes of the economic agents. Policy measures can help by establishing feasible organizational and behavioral institutions, but their success depends on a complex spontaneous process of societal interaction. The economics discipline can contribute to the analytical understanding of this process and its institutions.

At the moment, we are in the middle of the first decade of the 21st century. A High-Level Group on the Lisbon Strategy, under the chairman WIM KOK, has already given a mid-term review about the current achievements of this challenge. Because "some countries have managed and are still managing very well, whereas most others are lagging behind" (BMW Foundation Herbert Quandt & CESifo Group Munich, 2005, p. 6), the group made the following proposals (s. BMW, 7):

- Each member country will formulate its national action program ... these national programmes need to have a clear link with the Broad Economic Policy Guidelines and Employment Guidelines, but also with the legislative process in the member states.
- In addition, the open method of coordination can be approved. It has fallen far short of expectations. The central elements of this method – peer pressure and benchmarking – are clear incentives for the member states to deliver on their commitments by measuring and comparing their respective performance and facilitating exchange of best practice.
- And finally, a stronger focus on a limited number of top priorities is absolutely necessary.

The European Commission itself has brought its 6th Framework Program (FP) "Towards a European Research Area" under way, with a budget of €17.5 billion. The budgets of the first six four-year Research Programmes have been extraordinarily enlarged (s. Fig. 11.1). The average growth rate reached 6 % per year since the early eighties.

There seems to be no shortage of information about the current development of the European knowledge based economies. But if we look a little more closely at such publications, we only see "old wine in new bottles". For example, the OECD has just published the 7th issue of the biennial series *OECD Science, Technology and Industry Scoreboard*. The collected data are seen to explore "the growing interaction between knowledge and globalization at the heart of the ongoing transformations of

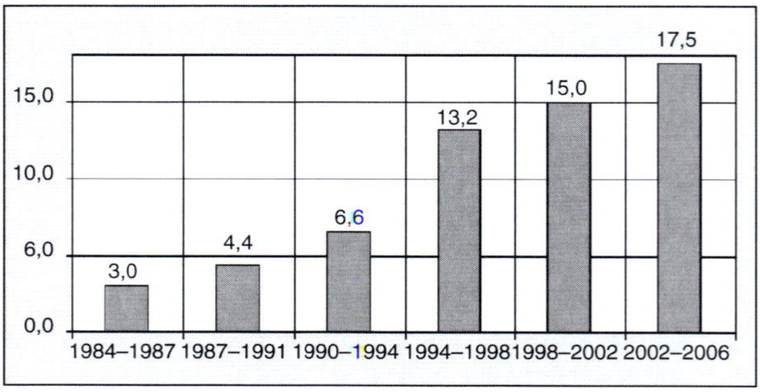

Fig. 11.1 The budget of the EU research programmes 1984–2006 in Mrd
Source: Bundesministerium für Bildung und Forschung, 6.

the OECD economies" (3). The titles of the six chapters of this volume all use the item "knowledge". But if we look at the so-called "knowledge flows" (Chap. 5, 132–155), we find nothing else but the usual trade figures. Indeed, the intention is to show indicators not for the *knowledge economy,* but for *globalization*. As the report states, even this objective is more difficult to date: "... the intensity and multiplicity of transactions have accelerated over the past decade, making the implications of globalization harder to quantify."

11.1.3 Globalization and Knowledge Society

Let us just have a short look at this new and often misunderstood item. *Globalization* is defined as the internalisation of *industrial production* (in contrast to the well-known internationalisation of *trade*). It is quite clear that the division of labor reaches a new quality in this way. Global players produce, or have their products produced, everywhere in the world. The preconditions are as follows:

- *low shipping costs* (see the giant container ships with their logistical efficiency),
- *modularisation* of composed products,
- application of the *same newest technology* around the world,
- strict *certification* of the output modules.

These four activities require advanced computer technology for telecommunication purposes and for the sensitive measuring and steering of automatic production. The produced modules must be exactly the same wherever they come from and wherever they get installed.

Surely globalization may be taken as an expression of the knowledge-based economy. The reason is to be seen in the intensive use of IT technology. With this in mind, however, other perspectives on the knowledge-based economy – and specifically that of the knowledge society – are left aside.

11.1.4 What is Missing in the Political Debate on the Knowledge Economy?

This debate understands knowledge too much as a subject that is feasible by management activities. But, we know from the experiences of business knowledge management (Beerli et al. 2003 a.os. 3–15) that the e-learning and the installation of telecommunication sets is not enough for efficient knowledge interaction. One must obey the individual interests that steer spontaneous knowledge searching and sharing. Exactly these problems are taken under consideration at various phases of the history of economic thought.

11.2 What Does the History of Economic Thought Tell Us About the Role of Knowledge?

In advance and in one phrase: the societal use of knowledge. We follow this basic idea from the *Classics* on until the *New Evolutionary and Institutional Economics*.

11.2.1 Adam Smith: Division of Labor Creates New Knowledge

Adam Smith (1723–1790) was not the first who wrote about the division of labor. Indeed it was already known to Greek philosophers like Plato and Xenophon that division of labor leads to increases in the productivity of labor. In his famous first chapter of the *Wealth of Nations* (1776/1961) has Adam Smith "dargethan, daß die Arbeitstheilung das Erzeugnis zugleich in hohem Grade vermehrt, und daß hierin ihr größter Nutzen besteht, weil sie dadurch eine Quelle des Überflusses an allen Hervorbringnissen der Arbeit wird." (Storch, 1819, Vol. 3, 6). Division of labor creates not only *efficiency* but particularly *development*! The way that it gets done leads over the growth of knowledge: "For Smith applies the division of labor to the growth of knowledge, pointing to the role of those specialised philosophers and men of speculation, 'whose trade is not to do any thing, but to observe everything; and who, upon that account, are often capable of combining together the powers of the most distant and dissimilar objects'. Moreover, because the division of labor also applies to the philosophers, 'Each individual becomes more expert in his own peculiar branch, more work is done upon the whole, and the quantity of science is considerably increased by it'" (Metcalfe & Ramlogan, 2005, 658) Adam Smith, describing all the utensils a men can dispose on in his "kitchen, all the furniture of his table, the knives and forks, the earthen or pewter plates upon which he serves up and divides his victuals, the different hands employed in preparing his bread and his beer, the glass window which lets in the heat and the light, and keeps out the wind and the rain", contributes all what even "the very meanest person in civilized country" is provided by "all the knowledge and art requisite for preparing that beautiful and happy invention" (Smith, 16, spreaded text E. H.). In any case, we can ascertain that Adam Smith sees the division of labor as a means for economic growth through increasing the use of knowledge.

However, an admirer of the work of Adam Smith criticized him for having applied the idea of division of labor only to *physical* labor but unfortunately not to *mental* labor. Heinrich Von Storch (1766–1835, 1819) and Charles Babbage (1791–1871, 1993) found, although with different arguments, that Adam Smith should have attributed the advantages of the "*division of knowledge*". Their own contributions have been the first attempts to formulate this type of knowledge oriented societal interaction, what Friedrich A. von Hayek (1899–1992, 1937) later demonstrated.

11.2.2 Schumpeter's View on the "Driving Force" of the Economic Development

Although the role of knowledge by Smith, Storch and Babbage has been set on the economic agenda, the economics discipline did not continue these first attempts to deal with this problem. However, during the first phase of the industrialization there must have been the prevailing view that the development was the result of a better use of knowledge. Joseph A. Schumpeter (1912), in his seminal work Die *Theorie der wirtschaftlichen Entwicklung*, contradicted this opinion: „Wir sind also jetzt imstande, die Unzulänglichkeit einer Auffassung nachzuweisen, welche ausgesprochen oder unausgesprochen zweifellos herrschend ist. Es ist die Auffassung, daß in diesem technischen und organisatorischen Fortschritt ein selbständiges Moment liegt, das sein Entwicklungsgesetz in sich selbst trägt und wesentlich auf dem Fortschritte unseres Wissens beruht. Die Kombination, die nach dem Stande unseres Wissens jeweils die beste ist, bildet demnach gleichsam das Gravitationszentrum und nach dessen Erreichung jedenfalls eine Tendenz besteht, eine automatische Tendenz, wie nach der Erreichung des Gleichgewichtszustandes in der statischen Wirtschaft nach unserer Auffassung. Das ist nicht richtig." (480) Thus, the mistake of the prevailing view consists in the hypothesis that new knowledge drives the economy automatically to a new state of the development.

Schumpeter's statement against the mainstream thinking of the discipline makes clear, that inventions and knowledge are not seen by him as the driving force of the economic development, and are therefore of no interest for Schumpeter's analysis. It holds even, that inventions are to be taken not as *cause* but even as *effect* of the development (479).

What Schumpeter had in mind when he denied the innovative influence of inventions and new knowledge, finds a very detailed expression in a lot of pieces of empirical evidence in his *Business Cycles* (1939), where he speaks about the first *Kondratieff cycle* 1767–1842 (Chap. VI). There he refers to two different types of new events. We may call them *original inventions* and *applied novelty*. It may be useful to add these notions in Schumpeter's following texts {included in parenthesis} just to illustrate what his meaning under the different connections seems to be:

- "We hold, ... , that the influence excerted was not in the direction of initiating new {*original inventions*, E. H.} – economic, political, artistic, and so on – creations." (233, footnote 1);
- "The cotton textile industry was the new leader, according to our terminology, but was not new {*original inventions*, E. H.} in the sense of common parlance." (271);
- "But if inventions is not the core of the matter, neither is objective opportunity. Study of our period shows us again that 'doing the thing' – the actual setting up new production functions {*applied novelty*, E. H.} – is a distinct phenomenon." (272);
- "The New Men and the New Firms stand out so well in this case because the industry itself was new {*applied novelty*, E. H.} as ... the industries that carry Kondratieff upswings ..." (273).
- We notice that everything depends on the new men and their new firms that things get done.

That is not all Schumpeter says with respect to economic development. In *Capitalism, Socialism and Democracy* (1942), the scenario changes from *men* to *processes*. If compared to the notion of *perfect competition* of the neoclassical theory, the entrepreneur is the actor in the process of a *new type of competition*. When Schumpeter introduces this process, he first speaks of a *process of qualitative change*, later of a process of *industrial mutation*, and finally of a *process of Creative Destruction*: "This process of Creative Destruction is the essential fact of capitalism." (83) As a further illustration of this process Schumpeter adds a time dimension: *the perennial gale of creative destruction* (84). Later on we only find once more the notion of *perennial gale* [88] and two times that of *process of creative destruction* (104, footnote 24, 1943[1]). In spite of the scarce mentioning of the notion of *creative destruction* in Schumpeter's original work, it has received a tremendous attention in the secondary literature. Presumably it is the most cited conceptional invention of Schumpeter!

In my view, however, it is an ill-formulated notion. By asking, what is cause and what is effect, we must ascertain that a creative innovation is obviously the *cause*, and the so-called destruction of old production functions is the *effect*. Thus, in fact, Schumpeter should have spoken of *destructive creation*! But this sequence of the dual notion does not sound as sexy as it does the other way around. However, what urges us at all to stigmatize the motive power of economic development by characterizing it as a kind of destruction? Innovations bring about better and more products; they increase productivity. *Destructions* by innovations should be understood as a normal outcome of selection by the competitive process. To give the specific type of competition Schumpeter had in mind a name, we mean it is enough to speak of *Dynamic Competition*.

[1]Here Schumpeter refers to the psychological ability of a socialist society to "*fulfill important functions in the process of creative destruction*".

There are some other processes mentioned in *Capitalism, Socialism and Democracy* which are hostile to the capitalist development: the obsolescence of the entrepreneurial function, the destruction of the protecting strata, and the destruction of the institutional framework of the capitalist society (131–142). Under the headline of a "*growing hostility*" (143–155), Schumpeter mentions the group of intellectuals as a specific phenomenon of the capitalist society: "Unlike any other type of society, capitalism inevitably and by virtue of the very logic of its civilization creates, educates and subsidizes a vested interest in social unrest." (146) Here one group of the knowledge sector appears on the wrong side of the capitalist development, the knowledge of intellectuals as the enemy, not as a power of development.

Where Schumpeter speaks of the *obsolescence of the entrepreneurial function* he also mentions a certain type of knowledge that influences technical progress: "Technical progress is increasingly becoming the business of teams of trained specialists who turn out what is required and make it workable in predictable ways" (132). What else can these specialists be other than *knowledge workers*? (to the role of knowledge in the Schumpeterian Economy see Helmstädter 2007)

To sum up Schumpeter's changing views on the economic role of knowledge we may say: he first refused the leading view that the progress of knowledge has to be seen as the driving force of the development and introduced the pioneer entrepreneur as the true heroes of the development. By this he had the first *Kondratieff cycle,* 1767–1842, in mind. Later, when he spoke of *trained specialists* as the researchers, he had the first half of the twentieth century in mind. To make such a difference seem plausible: In the first phase of industrialization, inventions (new knowledge) may have been numerous in relation to the innovators, in the second phase this relation had changed around with a shortage of new knowledge (inventions) instead of innovators.

In any case we see that Schumpeter withdrew his former reservation against knowledge as a source of the economic development. To differentiate between invention and innovation is still suitable. But, as we will see later on, we should better refer to two types of knowledge: scientific knowledge against the tacit knowledge of the competitive agents.

11.2.3 Hayek: The Division of Knowledge

The third author who has dealt with knowledge as a basic theme in the history of economic thought is Friedrich A. von Hayek (1899–1992). His article about *Economics and Knowledge* (1937) focuses on the notion of the *division of knowledge*. Fritz Machlup (1902–1983, 1977) remarks about it: „Eine der originellsten und bedeutendsten Ideen, die Hayek entwickelte, ist die Rolle der ‚Wissensteilung' in der Wirtschaftsgesellschaft" (41). This idea puts the division of knowledge on the side of the division of labor as a second basic societal process of interaction. And it is specifically regarded as a *competitive process*!

This fact led Hayek (1968) to the next question: Why is competition good for the society? His answer: it is a *search process*! Competition detects new knowledge. And: Economic competition on markets detects the system of prices and further knowledge about the market conditions. This type of knowledge sees Hayek in contrast to scientific knowledge as a specific type of competence for economic action under changing conditions. It is implicit or tacit knowledge, getting known only to the participating competitors fragmentarily. The knowledge of the society is not the pure sum of the individual knowledge's but a much bigger entity of its own, that nobody as a person or an institution can dispose on. That may be the cornel of Hayek's idea, and for the purposes of this paper a sufficient introduction into his concept.

Again, Hayek's idea of the division of knowledge was not accepted as part of the canon of economic themes and finally almost lost. One reason may have been that the article from 1937 was written as a concluding contribution to the famous debate about rational calculation under socialism. Thus, it was a little outside the mainstream situated. But its content was of great general importance. Indeed, it leads to the role of competitive interaction as a search process in the knowledge based economy. Its fruitfulness must be seen in its applicability for all spontaneous societal interactions under complex conditions.

The notion of competition as a search process should not get restricted to economic competition. There are many other forms of competition that are search processes as well. That holds particularly for status[2] or reputation competition, the basic interaction mode in the scientific sector. The usefulness of competitive processes depends on their efficient rules under which these processes run. Thus, we are led to the *New Institutional Economics*, and because of the competitive dynamics to the *New Evolutionary Economics*.

Hayek's thinking about the *division of knowledge* and the following interpretation of *competition as a search process* leads us, by the questions raised, to the *New Institutional* and *Evolutionary Economics*. These are the instruments for the analytical description of those processes that govern the knowledge-based economy (Brödner, P. a. os., 1999). Part III of this paper tries to make this proposal plausible.

11.3 About the Economics of Knowledge

11.3.1 The Starting Point: Knowledge as Good or Process?

Let us first add a few considerations about an appropriate starting point for the economics of knowledge. In his monograph "The Economics of Knowledge", Dominique Foray (2004) speaks of an original analytical discipline which "covers the properties of that economic good governing its production and reproduction as well as the historical and institutional conditions (such as information technology or patent rights) determining its treatment and processing in a decentralized economy" (1).

---

[2]There is also another economic interpretation of *status*. The game theory speaks of *status games* when goods are demanded for demonstrative purposes. (s. Reisch, 2003).

One may agree that Economics of Knowledge must be understood as a new discipline of its own. But there are many doubts whether the starting point for this new discipline should be the understanding of knowledge as a good. Look at a patent: is it a knowledge good? It is the property right to use what is written down in the patent article as the underlying invention. The good character is included in the right to use the invention in question. The patent itself becomes published, and thereby the included knowledge content gets publicly known.

A firm cannot buy a knowledge good, it must employ a knowledge worker and hope that he is able to apply the wanted knowledge. Explicit knowledge, published in any medium, to understand as knowledge good is economically worthless if one is unable to transform the text into a new competence of action. That requires itself competence and a context of interest to use the written knowledge. Those are our reservations against the understanding of knowledge as a good.

Furthermore, we do not find it useful to transpose all the instruments the economics discipline applies analogously to material goods to knowledge. Foray's remarks about the "production and reproduction" of knowledge and its "treatment and processing in a decentralized economy" remind us too much of material processing procedures.

Out of such reasons we prefer an alternative entry to the economics of knowledge. As David Hull (1988), evolutionary biologist, in his monograph "Science as a Process" we set in the center as the basic notion of a process. Thus, we speak of knowledge as a competitive process. First, that means that we concentrate on the creation of new knowledge. Second, we follow the lines of the history of economic thought, which now has to be shown in some further detail.

11.3.2 Knowledge: Created by Competitive Interaction Processes

Knowledge as a subject of economic use has always been seen as the result of societal interaction. When Adam Smith spoke about the advantages of specialization by the division of labor, he stressed at the same moment the need of the exchange of goods in competitive markets. Thus, one can say that the market exchange process has brought about the new knowledge of the specialized workers. The watching "philosophers" got their new ideas by observing this process from outside. Although Smith does not say very much about the societal knowledge searching process specifically, this interpretation fits well into the Smithian general picture of the market economy.

Schumpeter expressively speaks about different processes that drive – later also hinder – the economic development. Of the greatest interest for us is the process of Dynamic Competition. That means that in economic development, new markets compete continuously with older ones such that economic growth and development progress through restless structural changes caused by innovations. This perspective aims at a particular characteristic of the dynamic process. What we are missing in Schumpeter's analysis is a broader view to this process, specifically the preliminary stages of innovations: inventions and new knowledge. These questions are meanwhile taken under more and more consideration by the New Evolutionary Economics.

Hayek's view on the aspect of division of knowledge holds clearly under market competition as a process. He stresses continuously the importance of the reaction on the changes of the market conditions: "It is, perhaps, worth stressing that economic problems arise always and only in consequence of change." (Hayek, 1945, 523) The competitive process has not to find out one single price structure but the current change of the prices that follows out of the changes of the basic conditions. The knowledge that results is practical and implicit (tacit) knowledge, a kind of knowledge that seems to him underestimated. Another kind of knowledge, "... scientific knowledge, occupies now so prominent a place in public imagination that we tend to forget that it is not the only kind that is relevant." (521) The Hayekian kind of knowledge is "unorganized knowledge which cannot possibly be called scientific in the sense of knowledge of general rules: the knowledge of the particular circumstances of time and place" (521). It thus seems plausible to simply call this kind of knowledge *Hayek-*, *competition-knowledge* or *Knowledge I*. Under the perspective of innovations, this type of knowledge must be taken into account as well as scientific knowledge or *Knowledge II*. One of the intriguing problems of the knowledge-based economy is to find the right mixture of both kinds of knowledge to make innovation successful.

11.3.3 Competition as Search Process

In theoretical economics with competition played for a long time, the role of a mathematical algorithm to find a solution of a fully-specified static model of virtual societal interactions: *competition as a search instrument*. To understand, following Hayek (1968), *competition as a search process* is to regard the factual actions of economic agents as the subject of analysis. If one asks what competition as a general mode of free spontaneous interaction aims at, we answer: an ex ante unknown *ranking order*! That is evident in sports, politics and for example architectural competitions. And it is true also in economics.

Economic competition needs a ranking order of the supplier according to their supply prices and a second ranking order of the demanders according their willingness to pay. Only if such ranking orders on both markets are given, the market price and the amount of traded goods will result by competition. Another ranking follows out of this: The binary ranking of the participating and the excluded suppliers and demanders.

If we refer to the basic focus of competitive interaction as a procedure to find out a ranking order, we have the platform from which we can introduce several modes of competition under a unique aspect. Two kinds of competition are of great importance under the knowledge-based economy: *economic competition* and *reputation competition*. The first is valid in the economic exchange process, the second in scientific communities. In the frame of communities, not only reputation competition holds but of course *co-operation* as well. In any case, it is important to see that both

Table 11.1 Competition and knowledge

Economy	Scientific sector (communities)
Types of competition	
Economic competition	Reputation competition
Exchange of goods and services	Knowledge interactions (also co-operation)
Transactions	New knowledge

modes of interaction characterize communities of societal interaction. – Table 11.1 shows this terminology.

Asking about motivations of competitors in a search process one has to refer to a kind of intrinsic motivation. In any case: priorities matter. Under economic (dynamic) competition, the pioneer entrepreneur who reaches the market first with his innovation has the temporary advantage for extra profits. Under reputation competition in a scientific community, those scientists who lead the way with new methods and results which they found first, gain the greatest reputation. The community of competitors finds a subtle ranking order which gets known on a very informal but effective way. As it is generally known, priority plays again an important role.

All competitive search processes must be free and open for competitors. The efficiency of these processes depends on the spontaneity of the societal interactions. Of course, open processes need their rules. Thus, we must ask what the *institutions* are that govern these processes. Under institutions, we are referring to formal and informal rules of behavior, customs and organizational structures. Some formal institutions (law, public regulations) are the instruments of political measures.

11.3.4 The Institutions of Economic Competition

We introduce the institutional problem first with a few words about the institutions for competitive market exchange. This is the subject New Institutional Economics has addressed over the last two decades.

The New Institutional Economics (NIE) considers the institutions for societal interaction in a competitive market exchange economy as an economic issue of its own. Because the system of institutions creates opportunity costs, the neoclassical theory cannot tackle this problem, because important institutions are not the subjects of exchange. NIE particularly regards *transactions costs* as its genuine problem. Because of its success, NIE also attempts to expand this subject. Thus the question arises as to whether the instruments of NIE are also appropriate for the system of knowledge interactions? In general one may say yes, but that does not hold for the focal notion of *transaction*!

By the definition of Williamson (1996, 58) "a transaction occurs when a good or service is transferred across a technologically separable interface. One stage of activity terminates and another begins." Such an interface has to make the point clear as to where a *property* or a *responsibility sphere* (inside firms) ends and another begins. This is of importance for the exchange of tradable goods and services. They have their price and hold the balance of reciprocal benefits. By this the preconditions for *economic competition* are fulfilled. These basic institutions characterize all further considerations of NIE.

11.3.5 The Institutions of Reputation Competition in Scientific Communities

Reputation competition is the driving force in communities that search for new knowledge. We regard scientific communities (e.g., scientific societies, research institutes or groups and faculties) as a typical organizational unit searching for new knowledge. If a member of such a community has to offer new knowledge, he presents it to the community seeking recognition. The members of the community are not just demanders but are themselves also suppliers of new knowledge. So, the recognition of new knowledge must come from *competing* agents. Here we do not find two sides of a market.[3] Therefore the application of the market model fails. However, there is competition in action, one-sided competition, compared to the two-sided economic competition.

Reputation finds its expression particularly by citation. That is not the price to be paid for the use of new knowledge, but an institutional instrument that enhances knowledge sharing (Albert, 2004, 134–137). Citation is specifically efficient when articles (instead of monographs) are the usual instrument of a fast publication. In this case, citation makes it easier to signal the author's standpoint and to continue the state of affairs. In any case, it is not to compare with the function of a price, which has to balance reciprocal benefits.

Reputation is an indicator of the rank a scientist holds. In the long run and under normal conditions it will also influence the income position of a scientist. That is the price his knowledge output has brought about. But such a payment is not strictly reciprocal. Salary of a scientist is a *general payment*. It does not compensate current knowledge outputs but expected future knowledge outputs. The salary of a teacher is paid more for the future competence his or her students may have than for the current educational work. General payment is the mode knowledge workers are paid. It has not and cannot have the character of an exchange price, balancing current reciprocal benefits.

A creative scientific community needs free and open interactions between its members. To find new knowledge is not to be understood as a strict production function whereby output is the inevitable effect of inputs. Uncertainty makes it

---

[3]Walstad, 2002 represents this thesis, see Albert, 2004, 129.

impossible that input always brings about the wanted output. To tackle the risk of failure needs willingness to take risks and competence. But in principle, there is no secure way to find new knowledge, as Loasby's contributions (2003) to the knowledge issue show.

Co-operation among members is also a defining characteristic of a scientific community. Here, the balance of reciprocal benefits does not play a motivational role. Co-operation is fundamentally based on an altruistic motivation, while the motivational basis of competition is egoistic. Thus, the two modes of societal interaction oppose each other. Nevertheless they happen to appear in real life often more or less mixed. Competition can even be regarded under a formal societal aspect as a kind of societal co-operation or collective co-ordination. Seldom shall pure cooperation continue for a long time if the advantages of it are heavily unbalanced.

Our considerations about the institutions for economic and reputation competition are very short. Their only purpose was to draw attention to the institutional differences under which both modes of spontaneous societal interactions work. Referring to the publications under the reference section below can help further, but the New Institutional Economics of Knowledge Sharing is still very incomplete (see further institutional aspects of the knowledge society in Helmstädter 2001, 2003 and 2004).

11.4 Conclusions for a Science Policy

The German economic history shows two important examples of very successful political measures to advance the knowledge society. The first one happened at the beginning of the industrialization process. Around the year 1800, the industrial productivity of Prussia was only half that of the British (Kindleberger, 1955, 231). But the first World Exhibition of 1851 in London saw the German industry already on the same level as the British. "Among the most effective institutions advancing industrialization was the Gewerbeinstitut and the related Gewerbeschule, also known as Technische Hochschule, started by Peter Beuth. Beuth had become interested in industry while billeted during the Napoleonic wars with the Prussion army at William Cockerill's machinery plant at Liège. As head of the Department of Trade and Indsutry of the Ministery of Finance from 1816, he established what would today be called an industrial policy for Prussia. . .The Gewerbeinstitut also followed French practice in holding industrial exhibitions with prizes and honors. . . The passion for education was not limited to the middle and upper classes. An anthropologist has characterized the German workman as having 'an insatiable Lust for learning', and being 'culture greedy'" (235).

The second example is the German "Wissenschaftswunder" (Vereeck, 2001) of the late decades of the 19th century. It is the work of Friedrich Althoff (1839–1908), since 1882 "Vortragender Rat" in the Prussian Ministery of Culture in Berlin: "Die Performanz der Wissenschaftler im 'System Althoff' ist ganz und gar unbestritten. Die Reformen des preußischen akademischen Systems stellen daher eine Evidenz für die institutionellen Erfordernisse produktiver akademischer Aktivität dar"

(Vereeck, 26). The following institutions have been introduced through Althoff's system (Vereeck, 171–190):

- The policy of *openness* of the scientific community and the understanding that *transparency* of the reached results stimulates scientific performance.
- Statistical indicators are appropriate as *incentives*, but *not as binding targets.*
- Institutions matter in helping to enhance academic interactions.
- Institutional economics helps us understand the functioning of scientific communities.

Althoff's academic reforms resulted in a development of the German universities such that they became the master plan for the universities in the USA. Nelson (2002) mentions Althoff's successes in the case of the education of chemists. The development of the German Chemical Industry in comparison with the British has been "for a variety of reasons the supply of German chemists (has been, E. H.) much greater than the supply of British chemists" (Nelson, 26).

We see that there is some experience with how to bring the scientific system ahead through effective political measures, namely by institutional reforms. Today, political strategies refer to bringing more economic competition to universities and increasing research & development budgetary allocations.

There is no doubt that some services of the universities should experience more economic competition; this holds specifically for education and research for firms. Here reciprocal payment is possible and should be applied. This would enhance the financial situation of the academic sector. But it is not at all that kind of reform the German academic world could bring ahead.

The idea to make something of a scientific enterprise out of universities is an ill-defined strategy. Such a strategy would never function, because the exchange of tradable goods and services and the searching of new knowledge are two absolutely different cases of societal interaction. Knowledge must be seen as a process. That means that competitive processes which create new knowledge must get the targets of public support. The political conclusion can only be to look for appropriate and efficient institutions and for the self-steering of the complex spontaneous interaction process, which alone offers hope for the wanted results of scientific progress.

By the way, one must see that scientific communities are not bound by university walls. Openness of the science process means worldwide reputation competition. The financial preconditions (e.g., grants for attending international conferences, membership in scientific societies, distant co-operation) must be given. The restriction that the leading researcher (e.g., a professor) should not be paid for his or her research activities (e.g., in DFG projects) does not seem to provide an effective incentive today. The American model (the university pays for 9 months, three months are free for additionally paid research activities) seems to prevent a more fruitful model.

The most outstanding characteristic of the German science policy today is its establishment of ten pioneering universities. The reform idea behind this strategy is simply parting with the long-held opinion that public universities should present the same standard everywhere in Germany. This model of public universities and its misguided "openness" for everyone – whether sufficiently qualified or not – have been two underlying weaknesses of German scientific development.

References

Albert, M. (2004). Methodologie und die Verfassung der Wissenschaft – Eine institutionalistische Perspektive, In M. Held, G. Kubon-Gilke, & R. Sturn (Hrsg.), Normative und institutionelle Grundlagen der Ökonomik, Jahrbuch 3: Ökonomik des Wissens, (pp. 127–150). Marburg 2004: Metropolis Verlag.

Babbage, C. (1993). *On the economy of machinery and manufactures.* London 1835: Charles Knight, as reprint published London 1993: Charles Knight.

Beerli, A., Falk, S., & Diemers, D. (2003). *Knowledge management and networked environments: Leveraging inntellectual capital in virtual business communities.* New York a.os. 2003: Accenture LLP.

BMW Foundation Herbert Quandt & CESifo Group Munich, (2005, June 9–10). 4th Munich economic summit. Europe and the Lisbon goals: Are we Halfway there?, Munich.

Brödner, P., Helmstädter, E.,& Widmaier, B., (Hrsg.). (1999). Wissensteilung. Zur Dynamik von Innovation und kollektivem Lernen, Arbeit und Technik, Schriftenreihe des Instituts Arbeit und Technik, Bd. 13, München und Mering 1999: Rainer Hampp Verlag.

Bundesministerium für Bildung und Forschung. (2002). Chance für Deutschland und Europa. Das 6. Forschungsrahmenprogramm, Bonn 2002.

Deutscher Bundestag. (2002). Globalisierung der Weltwirtschaft – Herausforderungen und Antworten, 14. Wahlperiode, Abschlussbericht der Enquete-Kommission, 24. Juni 2002.

European Commission, Information Society Technologies. 2003–2004 Workprogramme, www.cordis.lu/ist.

European Commission, Towards a European research area. Science, technology and innovation. Key figures 2003–2004, Brussels 2003 (www.cordis.lu/indicators).

Foray, D. (2004). *The Economics of Knowledge*. Cambridge Mass. London: The MIT Press (original French edition: L'économie de la connaissance, 2000).

Fraunhofer-Institut für Systemtechnik und Innovationsforschung. (2004, August). Innovations- und Zukunftspotenziale Nordrhein-Westfalens. Struktur und Dynamik vonWirtschaft, Wissenschaft und Bildung, Karlsruhe.

Friedrich-Ebert-Stiftung. (2004). Deutschland in der globalen Wissensgesellschaft. Auswirkungen und Anforderungen, Berlin.

Hayek, F.A. von (1968). Der Wettbewerb als Entdeckungsverfahren, Kieler Vorträge Neue Folge 56.

Hayek, F.A. (1945). The use of Knowledge in Society, *The American Economic Review*, Vol. XXXV, Nr. 4, September, 519–530.

Helmstädter, E. (2001). Wissenteilung, Thünen-Vorlesung bei der Jahrestagung des Vereins für Socialpolitik Berlin 2000, in: Perspektiven der Wirtschaftspolitik, Bd. 2 Heft 4, 445–465.

Helmstädter, E. (Ed.). (2003). The Economics of Knowledge Sharing: A New Institutional Approach. Cheltenham, UK, Northampton, USA: Edward Elgar.

Helmstädter, E. (2004). Wirtschaft und Wissen. Die Institutionen der Wissensteilung als Aufgabe der Ordnungspolitik, in: ORDO. Jahrbuch für die Ordnung von Wirtschaft und Gesellschaft Bd. 55, 37–76.

Helmstädter, E., (2007). The Role of Knowledge in the Schumpeterian Economy, In Hanusch, H. and Pyka, A. (eds.), Elgar Companion to Neo-Schumpeterian Economics, Cheltenham, UK. Northampton, MA, USA: Edward Elgar, 296–315.

High Level Group on the Lisbon Strategy. (2004). Facing the challenge. The Lisbon strategy for growth and employment, November 2004. http://europa.eu.int/comm/lisbon_strategy/index_en.html

Hull, D.L. (1988) Science as a Process. An Evolutionary Account of the Social and Conceptual Development of Science, Chicago and London: The University of Chicago Press, Paperback edition 1990.

Kindleberger, C. P. (1955). Technological diffusion: European experience to 1850. *Evolutionary Economics, 5*(3), 229–242.

Loasby, B. (2003). The cognitive basis of institutions: an evolutionary perspective. In E. Helmstädter (Ed.), The economics of knowledge sharing: A new institutional approach (pp. 100–118). Cheltenham, UK, Northampton, USA: Edward Elgar.

Metcalfe, J. S., & Ramlogan, R. (2005). Limits to the economy of knowledge and the knowledge of the economy. *Futures, 37*, 655–674.

Nelson, R. R. (2002). Bringing institutions into evolutionary growth economics. *Evolutionary Economics, 12*(1–2), 17–28.

OECD Science, Technology and Industry Scoreboard, 2005.

Reisch, L. A. (2003). Statusspiele – Soziale Vergleichsprozesse und wirtschaftliches Verhalten. In: M. Held, G. Kubon-Gilke, & R. Sturn (Hrsg.), Normative und institutionelle Grundfragen der Äkonomik, Jahrbuch 2, Experimente in der Ökonomik (pp. 217–240). Marburg: Metropolis Verlag.

Schumpeter, J. A (1912). Die Theorie der wirtschaftlichen Entwicklung, Leipzig: Dunker & Humblot Verlag.

Storch, H. Von (1819). Handbuch der Nationalwirtschaftslehre, aus dem Französichen, mit Zusätzen von Karl Heinrich Rau, 3 Bde., Hamburg.

Technology Commercialization Group, Wettbewerbsfähigkeit im Forschungsbereich. Vergleichende Pilotstudie Deutschland – USA

UNESCO (2000). Science for the Twenty-first Century. A New Commitment Paris 2000.

Vereeck, L. (2001), Das deutsche Wissenschaftswunder: Eine ökonomische Analyse des Systems Althoff (1882–1907), Berlin: Duncker & Humblot.

Walstad, A. (2002). Science as a Market Process, The Independent Review, 7(1): 5–45.

Willaim, O.E. (1996). The Mechanisms of Governance, New York and Oxford: Oxford University Press.

Chapter 12
Technological and Growth Dynamics in the New Member States of the European Union

Andrea Szalavetz

12.1 Introduction and Overview

An impressive portfolio of theories facilitates the analysis of the impact of technical change on growth. Each one of them captures one aspect of the two factors' interrelation. Considered together, they contribute to gaining a broad view on technological and growth dynamics.

Out of a vast amount of theoretical approaches we select three key issues to address: the theories of

1. appropriate technology
2. technology accumulation, growth and development in stages
3. the direction of technical change

The question this paper is concerned with is how these theories apply to the technological dynamics and growth experience of new member states (NMS) at the EU-periphery. However, the results of our analysis can be put in a broader context, since NMS' experience has many overlapping elements with that of FDI-recipient high-performing developing countries.

Sections 12.2, 12.3, 12.4 take up each of the enumerated theories in turn and examine their explanatory power with respect to similarities and differences in technological and growth performance between both advanced and catching-up economies. Our explorations are characterized by a descriptive approach rather than an intention to test formal models. Section 12.5 summarizes.

12.2 Relevance of the Theory of Appropriate Technology

According to the theory of appropriate technology, technologies are specific to particular combinations of inputs. The choice of technology should therefore be in line with countries' factor endowments and the resulting capital-labor ratios with

A. Szalavetz (✉)
Institute for World Economics, Budapest, Hungary

P.J.J. Welfens, J.T. Addison (eds.), *Innovation, Employment and Growth Policy Issues in the EU and the US*, DOI 10.1007/978-3-642-00631-9_12,

relative factor prices. If the local processing and production organization skills do not match the chosen technology, total factor productivity and output per worker performance will be inferior to the ones in other locations with identical technology but no technology–skill mismatch. (Atkinson & Stiglitz, 1969; Basu & Weil, 1998; Caselli & Coleman, 2000; Acemoglu & Zilibotti, 2001)

This reasoning implies the existence of a wide variety of choices among possible techniques that differ in their required factor intensities. Another implication is that well-chosen technologies ensure local competitiveness even if we allow for significant inter-country differences in the level of particular techniques, provided that capital-labor ratios are in line with relative factor prices.

As opposed to the first conclusion, the variety of choice of technology proved to be rather limited in NMS and was only marginally influenced by local factor intensities.

The choice among alternative techniques, differing only in their relative factor intensities as a question of principle emerged in industries unable to attract foreign direct investment. Field investigations in Hungary (Artner, 2005) revealed, however, that local entrepreneurs' choice from a range of alternative techniques was rather limited. Limitation was only partly caused by lack of resources and/or institutional failure to provide entrepreneurs beneficial support schemes, credit and/or venture capital. Choice of technology was limited also because the applicability of particular technologies depended on the scale of production, rather than on relative local factor prices. Despite the availability (in principle) of up-to-date technology, the return on such investments was out of question for local entrepreneurs who did not manage to plug in dynamic global value chains and export large quantities of their products. In sum, it was not the deficiency of production and production organization skills that dampened technological upgrading, rather the lack of the capability to acquire a market large enough to ensure return on investment.

The other side of the coin was that on several occasions, these firms could assume orders of small batches while competitors with much superior technology could not, since their technology made the running small production batches prohibitively costly. In a way, this supports our second conclusion (that well-chosen technologies warrant competitiveness). Nevertheless, it is questionable whether a defensive strategy of keeping the outdated technology in order to be able to serve specific orders of small batches is sustainable.

Foreign investors could in principle also choose among a wide variety of production techniques to be located to their newly opened subsidiaries. The theory of appropriate technology suggests that the technological level of the production equipment to be located to the new subsidiaries and the resulting local capital-labor ratios are influenced by the availability of local skills and by relative factor prices. As opposed to this reasoning, it turned out that in several cases, even foreign investors' choice of technology was determined by factors other than local factor prices.

In emerging industries, dominated by foreign investors (e.g., in some ICT-related and in electronics industries), the technological level of the machinery and equipment at local subsidiaries is at the world technology frontier. The output of local subsidiaries in these industries is delivered to mother companies' global customers. Global production in these industries is extremely concentrated; therefore it is not surprising that the production facilities irrespective of their geographical location and of local factor proportions operate with frontier level technology. Note on the other hand, that production relocation to these countries concerned mainly the relatively most (unskilled) labor-intensive phases of the value chains.

In other more mature or traditional industries, foreign investors' choice of technology was indeed extensive. They could locate their own outdated machinery to the new production facilities or decide among a wide variety of alternative techniques when purchasing new machinery. Even in these cases however, the choice of technology was only partly influenced by local relative factor prices; instead, it was determined by the size of the local market.

The example of Poland clearly demonstrates the importance of market size in the choice of the technology. Some years ago, I investigated the existence of any possible differences among transforming countries, in market-seeking investors' investment behavior in the fast moving consumer goods industries (FMCG) – a typical recipient of market-seeking FDI[1] (Szalavetz, 2002). Privatization offered a unique opportunity for multinational companies (MNCs) competing in the saturated market of developed countries to gain huge new markets. Therefore, they established local subsidiaries with overlapping activity and product mix in several transforming countries. The technology level of the production equipment located in countries of a large market size like Poland turned out to be systematically higher, than that of the equipment in Hungary in these industries.[2]

To sum up, while in some industries an extensive spectrum of techniques may coexist with varying distance from the world technology frontier, in others the option of techniques inferior to the world technology frontier is excluded. Local factor endowment may influence the choice of technology in selected industries. However, the choice of technology is often explained by industry-specific features and the market size.

As for the growth impact of the choice of technology, our hypothesis is that the main difference between NMS and the advanced economies is the relatively higher importance in the former country group of the volume effect of technology upgrading. In NMS, the main channel through which the choice of technology has an impact on growth is through its volume effect, i.e. the upgrading of the technology is closely related to an increase in the scale of production. Compared to NMS, the quality change of capital input in advanced economies is more frequently related to the quality upgrading of goods; to switches to higher quality segments within the industry – and not necessarily to the increase of the scale of production. The growth

[1] Including products of food processing and household care industries, personal hygiene products, and tobacco and spirits.

[2] Wziatek-Kubiak and Jakubiak's (2003) case study results also found a much higher mechanization level in Polish industries than in Hungarian ones.

effect of an increasing share of quality competition is however ambiguous: often negative in the short run, switching to positive only with some time-lag.

There are two caveats to be mentioned when analyzing the growth effects of technology choice.

The first one is frequently emphasized in international literature. Although differences in technology figure high among the explanatory factors of inter-country income per capita differences – without the effective absorption and mastery of the acquired technology the growth impact of embodied technology accumulation is minimal.[3] Recent development has further intensified the need for endogenous capability accumulation in FDI-recipient countries where MNCs are the main source of technology. Investors' spectacularly enhanced mobility have made it possible for them to consider local capabilities one of the primary factor to influence their location choice. Therefore, the question of an effective assimilation of the transferred technology is preceded by another. Do foreign investors perceive local capabilities as strong enough to locate any productive activities here? The learning opportunity to achieve technical efficiency, facilitated by investors' know-how, transfers and technical assistance opens up only if the answer to this first question is positive.

As for the second caveat, while the quality level of technology continues to be an important explanatory factor of performance, and the rising capital- and research intensities of production technology are effective barriers to entry in several industries, the GDP share of activities with relatively little capital requirement is increasing rapidly. Activities with low (relative to manufacturing) capital-labor ratio e.g. knowledge-intensive business services are of mounting importance both in terms of their value added share, their productivity performance and their share in total profit[4] in determining competitive and growth performance. Overall growth performance is thus less tightly related to financing the accumulation of high quality level technology than before.

12.3 Technology Accumulation, Growth and Development in Stages

Many models describe latecomer technological catching-up as technology accumulation in well-definable stages. In the first phase, catching-up is based on technology acquisition and assimilation. In this phase reverse engineering and imitation boost growth and technological learning. In the second stage the upgrading of

---

[3]See the debate between "accumulationists" and "assimilationists" about the growth effects of physical capital accumulation in the analyses of the South-East Asian miracle – e.g. Kim and Nelson (2000) for a survey of the literature. Another debate in another related strand of the literature is between growth economists who have been trying to quantify the contribution of embodied and disembodied technological change to output growth. The classical representatives of this latter debate are Solow (1960) and Jorgenson (1966).

[4]The interested reader is referred to Wölfl (2005) for recent OECD-wide data.

manufacturing skills is accompanied by local firms' technological capability accumulation, which allows them to engage in incremental innovations both in terms of learning-by-doing process development and product design. The third stage is marked by intensifying indigenous research efforts forced by the obvious slowdown of technological progress as the technological level of activities gets closer to the world technology frontier. This process culminates in a locally emerging capability to generate new steps forward in world innovation and leads to a transition from investment-driven to innovation-driven growth. (e.g. Hobday, 1994; Kim, 1997).

While the empirical experience of development and technological upgrading of some Southeast Asian countries does in fact support this model, it is misleading in the case of NMS.

Firstly, despite NMS' follower status, development based on imitation of existing techniques is out of question. NMS' modernization has been based on their economic actors' re-integration into the global patterns of production, which was carried out mainly by foreign investors. MNCs have not only transferred the necessary technology, but also ensured market for the output of their local subsidiaries. Therefore, a stand-alone competition based on autonomous technological efforts along the path of transition from imitation of innovation and own brand building does not figure among the realistic development options of economic actors in NMS.

Secondly, re-integration has rapidly changed the production specialization in these countries. Countries, among others Hungary that specialized in high-tech goods, cannot follow the path of structural upgrading characteristic for industrializing Southeast Asian countries (the switching from traditional industries to mature ones and finally to technology-intensive industries). The beneficial coincidence of the process of their re-integration through FDI with the ICT revolution enhanced global integration, and the accelerating fragmentation of manufacturing value chains helped them acquire revealed comparative advantage ratios in industries, the technology-intensity of which is not in line with the predictions based on factor proportions theory. Of course, if it is the skill-intensity of operations and not the technology-intensity of industries that is taken into account, it becomes clear that these countries' RCA-profiles does conform to expectations based on factor proportions theory (see Feser (2003) for theoretical considerations). Nevertheless, in these countries structural and technological upgrading cannot follow a path along the ranking of industries (by technology-intensity). Upgrading is possible rather in a within-industry context.

Within-industry upgrading, i.e. specialization in segments and activities that require higher local skills and are technologically more complex, is an urgent task indeed. On the one hand, at present countries specialized in the *manufacturing* of technology-intensive products benefit more from the growth enhancing effects of the strong demand and productivity growth in these industries than countries specialized in related R&D and in other related knowledge-intensive activities. On the other hand, MNCs' first divestment steps that recently hit "well-specializing" NMS demonstrated that pure price-competitiveness makes them vulnerable irrespective of the industries they are specializing in.

As for within-industry upgrading in the context of FDI-dominated industries, the theory of intrafirm competition and subsidiary charter evolution of Birkinshaw & Lingblad (2005) applies also to subsidiaries in NMS. Subsidiaries engage in "allocational competition" within the organization of the multinational owner for additional resources and for tasks that are technologically more demanding and more human capital intensive. This type of upgrading not only ensures more local value added, but also augments investors' commitment and reduces the risk of divestment.

12.4 Direction of Technical Change

12.4.1 Rapid Increases in the Capital Stock and in Capital-Intensity

Advanced economies' spectacular capital deepening experience in the second half of the 1990s and the puzzling divergence of productivity performance among them have motivated a large number of theoretical and empirical contributions to explain the underlying factors of productivity performance. In a sources-of-growth context, several papers computed various factors' (ICT capital, non-ICT capital, human capital, labor etc.) relative contributions to productivity growth. (For a recent paper and overview see Jorgenson (2005); see also OECD (2003); Timmer, Ypma, and Van Ark (2003)). Productivity differentials were related to factor accumulation differentials and to differences in the weights of individual factors' growth and productivity contribution.

At the same time, another impressive body of literature is concerned with the direction of technical change and its macroeconomic and distributional effects (see e.g. Acemoglu, 2003). Analyzing the direction of technical change is particularly important in an era of rapid physical capital accumulation. Unless technical change increases the marginal product of capital, accelerating physical capital accumulation will be characterized by falling output elasticity of capital – this textbook thesis is deeply rooted in analysts' memory in NMS regarding the economic history of socialism, marked by forced accumulation and poor productivity performance.

As for the past decade, international growth accounting data indeed show a rapid increase in both the total capital stock and in overall capital-intensity. A large part of both can be explained by the "ICT-factor". Table 12.1 gives an overview of selected advanced economies' most recent experience concerning the growth of the capital stock in the "new economy" era.

ICT capital is measured as the sum of gross fixed stocks of ICT-equipment, communication equipment and software. Non-ICT capital includes only machinery and equipment (gross fixed stocks of structures, and transport equipment are not included). Data are calculated from the Groningen Growth and Development

Table 12.1 Rapid increase in total and in ICT capital stock, moderate increase in non-ICT capital stock, 1995–2004 (%)

Country	Δ gross fixed capital stock (%)	Δ ICT capital stock (%)	Δ non-ICT capital stock (%)
EU 15	123.0	260.5	104.3
United States	130.2	292.8	124.6
Austria	125.3	232.9	101.8
Denmark	134.8	304.2	126.1
Finland	111.7	373.6	66.6
France	124.9	259.3	120.5
Germany	110.7	197.2	91.4
Ireland	168.2	757.0	110.6
Netherlands	118.9	334.1	93.8
United Kingdom	126.7	322.5	100.4

Source: Timmer et al. (2003) Appendix Tables, updated June 2005, own calculations

Center's Total Economy Growth Accounting Database (mid-year, in 2000 prices for capital stock, in euro, USD and GBP respectively).

The data in Table 12.1 give an idea about the structural change in the capital stock of advanced economies: the rapidly increasing weight of ICT capital at the expense of non-ICT capital. They make it clear that the accumulation of physical capital stock was mainly due to a rapid increase in the stock of ICT capital in most countries.

Although accumulation trends exhibit really large asset-specific differences, we cannot speak of a resulting "massive" structural change in the capital stock, since ICT capital (and especially IT capital) is still only a fragment of the total capital stock. The share of IT equipment and software has doubled, tripled and even quadrupled in the past decade (Table 12.2). Nevertheless, the share of these two types of assets seems statistically significant only because it is related to the total *equipment* stock, i.e. because we omitted non-residential structures from the denominator. We calculated with the *equipment stock* and not with *productive capital stock,* because

Table 12.2 The share of IT equipment and software in total equipment stock (%)

	EU (15)	Austria	Denmark	Finland	France	Germany
1995	3.68	2.02	7.18	5.74	2.68	3.05
2004	11.32	9.45	19.05	10.84	8.44	9.92
1995 = 100	307.8	467.0	265.4	188.9	315.3	325.7
	Ireland	Italy	Netherlands	Sweden	UK	USA
1995	1.66	3.9	4.55	5.91	5.79	6.97
2004	8.96	7.84	15.90	17.7	16.42	18.36
1995 = 100	541.2	253.8	349.8	299.6	283.6	263.2

Source: GGDC Total Economy Growth Accounting Database, author's calculations

Table 12.3 Rapid increase in total and in ICT capital intensities, reduction of non-ICT capital-intensity, 1995–2004 (constant 2000 prices, %)

Country	Δ total capital-intensity (%)	Δ ICT-capital-intensity (%)	Δ non-ICT capital-intensity (%)
EU 15	115.6	244.8	98.1
United States	120.2	270.3	115.0
Austria	128.8	239.3	104.6
Denmark	131.3	296.4	122.9
Finland	105.6	353.1	63.0
France	124.4	258.2	120.0
Germany	116.7	207.9	96.4
Ireland	135.2	608.6	88.9
Netherlands	102.5	288.0	80.8
United Kingdom	118.7	302.3	94.1

Source: Timmer et al. (2003) Appendix Tables, updated June 2005, own calculations

despite the IT-investment boom, non-residential structures still account for more than two-thirds of the physical capital stock![5]

An overview of capital deepening i.e. an increase in the capital/labor ratio, in Table 12.3 supports the claim of the outstanding role of the "ICT-factor". While non-ICT stock data, with a few exceptions, exhibit a moderate increase compared to ICT capital accumulation performance, we can observe an absolute reduction in non-ICT capital-intensity in many of the surveyed countries, parallel to substantial increases in overall and especially in ICT capital intensities. Capital-intensities are calculated as gross fixed capital stock; ICT capital stock and non-ICT capital stock per total hours worked, respectively.

A key message of Tables 12.1 and 12.3 is that technical change in the past decade was ICT-capital using in a number of countries and industries. Quantifying factor inputs' growth contribution an extensive literature emphasized this feature of (productivity) growth together with other important aspects such as the skill-biased nature of capital accumulation and embodied technical progress. (OECD, 2003; O'Mahony & Van Ark, 2003; Acemoglu, 2002a,b; Caselli, 1999; Goldin & Katz, 1998) However, this literature fails to tackle the other key message of these tables, the non-ICT capital saving nature of technical change. Since ICT-taxonomy, and in particular the ICT-using – ICT-producing dichotomy is the key analytical tool of this literature, it neglects the significant sector- and industry-specific differences in both capital accumulation and deepening.[6] These differences are related to technical progress driven structural change rather than to the variations in industries' absorptive capacity for ICT. This issue is addressed in the next section.

[5]Their share was 67% in the US' and 72% in EU(15)'s total non-residential capital stock in 2004. (Author's calculations from GGDC's Total Economy Growth Accounting Database)

[6]See Colecchia and Schreyer (2002); Timmer et al. (2003); O'Mahony and Van Ark (2003)

12.4.2 Direction of Technical Change – A Case for Capital-Saving Technical Progress in the Manufacturing Sector

The stylized evidence presented in Tables 12.1 and 12.3 suggests that the direction of technical progress may differ in individual economic sectors. Recently we can observe a revival of the debate whether technical progress is purely labor augmenting or capital augmenting,[7] whether the sum of elasticities of substitution for capital and labor add up to unity, exceed unity or remain below it. (For an overview of the debate see Klump, McAdam, and Willman (2004), for a conceptual approach and a specification of the way to distinguish between different types of technological progress, see Welfens, 2005.)

Several papers in this literature share the view that recent technical progress is capital saving and not (or not only) labor saving at least in specific sectors of the economy (e.g. Boskin & Lau, 2000; Welfens, 2005). Boskin and Lau have come to this conclusion as a result of their estimates of the elasticities of substitution between composite capital and labor. Welfens (2005) points to case study evidence for capital-saving technical progress in the ICT sector, referring to the case of computer chips and fiber-optic cables.

In our view, capital-saving technical progress can be detected in several manufacturing industries, not only in the ICT sector.

Our point of departure, however, is not the calculation of elasticities of substitution rather the analysis of data on capital stock and capital intensities.

In this section we present the first results of ongoing[8] capital stock and capital-intensity analyses. The focus of this paper is the manufacturing sector – later we will shortly tackle other sectors' capital accumulation performance as well. We also investigate the relevance and the consequences of our results to NMS.

The American and German data presented below intend to provide stylized evidence for the claim, that technical change is capital saving in several industries of the manufacturing sector. Table 12.4 `shows a noticeable deceleration of growth of the average stock of machinery and equipment – a trend that already started in the 1980s, and Table 12.6 presents a similar trend for capital-intensity.

US data are taken from the NBER Manufacturing Industry Productivity Database (Bartelsman, Becker, & Gray, 2000) The cited NBER database provides highly disaggregated data for 459 four-digit industries in the manufacturing sector from 1958 up to 1996. Table 12.4 gives an overview of capital stock increases and demonstrates

---

[7]Without an intention to take a stand in this debate, the author of this paper recalls Boskin and Lau's (2000) insightful paper, in which the authors argue that technical progress is on the one hand purely capital augmenting (Solow-neutral) but at the same time also purely human capital augmenting. They refer to this form as "generalized Solow-neutral technical progress". Their approach, according to which tangible capital and human capital form a composite capital input separated from labor input, is particularly relevant in the age of the "new economy" marked by exceptionally strong capital-skill complementarity (Acemoglu, 2002a; Goldin & Katz, 1998).

[8]My research is sponsored by the Hungarian National Fund for Social Sciences Research (OTKA K61950)

Table 12.4 Growth of the average stock of machinery and equipment in 459 manufacturing industries in the US (1987 prices, %)

1968–1976	1972–1980	1980–1988	1988–1996
141.4	146.6	119.2	125

Source: NBER Manufacturing Industry Productivity Database – Bartelsman et al. (2000), own calculations

that the rate of increase was the highest in the 1970s, while decelerating considerably in the 1980s and early 1990s.

We could obtain recent data from the Survey of Current Business (SCB) statistics[9] SCB data are not compatible with NBER's cited database (in SCB current data are provided, and the value of software is also included into the stock of machinery and equipment – reflecting the results of capitalization of software in the US National Income and Product Accounts). Nevertheless capital stock calculations based on these data are useful complements to the NBER-data based calculations presented in Table 12.4, since they provide recent data reflecting tendencies in the volume of manufacturing capital stock during the key years of information technology revolution (1996–2004). According to our calculations based on SCB data the increase in the net stock of machinery, equipment and software of the American manufacturing sector was 18.7% in the period of 1996–2004 (1996–2004 = 118.7).

Considering that (i) this latter figure includes also capitalized software and that (ii) the period of 1996–2004 was marked by an outstanding accumulation of physical capital and software, the increase of a mere 18.7% in the equipment stock of the manufacturing sector seems puzzlingly low.

The reduction of the speed of capital accumulation at a time of rapidly increasing output suggests an increase in the productivity of capital. This is confirmed by the falling of the equipment-output ratio of the American manufacturing sector (Table 12.5).

The trend of the evolution of capital intensity also suggests a deceleration instead of acceleration. Capital-intensity is calculated in Table 12.6 as machinery and equipment stock per production worker. As opposed to usual approaches of calculating

Table 12.5 Evolution of the equipment-output ratio of the US manufacturing sector (previous year = 100)

1995	1996	1997	1998	1999	2000	2001	2002	2003	2004
99.5	101.9	101.1	102.8	99.7	98.6	106.9	101.4	96.9	92.7

Source: Author's calculations from data of SCB op.cit and the Statistical Abstracts of the United States (various years)

[9]Fixed Assets and Consumer Durable Goods for 1994–2004. Survey of Current Business, September, 2005 pp. 19–30

Table 12.6 Growth of average capital-intensity in 459 manufacturing industries in the US (1987 prices, %)

1976/1968	1980/1972	1988/1980	1996/1988
155.5	144.8	142.5	124.5

Source: NBER Manufacturing Industry Productivity Database – Bartelsman et al. (2000), own calculations

capital-intensity as equipment per total employment or hours worked, using the number of production workers in the denominator will better reflect the capital intensity of processing activities. At the same time, we can control for the shifts in the structure of manufacturing employment away from tangible processing activities towards production-related service activities (characterized by smaller than the industry average capital intensity).[10] NBER data demonstrate a deceleration in the increase of capital-intensity in the first half of the 1990s. However, as for the most recent period, SCB data point to another wave of accelerating capital-intensity increase.

According to the referred SCB statistics, average capital-intensity increase in the period of 1996–2003 was 43.5% (2003/1996 = 143.5) in the US manufacturing sector. This pick-up in the pace of capital-intensity increase can partly be explained with the inclusion of software in equipment stock calculations.

It is worth recalling Corrado, Hulten, and Sichel (2006) calculations. According to their estimates, total business investment in intangibles was approximately one trillion dollars in 1999, an amount that is roughly the same as investment in tangible capital at that time. Although software spending was less than 15% of intangible business investments – the capitalization of software into US National Income and Product Accounts (NIPA), i.e. the changed calculation method has in itself greatly contributed to the observed apparent acceleration of capital-intensity. As for 2000–2003, the cited authors calculated software investment amounting to a total of USD 172.5 billion. In NIPA series USD 169.6 billion have been officially recognized, which, together with USD 83.6 billion for the period of 1990–1999 contributes to the measured higher numerator of capital-intensity. (Corrado et al. (2006) p. 40)

Another explanatory factor is a sizable change in the denominator of the indicator. The number of production workers in manufacturing decreased markedly in this period. Compared to 1996 (1996 = 100) the value of the indicator was 80.5 in 2003[11] as a consequence of the offshoring of various processing activities.

We have also examined capital-intensity tendencies of the 459 industries included in NBER's cited database to see how many of these industries can be characterized with decelerating capital-intensity increase in the period of

[10] According to Pilat and Wölfl's (2005) estimates, the share of employment in service-related occupations within the manufacturing sector was as high as 40% on the average in 2002, in a sample of OECD economies (the U.S. where this share is probably much higher was not included in their sample. (see Fig. 6, p. 19)

[11] Calculated from the Statistical Abstract of the United States (various years)

1988–1996 compared to the period of 1976–1987. There were 280, out of 459 industries, the capital-intensity of which showed a slower increase between 1988 and 1996 than in the period of 1976–1987. We have found 93 industries (out of the above 280) for which the capital-intensity showed an absolute decrease between 1988 and 1996.

German data also show a slow and decelerating growth of manufacturing capital (machinery and equipment) stock in the 1990s (Table 12.7).

The relative deceleration of capital accumulation and capital intensity increase in the manufacturing sector has been caused by three factors. Firstly, the emergence of a new general-purpose technology generated sizeable structural change in the composition of output, which has led to changes in the aggregate factor intensities. Secondly, technical change triggered organizational changes have led to *the substitution of intermediate input for capital.* Production fragmentation, the offshoring of capital-intensive activities, and the specialization of manufacturing companies away from tangible processing activities towards production related service activities characterized by smaller than the industry average capital intensity[12] were facilitated by the revolutionary advances in ICT that contributed to the declining transaction and monitoring costs (Arndt & Kierzkowski, 2001; Feenstra, 1998).

The third reason, to be examined in the next section is the effect of capital-saving technical progress, which has also contributed to the reduction of the speed of both capital accumulation and capital intensity increase. Before turning to this issue, we shortly summarize other sectors' capital accumulation experience.

As opposed to the manufacturing sector that displayed a noticeable deceleration of capital accumulation and capital intensity increase, technical change in advanced

Table 12.7 Chain-type quantity indexes for machinery and equipment (gross) stock in the German manufacturing sector (1995 = 100)

Year	1991	1992	1993	1994	1995	1996	1997
1	84.9	91.6	97.1	99.3	100	100.7	101.2
2	91.1	95.1	98.3	99.4	100	100.2	100.5
Year	1998	1999	2000	2001	2002	2003	
1	102.2	102.3	103.0	105.1	106.1	103.9	
2	100.7	101.2	102.1	103.0	103.6	103.0	

1 = at replacement costs 2 = at 1995 prices
Source: Statistical Offices of the Länder and the Federal Statistical Office: Fixed Assets in the Länder and East-West Large-Scale Regions1 of Germany, 1991–2003 http://www.statistik-portal.de/Statistik-Portal/en/en_publ.asp#Gesamtrechnungen

[12] According to Pilat and Wölfl's (2005) estimates, in 2002, the share of employment in service-related occupations within the manufacturing sector was as high as 40% on the average, in a sample of OECD economies (the U.S. where this share is probably much higher was not included in their sample. (see Fig. 6, p. 19)

Table 12.8 The share of manufacturing (equipment and software) in total private net stock of equipment and software in the USA (at current costs, %)

1994	1995	1996	1997	1998	1999	2000	2001	2002	2003	2004
27.8	27.9	27.9	27.7	27.4	26.5	25.6	25.0	24.5	23.7	23.2

Source: Fixed Assets and Consumer Durable Goods for 1994–2004. Survey of Current Business, September, 2005 p. 25, own calculations

Table 12.9 The share of manufacturing (machinery and equipment) in total machinery and equipment net stock of Germany (at replacement costs, %)

1991	1992	1993	1994	1995	1996	1997	1998	1999	2000	2001	2002	2003
48.4	48.0	47.5	46.9	46.2	45.8	45.3	44.5	43.6	42.9	42.0	41.4	40.7

Source: Statistical Offices of the Länder and the Federal Statistical Office: Fixed Assets in the Länder and East-West Large-Scale Regions1 of Germany, 1991–2003 http://www.statistik-portal.de/Statistik-Portal/en/en_publ.asp#Gesamtrechnungen

economies generated sizeable capital accumulation in the other two sectors of the economy: agriculture and in services. A detailed analysis of the latter two sectors does not figure in the ambition of this paper, only two tables are provided to show the shrinking importance of manufacturing in the total gross machinery and equipment stock of the American and the German economies (Table 12.8 and 12.9).

12.4.3 Capital-Saving Technical Change in the Manufacturing Sector

An important question related to capital-saving technical change is the way it becomes manifest. Is it apparent in the new and superior quality (various performance parameters) of capital goods that make a unit of capital good more productive than before? Does it become manifest in changed technologies that make specific highly capital-intensive processing tasks less capital-intensive than before?

We have a lot of recent evidence concerning the former case – see Moore's law in the case of computer chips (Welfens, 2005), the dramatic efficiency improvement of information processing equipment, telecommunication equipment,[13] diagnostic instruments, photographic equipment, measuring, testing and optical instruments, as well as the effects of the coming nanotechnology revolution.[14]

[13]Examples include digital switching, fiber optics, satellite and cellular transmission equipment, local area network equipment etc.

[14]Nanotechnology – the ability to generate structures, devices and systems with novel properties due to their nanoscale dimensions – is predicted to bring about orders-of-magnitude efficiency increases information storage and data processing. Being a major, general-purpose technology that will define the next technological paradigm it will radically enhance the performance of both capital equipment and the quality of final goods in a wide range of industries.

We have anecdotal evidence for the latter case but also against it. Consider, for example, the case of the steel industry where the capital requirement as well as the operational costs of minimills are inferior to the ones of integrated mills or the fact that information technology brought great reductions in the capital requirement of the printing and publishing industry. As for anecdotal evidence against the capital-saving effect of recent technical progress, we can have in mind the incremental development of industrial machinery in several industries. Combining precision machinery with computerized control, new equipment among others in the textile (spinning, knitting etc.) industry, in the pulp and paper industry, or in metal processing industry have large investment requirement. Production facilities equipped with the most up-to-date machinery in these industries are considerably more capital-intensive than before. In these cases technical progress is labor-augmenting; it is energy- and raw material saving rather than capital-saving.

We have tried to find indirect evidence for the case of technical change that makes some highly capital-intensive processing tasks less capital-intensive than before[15] by elaborating on Caselli's (1999) reasoning. In a stimulating paper, Caselli demonstrated that the inter-industry dispersion of equipment/employment ratio, i.e. the inequality of capital-intensity increased sharply in the 1980s. Investigating possible evidence for skill-biased technical change, Caselli related this phenomenon to the increasing inter-industry inequality in wages. Our point of departure is however different. Our working hypothesis is that with capital-saving technical change, the sharply increasing trend of the inter-industry dispersion of capital intensity will decelerate and the trend may even be reversed.

We undertake a test similar to Caselli's referred one, but we compute the evolution of the inter-industry dispersion of capital-intensity using the number of production employees in the denominator. While Caselli's surveyed period is from 1958 to 1991, we focus on the already available, somewhat more recent data to see whether this trend continues and how the evolution of dispersion is related to the evolution of average capital intensity. Our results are summarized in Fig. 12.1 and Table 12.10.

In fact, the sharp increase in the inter-industry dispersion of capital equipment per production worker that was obvious between 1975 and 1985 began to decelerate afterwards. Given the relative shortness of the period that has elapsed since Caselli's calculations and for which data are available, we still cannot provide conclusive evidence at present for the hypothesis of a reversal in the trend of inter-industry dispersion of capital-intensity.

To find out, whether the deceleration in the dispersion of capital-intensity increase was influenced by technical change that makes the processing tasks of industries with above-the-manufacturing-average capital intensity less capital-intensive than before, we examined the 93 industries within which capital-intensity showed an absolute decrease in between 1988 and 1996 to see if their capital intensities in 1988 were above the manufacturing average. We found 31 industries that

[15] Note that the capital-saving effect of organizational innovations that accompanied the recent ICT revolution may even exceed that of the technical innovations.

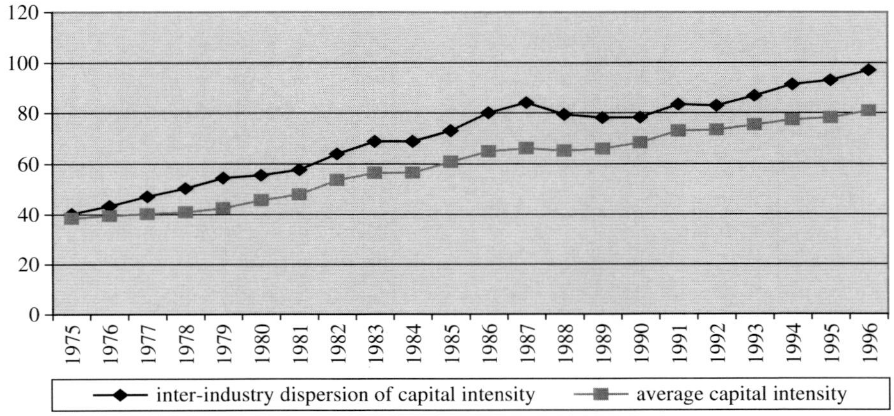

Source: NBER Manufacturing Industry Productivity Database – Bartelsman et al. (2000), own calculations

Fig. 12.1 Evolution of inter-industry dispersion of capital-intensity (459 manufacturing industries[16] in the US), in 1,000 USD, 1987 prices

Table 12.10 A decelerating growth of inter-industry dispersion of capital-intensity (%)

1980/1975	1985/1980	1990/1985	1995/1990	1996/1991
138.1	131.4	107.1	118.9	116.3

Source: NBER Manufacturing Industry Productivity Database – Bartelsman et al. (2000), own calculations

fit this assumption. This lack of a more significant overlapping makes us reject the hypothesis that capital-saving technical change affected *mainly* highly capital-intensive industries. Nevertheless, a weaker conjecture still holds: we can claim that recent capital-saving technical change affected quite a significant number of industries with high, i.e. above-the-manufacturing-average capital-intensities.

We have hitherto documented some facts and trends that support the claim of a partial and selective capital-saving technical progress – in advanced economies' specific manufacturing industries. The next section investigates whether these results apply also to the NMS of the European Union.

12.4.4 What About the New Member States of the European Union?

The FDI-driven rapid capital accumulation experience of NMS features and benefits from tendencies that are different from the ones in advanced economies.

[16] Following 1991 we calculated with 458 industries. We omitted asbestos products (NACE 3292) since its production was quickly phased out in the 1990s.

In advanced economies' manufacturing firms, part of the sizable structural change in the composition of physical capital derives from offshoring labor and capital-intensive activities. In FDI-recipient NMS[17] it is FDI and the accompanying embodied and disembodied technology transfer that have triggered a rapid structural change both in the industry mix and in the composition of manufacturing companies' equipment stock. This structural change is thus driven by their *receiving* the offshored physical capital. In line both with Vernon's (1966) original product cycle model and with Krugman's (1979) model marked by exogenous technology transfer that contributes to narrowing the income gap between rich and poor countries, these transfers of embodied and disembodied technology contributed to an increase in recipient countries' capital-intensity, as well as to the upgrading of their capital stock.

In industries privatized and technologically upgraded by foreign investors, the transfers of embodied and disembodied technology have initiated capital-augmenting technical progress. Transfers have been preceded by defensive restructuring steps that affected both the capital stock and the capital intensity of production: asset stripping and downsizing. Both the offensive and the defensive restructuring steps increased the efficiency of capital; that is they had capital-augmenting effects.

On the other hand, massive structural shifts within manufacturing[18] have also contributed to both the observed capital-saving and capital-augmenting changes. One feature of structural change was a rapid decline in the GDP share of capital-intensive industries, like chemicals or basic and fabricated metal products. This decline took place not only in a relative sense. In other words, it was not only caused by the rapid expansion of new industries, but also in the absolute sense. Since capital-intensive industries were the most hardly hit by transformation recession and several companies have exited from the market in these industries, their output and value added declined in absolute terms as well. The ones that managed to survive did so due to considerable asset stripping, downsizing and production reorganization, which eventually increased the marginal product of both capital and labor.

The proposition to be argued for – though not formally tested – in this section is that as opposed to advanced economies' recent experience marked by a relatively strong role of Solow-neutral technical progress, the role of Harrod-neutral technical progress has so far been relatively more important in NMS' manufacturing sector.

The main obstacle to performing a formal or at least a descriptive statistical analysis is the lack of reliable time series on capital stock in NMS. Transformation has led to radical changes in the capital stock of manufacturing companies that makes the survey of time series impossible. After the change of the regime, a substantial part of fixed assets has become obsolete under the new market economy conditions. The previous trend of exit mortality function (i.e. the distribution

---

[17]The case of other developing countries where it is also FDI that accounts for the rapidly improving performance is similar.

[18]See Gács (2003) about the extent of structural change in the ten candidate countries.

of retirements around the average service life of physical assets) which was more or less characterized by a lognormal pattern drastically changed after the change of the regime. There was a significant volume of unplanned, simultaneous discard of obsolete assets.

As far as Hungary is concerned, the Central Statistical Office stopped publishing data on capital stocks in the beginning of the 1990s. The book value of the capital stock was already at that time highly distorted, since the last comprehensive valuation of the national capital stock was undertaken in 1968 (Fazekasné, Imre, Nádudvari, & Nagy, 2003).

Capital stock changes can in principle be traced back from the balance sheets of corporations. These data bear however seriously distortions, because book values reflect the *historic acquisition price* of the assets. Data have become distorted not only because of the lack of reliance on perpetual inventory methods (PIM), but also because in the 1990s, the revaluation of assets depended on the ad hoc decision of the owners. Some firms have revalued their assets already in the early transformation years, others revalued half a decade later, while some of them haven't revalued their asset stock at all. The Hungarian Central Statistical Office initiated a project on the elaboration of a new methodology of the valuation and the measurement of capital assets in line with the European requirements. The survey and data collection to estimate the PIM based capital stock of Hungarian companies in the private sector was recently completed. The revaluation will serve as a good basis for future analyses concerning changes after 2000. Information concerning the changes in the capital stock during the first decade of transformation is however lacking.[19]

Labor-augmenting technical change in NMS has been driven on the one hand by labor-saving embodied and especially disembodied technical progress. The upgrading of FDI-recipient industries' equipment stock, accompanied by both organizational innovations and technical assistance that improved factory practices have all brought about considerable labor shedding,[20] which resulted in non-negligible productivity gains. Both the embodied and the disembodied sources of productivity growth increased the rate of labor- augmenting technical change and permitted a shift to a higher production function with significantly increased output per worker.

Structural change has brought about the substantial quality upgrading of NMS' aggregate machinery and equipment stock. Capital accumulation took place mainly in the new and quickly expanding industries set up by foreign investors. Although the transferred capital embodied high technology, the new industries can be considered technology- and skill intensive only according to the OECD taxonomy[21] that uses industry as a basic unit of analysis. In reality, however, the high-tech and medium high-tech industries located in NMS are mobile Schumpeterian industries

[19]This paragraph draws on Szalavetz (2005).

[20]It was of course defensive restructuring marked by downsizing that accounted for dominant part of labor shedding.

[21]The OECD taxonomy classifies industries by technology-intensity (see Hatzichrinoglou, 1997). This taxonomy is widely used in international statistics, e.g. in Eurostat, while alternative taxonomies (see Peneder (2003) for survey) are used in various analytical papers.

(Klodt, 1992) with local value added composed by mainly unskilled labor-intensive activities.[22]

The most puzzling feature of NMS' experience with mobile Schumpeterian industries is that these industries have been an exception to the consensus thesis of capital-skill complementarity, i.e. a significant complementarity between new technological capital and human capital (e.g. Acemoglu, 2002a; Machin & Van Reenen, 1998). As opposed to this consensus finding, transferred physical capital that embodied high technology in NMS *was complementary to low-skilled labor*!

Although labor-augmenting technical change and the upgraded technology that significantly increased the efficiency of using unskilled labor has ensured rapid catching-up in NMS, it is questionable how long these countries can remain competitive in an era of quickly increasing demand for products (and services) produced applying skilled labor using technologies.

The transition to a technical change with a direction and features more in line with those of advanced economies is an urgent necessity for NMS. Of course, first it is indispensable to create the necessary preconditions of this transition and later, to manage its painful employment-related trade-offs.

References

Acemoglu, D. (2003). Labor- and capital-augmenting technical change. *Journal of the European Economic Association, 1*, 1–37.

Acemoglu, D. (2002a). Technical change, inequality, and the labor market. *Journal of Economic Literature, 40*, 7–72.

Acemoglu, D. (2002b). Directed technical change. *Review of Economic Studies, 69*, 781–809.

Acemoglu, D., & Zilibotti, F. (2001). Productivity differences. *Quarterly Journal of Economics, 116*, 563–606.

Arndt, S. W., & Kierzkowski, H. (2001). *Fragmentation: New production patterns in the world economy*. Oxford and New York: Oxford University Press.

Artner, A. (2005). Production technology and competitiveness in the hungarian manufacturing industry. *Acta Oeconomica, 55*, 317–340.

Atkinson, A. B., & Stiglitz, J. E. (1969). A new view of technological change. *Economic Journal, 79*, 573–578.

Bartelsman, E. J., Becker, R. A., & Gray, W. B. (2000). NBER-CES manufacturing industry database. New York: National Bureau of Economic Research.

---

[22]This calls for analyzing occupations rather than industries when assessing the quality of structural upgrading (Thompson & Thompson, 1985; Feser, 2003). Industries as the basic unit of analysis may lead to erroneous conclusions. Hungary for example shows a large share of high-tech exports in total exports – in 2004 this share was 24.6% in Hungary as compared to 17.7% in EU 15 (Götzfried, 2005, p. 6). Hungary features large revealed comparative advantage in technology- and skill intensive industries (Eichengreen & Kohl, 1998; Kaminski, 1999; Zaghini, 2005). According to recent Eurostat data however, value added per person employed in high-tech manufacturing was € 19,000 in Hungary in 2002, as compared to e.g. 63,000 in Germany, 127,000 in Finland, and 70,000 in EU 15. (Götzfried, 2005, p. 4) See also Borbély's (2004) calculations about the inter-country differences in the quality composition of high-tech goods.

Basu, S., & Weil, D. N. (1998). Appropriate technology and growth. *Quarterly Journal of Economics, 113*, 1025–1054.

Birkinshaw, J., & Lingblad, M. (2005). Intrafirm competition and charter evolution in the multibusiness firm. *Organization Science, 16*, 674–688.

Borbély, D. (2004). Competition among cohesion and accession countries: Comparative analysis of specialization within the EU market. EIIW Discussion Papers, No. 122, University of Wuppertal.

Boskin, M. J., & Lau, L. J. (2000). Generalized Solow-neutral technical progress and postwar economic growth. NBER Working Paper, No. 8023.

Caselli, F. (1999). Technological revolutions. *American Economic Review, 89*, 78–102.

Caselli, F., & Coleman W. J. (2000). The world technology frontier. NBER Working Papers, No. 7904.

Colecchia, A., & Schreyer, P. (2002). ICT investment and economic growth in the 1990s: Is the United States a unique case? A comparative study of nine OECD countries. *Review of Economic Dynamics, 5*, 408–442.

Corrado, C. A., Hulten, C. R., & Sichel, D. E. (2006). Intangible capital and economic growth. NBER Working Papers, No. 11948.

Eichengreen, B., & Kohl, R. (1998). The external sector, the state and development in Eastern Europe. BRIE Working Paper No.125, San Francisco: Berkeley Roundtable on the International Economy.

Fazekasné, K. K., Imre, J., Nádudvari, Z., & Nagy, A. (2003). Az állóeszköz-statisztika fejlesztésének eredményei. (The results of the development of real asset statistics) *Statisztikai Szemle, 81*, 973–990.

Feenstra, R. C. (1998). Integration of trade and disintegration of production in the global economy. *The Journal of Economic Perspectives, 12*, 31–50.

Feser, E. J. (2003). What regions do rather than make: A proposed set of knowledge-based occupation clusters. *Urban Studies, 40*, 1937–1958.

Gács, J. (2003). Structural change and catching up: The experience of the ten candidate countries. In G. Tumpel-Gugerell & P. Mooslechner (Eds.), *Economic convergence and divergence in europe: Growth and regional development in an enlarged European union* (pp. 131–167). Cheltenham (UK), Northampton (US): Edward Elgar. .

Goldin, C., & Katz, L. F. (1998). The origins of technology-skill complementarity. *Quarterly Journal of Economics, 113*, 693–732.

Götzfried, A. (2005). High-technology: Enterprises and trade. Eurostat, Statistics in Focus 9/2005.

Hatzichrinoglou, T. (1997). Revision of the high-technology sector and product classification. STI Working Papers No. 2, OECD, Paris.

Hobday, M. (1994). Export-led technology development in the four dragons: The case of electronics. *Development and Change, 25*, 333–361.

Jorgenson, D. W. (2005). Information technology and the g7 economies. Mimeo, http://post.economics.harvard.edu/faculty/jorgenson/papers/IT_G7_economies_05012005.pdf

Jorgenson, D. W. (1966). The embodiment hypothesis. *Journal of Political Economy, 74*, 1–17.

Kaminski, B. (1999). *Hungary: Foreign trade issues in the context of accession to the EU*. Washington, DC: World Bank Publications.

Kim, L. (1997). *Imitation to innovation: The Dynamics of Korea's technological learning*. Boston, MA: Harvard Business School Press.

Kim, L., & Nelson, R. R. (Eds.). (2000). Technology, learning and innovation. Cambridge, UK: Cambridge University Press.

Klodt, H. (1992). Technology-based trade and multinationals' investment in Europe: Structural change and competition in Schumpeterian goods. In M. W. Klein, & P. J. J. Welfens (Eds.), *Multinationals in the New Europe and global trade* (pp. 107–121). Dordrecht: Springer.

Klump, R., McAdam, P., & Willman, A. (2004). Factor substitution and factor augmenting technical progress in the US: A normalized supply-side system approach. ECB Working Papers, No. 367, ECB, Frankfurt am Main.

Krugman, P. (1979). A model of innovation technology tansfer and the world distribution of income. *Journal of Political Economy, 87*, 253–266.

Machin, S., & Van Reenen, J. (1998). Technology and changes in skill structure: Evidence from seven OECD countries. *Quarterly Journal of Economics, 113*, 1215–1244.

OECD (2003). *The sources of economic growth in OECD countries*. Paris: OECD.

O'Mahony, M., & Van Ark, B. (2003). EU Productivity and competitiveness: An industry perspective. Can Europe resume the catching-up process? European Commission, Enterprise Directorate General. www.ggdc.net/pub/EU_productivity_and_competitiveness.pdf

Peneder, M. (2003). Industrial structure and aggregate growth. *Structural Change and Economic Dynamics, 14*, 427–448.

Pilat, D., & Wölfl, A. (2005). Measuring the interaction between manufacturing and services. OECD STI Working Paper, No. 5.

Solow, R. (1960). Investment and technical progress. In K. J. Arrow, S. Karlin, & P. Suppes (Eds.), *Mathematical methods in the social sciences* (pp. 89–104). Stanford: Stanford University Press.

Szalavetz, A. (2005). Physical capital stock, technological upgrading and modernisation in hungary. *Acta Oeconomica, 55*, 201–221.

Szalavetz, A. (2002). Some neglected effects of EU enlargement: Rationalization and Specialization. IWE Working Papers, No. 129, MTA VKI, Budapest.

Thompson, W., & Thompson, P. (1985). From industries to occupations: Rethinking local economic development. *Economic Development Commentary, 9*, 12–18.

Timmer, M., Ypma, G., & Van Ark, B. (2003). *IT in the European Union: Driving productivity divergence*? Groningen: Groningen Growth and Development Centre, Research Memorandum GD-67.

Vernon, R. (1966). International investment and international trade in the product cycle. *Quarterly Journal of Economics, 80*, 190–207.

Welfens, P. J. J. (2005). A Quasi-CobbDouglas production function with sectoral progress: Theory and application to the new economy. EIIW Discussion Paper, No. 132, University of Wuppertal.

Wölfl, A. (2005). The service economy in OECD countries. OECD/CEPII, *OECD Science,Technology and Industry Working Papers*, 2005/3, OECD Publishing.

Wziatek-Kubiak, A., & Jakubiak, M. (2003). Results of field work on firm specific determinants of productivity gap among Polish, Hungarian and West German Enterprises: Furniture and Investment Goods Industries. http://www.iwh-halle.de/projects/productivity-gap/prelim_results/WP6_CASE_02.pdf

Zaghini, A. (2005). Evolution of trade patterns in the new EU member states. *Economics of Transition, 13*, 629–658.

Chapter 13
Benchmarking Innovation Performance on the Regional Level: Approach and Policy Implications of the European Innovation Scoreboard for Countries and Regions

Bernhard Iking

13.1 A Short Introduction

Benchmarking is a methodology which measures processes, performances and results of the own region (firm) and compares the gathered data with those of other (better) regions (firms) by using indicators (benchmarks). The goal is to learn from the "Best". Hence, benchmarking can be understood as a continuous systematic process to compare the success of organisations, functions, processes, regions, policies or firms with the "best" aiming at adapting success experiences and improving own procedures in order to exceed the best performers.

In the given context we should understand the function of a regional benchmarking as follows: it aims at finding out the strengths and the weaknesses of certain regions and/or countries in order to initiate a process of change in the sense of closing identified gaps, or maintain respectively extend competitive positions.

Innovation in an economic sense is here understood as the successful implementation of an idea into a marketable or improved product, process or service. Following this interpretation, benchmarking innovation is the comparison of specific criteria concerning innovation base (mainly input oriented), innovation results (mainly quantative innovation outcome) and innovation performance (qualitative innovation outcome).

For this reason, benchmarking innovation performance in regions is a suitable instrument

- to assess a country's/region's performance in terms of innovation inputs and outputs
- to analyse a country's/region's performance related to the factors that drive innovation and innovation processes

Based on the identified strengths and weaknesses future opportunities and threats can be derived and the appropriateness of current innovation policies assessed. These analyses may initiate policy changes in order to improve regional innovation performance in future.

B. Iking (✉)
ZENIT - Zentrum für Innovation und Technik mbH, Mülheim/Ruhr, Germany

P.J.J. Welfens, J.T. Addison (eds.), *Innovation, Employment and Growth Policy Issues in the EU and the US*, DOI 10.1007/978-3-642-00631-9_13,

13.2 The European Innovation Scoreboard (EIS): Approach and Key Results 2006

The European Innovation Scoreboard is an instrument developed by the European Commission in order to evaluate and compare the innovation performance of the EU member states. The decision goes back to the Lisbon European Council from 2000 where the EU-leaders pronounced the strategic goal for the European Union to become the most competitive and dynamic knowledge-based economy in the world by 2010.

The EIS 2006 includes 25 different indicators which assess five key dimensions of innovation:

1. Human Resources (5 indicators)
2. Knowledge creation (4 indicators)
3. Innovation and Entrepreneurship (6 indicators)
4. Output -Applications (5 indicators)
5. Output – Intellectual Property Rights (5 indicators)

These indicators are used to derive trend analyses for all 25 European Union Member States, as well as for Bulgaria, Romania, Turkey, Iceland, Norway, Switzerland, the USA and Japan.[1]

13.2.1 Overall Country Trends

The underlying indicators are combined into a composite index (SII), which provides an overview of the relative national innovation performance (see Fig. 13.1)

Calculation of the Summary Innovation Index (SII):

$$SII_i^t = \frac{\sum_{j=1}^{m} q_j y_{ij}^t}{\sum_{j=1}^{m} q_j} \qquad y_i^t = \frac{x_{ij}^t - Min\left(x_j^t\right)}{Max\left(x_j^t\right) - Min\left(x_j^t\right)}$$

[1]National data sources were used whenever Eurostat and OECD data were not available.

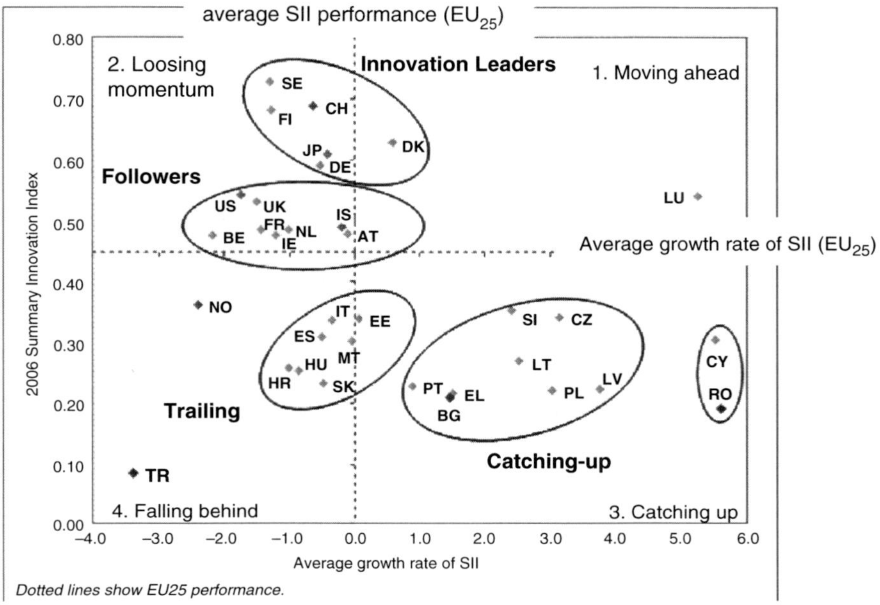

Fig. 13.1 Average country trend by Summary Innovation Index (SII)
Source: European Commission 2006, p. 4.

q_i is the weight given to indicator j in the composite index.
x_{ii} is the value of indicator j for country i.
y_{ij} equals the value of the transformed indicator for country i

Due to data restrictions not all indicators were available for all countries in EU_{25}. Hence, the number of indicators that entered in the Summary Innovation Index (SII) of each country vary between 14 to 25 indicators. For this reason, innovation classification based on the SII index need to be interpreted with caution. However, it gives a slight idea of the relative innovation positions by country (see Fig. 13.1). The current performance is represented by the vertical axis against the average-trend performance of SII (horizontal axis). This leads to the differentiation between countries that move ahead[2] and those that are losing momentum[3] on one side and those that are catching up[4] and other, that are falling behind.[5]

[2]Countries above both the average EU_{25}-trend (SII growth rates) and the current average SII_{EU25}performance (Innovation leaders).

[3]Countries above the average SII_{EU25} performance but below the current average SII_{EU25} performance (Innovation followers).

[4]Countries below the average EU_{25}-trend (SII growth rates) but above current average SII_{EU25} performance.

[5]Countries below both the average EU_{25}-trend (SII growth rates) and the current average SII_{EU25} performance.

Countries like Luxembourg and Denmark are moving ahead with above average values for current performance and trends. Also Switzerland, Germany, Finland, Sweden and Japan are performing on average better than most of the other EU countries do (Innovation leaders). Finland and Sweden are the top performers on the SII but with below average trends.

The situation in the Netherlands, France, Ireland, Belgium, United Kingdom, Island and Austria seems to be less advantageous as trend rates in these countries are quite far below the EU_{25}-average. Their SII scores are below thoses of the innovation leaders but above that of the EU_{25} and the other countries. The EU calls this group "Innovation followers". The EU states that the gap of the innovation followers with the innovation leaders has on average slightly increased.[6]

However, only the comparison of innovation performance at indicator level gives more detailed insights on strengths and weaknesses of the compared countries.

Table 13.1 shows the innovation performance for EU_{15}, EU_{25} and selected EU-countries for 25 indicators including the US and Japan.[7] The leading positions in trend performances are taken up to 60% by the innovation leaders (SE, FI, CH,

Table 13.1 Innovation performance of selected EU-countries

No.	Indicator	EU_{25}	EU_{15}	European "innovation leaders" EIS 2006			DE	US	JP
1.1	S & E graduates/20–29 years	12.7	13.6	23.1 (IE)	22.0 (FR)	18.1 (UK)	9.0	10.2	13.4
1.2	Population with tertiary education/25–64 years	22.8	24.0	34.6 (FI)	33.5 (DK)	33.3 (EE)	24.6	38.4	37.4
1.3	Broadband penetration rate (≥ 144 Kbits/s)	10.6	12.0	22.5 (IS)	22,4 (NL)	22.0 (DK)	10.2	14.9	16.3
1.4	Life-long learning	11.0	12.1	34.7 (SE)	29.1 (UK)	27.6 (DK)	8.2	–	–
1.5	Youth education	76.9	74.1	96.3 (NO)	91.5 (SK)	90.6 (SI)	71.0	–	–
2.1	Public R&D/GDP GERD	0.65	0.66	1.17 (IS)	0.99 (FI)	0.92 (SE)	0.76	0.68	0.74
2.2	Business R&D/GDP BERD	1.20	1.24	2.92 (SE)	2.46 (FI)	2.16 (CH)	1.76	1,87	2,39
2.3	Share of medium-high/high-tech R&D	–	89.2	92.7 (SE)	92.3 (DE)	92.0 (CH)	92.3	89.9	86.7
2.4	Share of firms receiving funding	–	–	39.3 (LU)	27.8 (IE)	17.8 (AT)	9.2	–	–
3.1	SMEs innovating in-house	–	–	47.2 (IE)	46.5 (IS)	46.2 (DE)	46.2	–	15.3

[6]European -Commission, 2006, p. 3.

[7]For the indicators 1.3 (broadband penetration rate), 1.4 (life long learning), 2.4, 3.1, 3.2, 3.3, 3.6, 4.3, 4.4, and 5.5 (high-tech manufacturing value added) time series are only available for a very small number of new member states and just a few years.

Table 13.1 (continued)

No.	Indicator	EU_{25}	EU_{15}	European "innovation leaders" EIS 2006			DE	US	JP
3.2	Innovative SMEs co-operating with others	–	–	20.8 (DK)	20.0 (SE)	17.3 (FI)	8.6	–	6.9
3.3	Innovation expenditures	–	–	3.47 (SE)	3.08 (EL)	2.93 (DE)	2.93	–	–
3.4	Early-stage venture capital (% of GDP)	–	0,023	0.068 (DK)	0.067 (SE)	0.048 (UK)	0.015	0.072	–
3.5	ICT expenditures (% of GDP)	6.4	6.4	9.8 (EE)	9.6 (LV)	8.6 (SE)	6.2	6.7	7.6
3.6	SMEs using organizational innovation	–	–	63.0 (CH)	58.4 (LU)	57.1 (DK)	53.2	–	–
4.1	Employment in high-tech services	3.35	3.49	5.13 (SE)	4.97 (IS)	4.69 (DK)	3.36	–	–
4.2	Share of High-tech exports	18.4	17.7	55.9 (MT)	29.5 (LU)	29.1 (IE)	15.4	26.8	22.4
4.3	Sales share of new-to-market products	–	–	13.6 (MT)	12.8 (SK)	10.8 (PT)	7.5	–	–
4.4	Sales share of new-to-firm products/turnover	–	–	15.1 (PT)	10.0 (DE)	10.0 (ES)	10.0	–	–
4.5	Employment in medium-high/high-tech manufacturing/turnover	6.66	6.71	10.43 (DE)	9.63 (SI)	9.42 (CZ)	10.43	3.84	7.30
5.1	EPO patents/per million population (pmp)	136.7	161.4	425.6 (CH)	311.7 (DE)	305.6 /FI)	311.7	142.6	174.2
5.2	USPTO patents/pmp	50.9	60.2	168.4 (CH)	123.0 (DE)	109.7 (SE)	123.0	277.1	304.6
5.3	Triad patents /pmp	32.7	38.9	108.9 (CH)	101.7 (FI)	85.2 (DE)	85.2	47.9	102.1
5.4	Community trademarks	100.7	115.7	782.7 (LU)	225.2 (CH)	187.0 (AT)	140.5	33.8	11.7
5.5	Community designs	110.9	127.6	377.6 (LU)	243.2 (DK)	210.0 (CH)	186.5	17.5	13.2

Source: European Commission (2006), p. 13; own adaptations

JP, DE, DK, LU). The innovation leaders are particularly dominant in the fields of knowledge creation, innovation & entrepreneurship and intellectual property whereas the innovation follower (US, UK, FR, NL; BE, IE, IS, AT) occupy the best slots in the area of "Innovation driver".

The European innovation leaders are ahead of the US in 12 out of 15 comparable indicators and ahead of Japan in 13 out of 17 comparable indicators. On the other hand, compared with the European average performance the relation to the US is negative: The EU_{25} (EU_{15}) does still significantly better in four (five) indicators against nine (eight) where EU_{25} (EU_{15}) does significantly worse.

The trend balance in relation to Japan is disastrous. Just in the field of community trademarks and designs EU_{25} (EU_{15}) does better than Japan. In ten indicators

EU_{25} (EU_{15}) does significantly worse. Especially in terms of Business R&D the EU as well as the US are losing ground in relation to Japan. This trend might find its reflection already in the trend performance of patents at the European Patent Office where European countries and the USA lose remarkable ground in relation to Japan too.

The comparison of trends in innovation performance does not offer many positive future scenarios from the European point of view in relation to the US and Japan. In the static view the US outperforms the EU_{25} (EU_{15}) average performance in 9 (8) out of 11 (12) indicators and Japan is better in 11 (10) out of 13 (14) indicators.[8]

According to EIS 2006 the main factors underlying this gap are: USPTO and Triad patents [–49% (US); –38% (JP)] of the gap, working population with tertiary education [–38% (US), –29% (JP)], BERD [–15% (US); –20% (JP)], high-tech exports [–12% (US)], Broadband penetration [–11% (US), –12% (JP)].

13.2.2 Results for Germany

Germany's innovation performance in relation to EU_{25} average is demonstrated in Fig. 13.2. The strength of Germany lies in the fields of "Knowledge creation" and "Intellectual Property Rights". Public and especially private R&D expenditures (input side) in Germany are broadly above European average. Even more noticeable is the patent output, both for medium and high-tech patents, at European and US patent offices as well as in relation to triad patents. This reflects the high inventory potential of the German innovation system.

Even the transformation of the inventory potential in new products and services through SME's is much better than EU_{25} average and public discussion in Germany suggests. However, there is still a performance gap between "knowledge creation" and "knowledge application" in the sense that performance potential is by far not exploited as it could be.

The most alarming signal in Germany's performance is the extremly poor development of graduates in Sciences and Engineering studies. This development threatens to undermine the inventory potential in industrial sectors where Germany has actually its strength as the high employment base in medium and high tech manufacturing (indicator 4.6) suggests.

The weak performance in terms of Lifelong learning in relation to EU_{25} can be partly explained by the so-called "Duales System" as the qualification system in Germany is named. This training system focuses on a intensive two to three year lasting qualification period in which job starters leaving secondary school are offered a training on the job accompanied by study periods of two to six weeks.

[8] cf. European Commission (2006), p. 17.

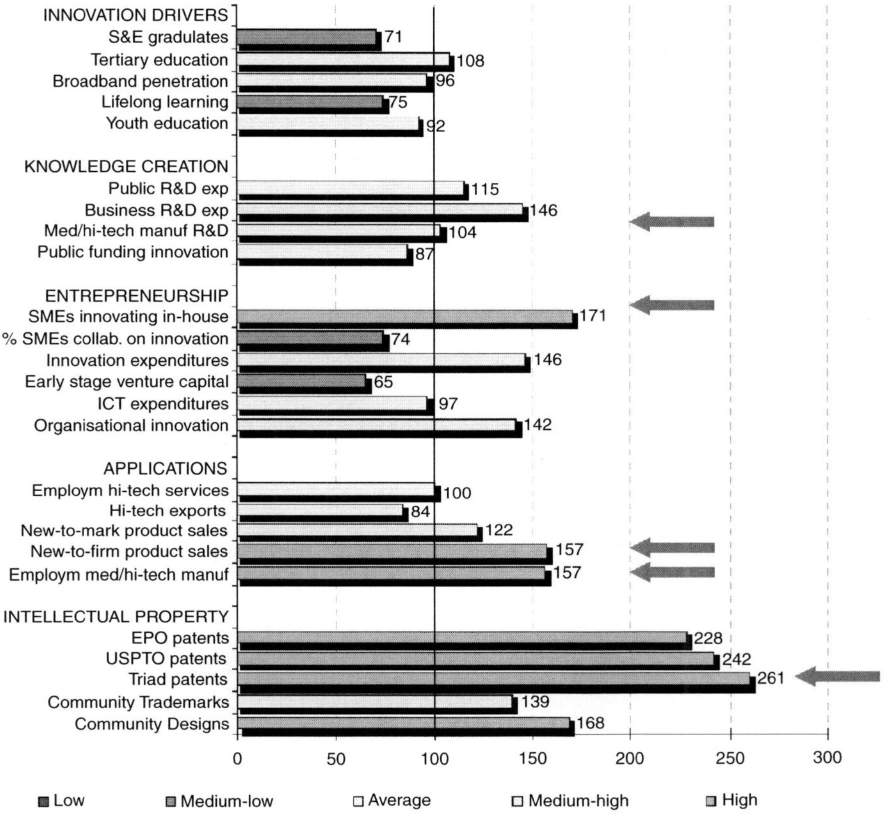

Fig. 13.2 EIS 2006 Innovation performance relative to EU_{25}- Germany
Source: European Commission (2007a), p. 35

Figure 13.2 also demonstrates that the finance sector is underdeveloped when it comes to early stage investments in know how intensive and technology based start ups. This is an important obstacle to speed up innovation performance especially for SME's which are marked by a low capital base.

Figure 13.3 visualises the relative performance of Germany for selected indicators in a dynamic perspective. All indicators to the right of the vertical axis have shown growth in the country, while all indicators above the horizontal axis are above the EU_{25} average.

Figure 13.3 underlines the strong patent performance of German industry (indicator 5.1 to 5.3) and the strong know-how base incorporated in manufaturing employees (indicators 4.5). The graph also underlines that German industry invests above-average in R&D. On the other side Fig. 13.3 also pinpoints the highest weakness of the German innovation system which is the below average development of graduates in science and engineering (indicator 1.1). Germany is threatened to lose its creative and industry related know-how base and hence its future innovation ability.

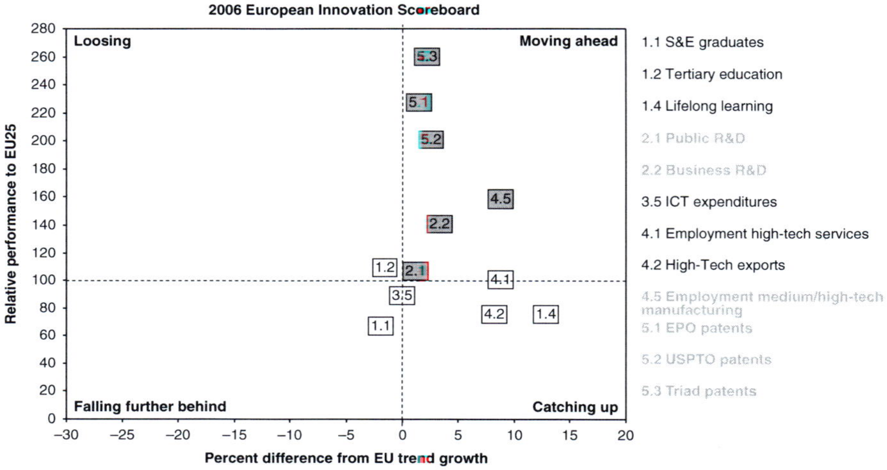

Fig. 13.3 Relative trends for Germany in 2006 in relation to EU_{25}
Source: Illustration is based on EIS 2006 database; European Commission (2007b)

13.2.3 Conclusions

The set of indicators used by the European Innovation Scoreboard gives first insights to the innovation potentials and capabilities of European countries in relation to each other and the United States and Japan. The EIS 2006 pinpoints strengths and weaknesses of the EU as such and of single countries. In a dynamic perspective it becomes clear that the EU_{25} countries perform very differently. Some countries seem to make their homework better than others.

However, the EU in general seems to improve its know-how base as the increasing number of graduates in science and engineering shows while the US falls further behind. Against the European trend highly industrialised Germany is facing an alarming below average development of graduates in science and engineering which could threaten its future innovation potential and capabilities. However, employment performance in high and mediumtech sectors in Germany is still stable whereas EU_{25} as such and the US in particular are facing extreme job losses of qualified workers in high and medium-tech manufacturing.

The level of US public R&D expenditures (mainly in the military sector) is more or less the same as in Europe. On the other hand US and especially Japanese firms invest on average much more in their future as European firms do. However, in Sweden, Finland, Switzerland, Denmark and Germany the Business expenditures in R&D are quite comparable to those in US and partly to Japanese firms as well. This is remarkable since in the end firms compete with each other and not states. The competitiveness of firms is much more decisive for economic welfare as the "indicator related performance" of states. Hence, it is an alarming signal that European firms invest in average much less in their future competitiveness than US and especially Japanese firms do. Especially Japanese firms have intensified their research activities at an impressive rate. Despite high salarys in Japan Japanese firms remove their research activities back "home". Business Enterprise Expenditures increased by more than 10% per year during the last 5 years. The main reasons are to maintain technological leadership and to avoid know-how and technology exports, which imply always the danger of imitation and know-how losses, especially when R&D capacities are once sourced out to other countries.

13.3 The Regionalised Innovation Scoreboard (RegIS): Approach and Key Results 2006

The RegIS 2006 largely follows the definition of the indicators used in the EIS 2006. The main aim of the RegIS 2006 is to regionalize the results gathered on country level from EIS 2006. However, when gathering the data at regional level available data do not always meet the indicator definitions of EIS. Twelfe of the 25 indicators used in EIS 2006 could be applied at a regional level within the RegIS 2006. In particular, indicator area 3 "Transmission and Application of Knowledge" could not be established at regional level, since the indicators were established in the EIS 2006 by an innovation survey conducted throughout Europe on the basis of a questionnaire whose results could not be broken down to the NUTS-I-level in a scientifically appropriate way. For those indicators where comparable data on regional level were available calculations had to be linked logically to those obtained in the European Innovation Scoreboard 2006.

The underlying mathematical calculation reads as follows:

$$Index\ value_i^t = \left[\frac{\left[proportional\ value_i^{\mathrm{Reg}IS}\right]^t * \left[\dfrac{proportional\ value_{BRD}^{EIS}}{proportional\ value_{EU15}^{EIS}}\right]^t}{\left[proportional\ value_{BRD}^{\mathrm{Reg}IS}\right]^t} \right] * 100$$

$$Proportional\ value_i^{\mathrm{Reg}IS} = \left[\frac{X_i^{\mathrm{Reg}IS}}{Z_i^{\mathrm{Reg}IS}}\right] * \alpha$$

t	respective year
i	respective federal land
X	feature to be examined
Z	reference value
α	percentage or thousandth part depending on the definition
RegIS	survey on the basis of regionalized data
EIS	survey on the basis of data specific for a federal

Seventeen new indicators[9] have been added to RegIS 2006 which have been ascertained at German level. Although they are not included in the EIS 2006, these indicators complete the picture of the existing innovation potential and skills within Germany in the regionalized analysis.

13.3.1 The German Federal Lands in Focus

The present analysis aims at regionalizing the results of the "European Innovation Scoreboard" which allows a comparison of the performance of the Federal Lands of North Rhine-Westphalia (NRW), Bavaria and Baden-Württemberg within Germany. A comparison with the EU_{15} performance puts the relative strengths and weaknesses of the Federal Lands and the Federal Republic of Germany as a whole in an European context. In particular, a comparison of the German Lands, on which the focus is laid here, with the Netherlands, which are comparable in terms of size and population with NRW, is to put the analysis of the innovation potential on a strong basis.

13.3.1.1 Key Results for North Rhine Westphalia, Bavaria and Baden-Württemberg

The Figs. 13.4, 13.5 and 13.6 illustrate the survey results. In each figure you find information about the innovation performance of the Federal Lands Baden-Württemberg, Bavaria, North-Rhine Westphalia and The Netherlands, in comparison to the EU_{15} average and the average performance of Germany. Innovation performance is demonstrated for two different years for every indicator in order to get an idea about the performance trend over time.

Figure 13.4 indicates the results for the innovation area "Human Resources". Compared with the EU_{15}, the number of graduates produced by the German Federal Lands in natural science and engineering is below average (see indicator 1.1). Even worse is the situation in The Netherlands which is, due to its weaker industrial base, less dependent on a high rate of graduates in engineering and natural sciences. Between 1999 and 2006, only the new German Federal Lands could maintain or even improve their relative position within the EU_{15}.[10] All other Federal Lands continue to lose ground as compared to the EU_{15} average. Even Baden-Württemberg, which had still a performance above the EU_{15} average in 1999, fell significantly back in 2006. This development causes serious concerns, since the competence in natural science and engineering is relevant for the technological capacity of the manufacturing industry. A new study suggests that political counter-actions are already required at secondary school levels I and II, since there is a high correlation

[9]These are the indicators 1, 2, 2.3, 2.4, 2.6 to 2.13, 3.1, 4.2 to 4.7; see also Figs. 13.4, 13.5 and 13.6.

[10]cf. Iking, 2007, 83 ff.

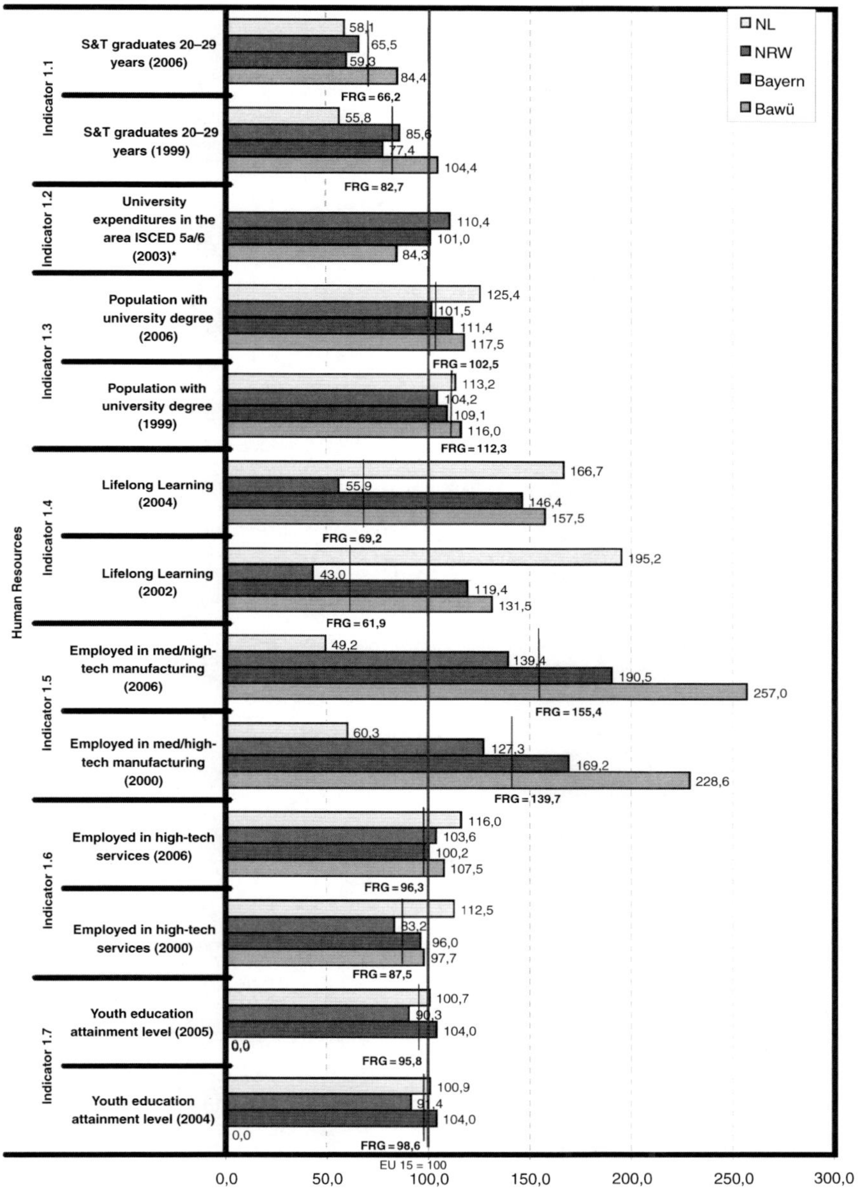

*Base: FRG = 100

Fig. 13.4 Regional Innovation Scoreboard 2006 – Results for North-Rhine Westphalia, Bavaria, and Baden-Württemberg (EU_{15}=100) – Area Human Resources
Source: Iking (2007), p. 11

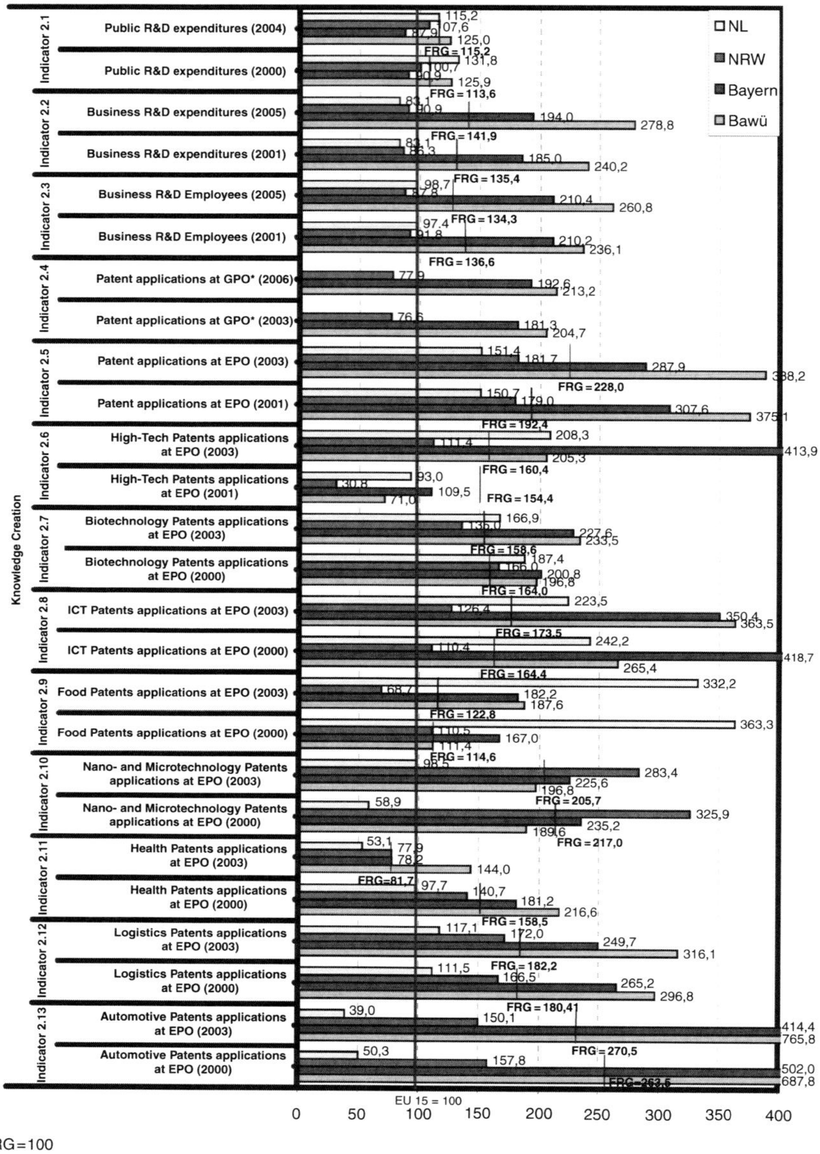

Fig. 13.5 Regional Innovation Scoreboard 2006 – Results for North-Rhine Westphalia, Bavaria, and Baden-Württemberg ($EU_{15}=100$) – Area Knowledge Creation
Source: Iking (2007), p. 12

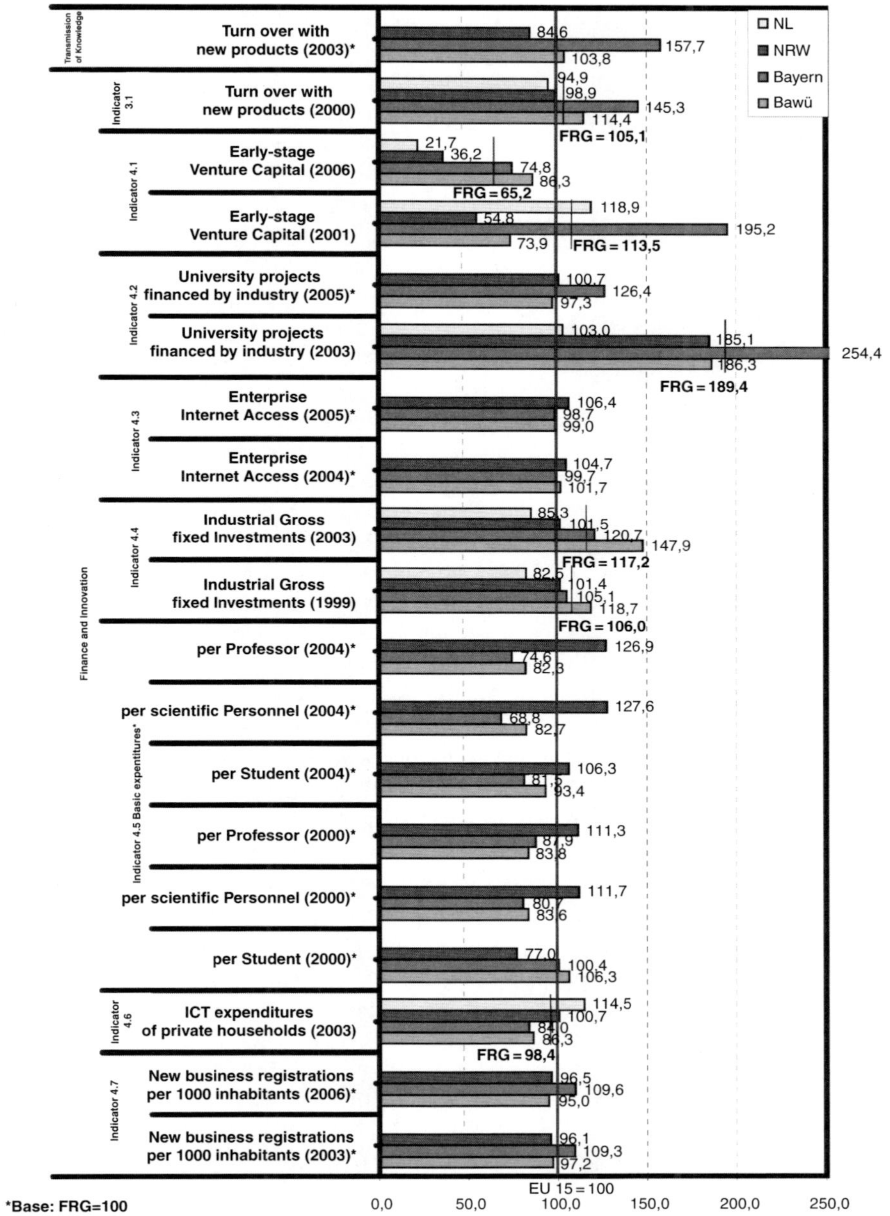

Fig. 13.6 Regional Innovation Scoreboard 2006 - Results for North-Rhine Westphalia, Bavaria, and Baden-Württemberg ($EU_{15}=100$) – Area Knowledge Transmission, Finance and Infrastructure Source: Iking (2007), p. 13

between the chosen subject of study and the special subject chosen in grammar school education.[11]

The educational level of Germany remains at an unchanged high standard (see indicator 1.3). In 2006 more than 7,5 million persons held a university degree which is one million more than 1999. However, in 2006, five Federal Lands were below the EU_{15} mean. In 1999, 6 Federal Lands had been below the mean.[12] Nevertheless, the relative position of some Federal Lands in 2006 is worse than in 1999.

Looking back on the development of indicator 1.1, the following pithy conclusion can be drawn: German students are still prepared to enroll for degree courses at universities, but the tendency in the chosen subjects has unfortunately shifted to the disadvantage of engineering and natural sciences.

Besides the shifted tendency in the chosen subjects of study, the access to university is a serious problem. The number of persons from "lower" social groups ("low" educational level of parents, low income) entering grammar school (33 of 100) or subsequently taking up studies (8 of 100) is by far too low.[13] These differences do not express the different talents specific for a social group, but rather the deficiency of the German school system in making up for the different preconditions related to the parental home by a corresponding (financial) promotion.[14]

In the area of "Lifelong learning" (indicator 1.4), Germany is clearly lagging behind its European neighbors within the EU_{15}. In the regionalized detail assessment, Baden-Württemberg and Bavaria are above the German and the EU_{15} mean. North Rhine-Westphalia fails to meet the German and European mean to an unacceptable clear extent (data from 2002 and 2004). Unfortunately, the tight economic situation leads to the cutting of expenses in the education area both in the private and in the industrial sector. It is not understood that more investments in the development of knowledge and additional qualifications are an investment into the future in the sense of providing a forward strategy for the employed and the companies. It is necessary to raise awareness among the public, so that increased importance is attributed to lifelong further training in order to achieve more dynamics. It is moreover necessary to design and implement concepts for vocational training which enable a successful career despite the enormous changes taking place in the economy within short periods. This necessity implies significant consequences for the educational demands on our universities. They will have to focus their education and training on transmitting the ability to quickly acquire the latest developments in knowledge! This means that the main educational goal will be to convey methodical fundamental knowledge, even if application skills of course continue to be important.[15]

[11] cf. BMBF (2005b), 7 ff.

[12] cf. Iking, 2007, 86 ff.

[13] BMBF (2005a), p. 78 and Table 10-1 on p. 81.

[14] ZEW (2003), p. 2.

[15] cf. Mathias Göbel in www.lv1.ifkomhessen.de/halbwert.htm

In Germany, the share of employed in medium-/high-tech manufacturing (indicator 1.5) in 2006 is above average in comparison to the European neighbouring countries of the EU_{15}. This indicates a still high innovative ability of the German economy as a whole. This area remained a German domain even in phases of economic slowdowns in Europe. The above average development of Germany regarding the index values, however, is not based on the fact that Germany experienced an absolute growth in employment. This relatively good ranking inside EU is rather caused by the reduction of jobs losses in the other EU_{15}-member states. When compared at German level but also at European level, the Federal Lands of Bavaria and Baden-Württemberg reach index values of 191 and 257 points in 2006.

Germany showed a strong growth in the number of employees in the high tech service sector in the period between 2000 and 2006. In absolute figures, the number of employees increased by approx. 81,000 persons to 1,290,000 persons from 2000 to 2006.[16] Thus the number of employees has grown more in Germany (6.7%) than in the EU_{15} mean (6%). Nonetheless, the rate of the employees in high tech services obtained only 96 index points in comparison with the EU_{15} performance in 2006.

On regional level Baden-Wuerttemberg experienced growth rates of 8.8% in terms of high tech service jobs and North Rhine Westphalia even by 21.3%. On contrary Bavaria just had 2.3% more jobs in 2006 compared to 2000.

While the indicator group 1 "Human Resources" (Fig. 13.4) tries to evaluate the basis for successful innovative efforts, the analysis of the indicator group "Knowledge Creation" (Fig. 13.5) is concentrated on measurable inputs (2.1, 2.2, 2.3) and quantifiable results and outputs (2.4 to 2.13). In particular, the Business R&D expenditures (indicator 2.2) can be interpreted as an indicator for the businesses′ willingness to innovate and their preparedness to take risk.

At the same time this indicator is indirectly also an indicator for the attractiveness of a location. The preparedness to take risk is particularly determined by the capacities described in the Human Resource area. There is empirical evidence for the significance of R&D capacities (indicators 2.1 to 2.3) for economic growth. In the past decade, economic growth was especially strong in those countries where the R&D capacities increased most.

Figure 13.5 shows that Germany's public R&D expenditures in 2004 continue to be above the EU_{15} mean. It is striking that the smaller Federal Lands make relatively more investments in R&D than the bigger Federal Lands.[17] In relation to their economic power, the R&D expenditures of the Federal Lands of Baden-Württemberg (0.81%) and Saxony (1.15%) are fair above average. The R&D expenditures of Bavaria (0.57%), Rheinland-Pfalz (0.5%) and Schleswig-Holstein (0.55%) are far below national average. NRW reports a public investment in R&D of 0.7% which is in national average and a remarkable increase compared to 2000. In absolute terms the public investment amounts to 500 million Euros being invested in public research projects. No other Federal Land invested as much as NRW in 2004.

[16] Iking (2007), p. 92.

[17] cf. Iking (2007), 94 ff.

Against that, the public R&D expenditures in other European countries increased more strongly than those of the German Federal Lands. However, in relation to its economic power the Netherlands faced a decrease of public R&D expenditures between 2000 and 2004 from 0.87% to 0.76%. In absolute terms the public R&D investments increased by more than 600 million Euros to 3,7 million Euro in 2004 which were 350 million Euro more than in NRW.[18]

Absolute or relative figures of invested capital do of course not tell anything about the quality of the output. Nevertheless, available financial resources significantly influence a country's innovation ability. The expansion of public R&D expenditures is above all important for Germany, since the German economy traditionally maintains intensive cooperation with research institutions.

In the period between 2001 and 2005, the business R&D expenditures slightly increased in Germany and most Federal Lands on a relatively high level. In 2005 North-Rhine Westphalia failed to meet the national mean as well as the EU_{15} mean, showing a declining trend. This trend is particularly alarming since the R&D investments made by enterprises can be regarded as investments into the future and can also be interpreted as the enterprises′ "confidence in the location". Against that, Germany as a whole maintained its position in comparison with the EU_{15} performance and the Federal Lands of Bavaria, Baden-Württemberg and Berlin but also Hesse are clearly at the top. The big differences established in the levels of R&D expenditures in Bavaria and in Baden-Württemberg on the one hand and North-Rhine Westphalia on the other hand refer partly to the different industrial cultures. In the case of NRW, e.g., the industrial culture is inseparable from the history of the coal, iron and steel industry. NRW has not yet been able to change into a location for R&D-intensive industries.[19]

Sector-specific studies of the industrial R&D development moreover suggest that the German R&D strengths are in particular focused on medium-/high-tech areas, in particular the chemicals industry, the electrical engineering industry, mechanical engineering and vehicle construction. The German innovation system is increasingly depending on automobile construction in particular.[20]

It was furthermore found that since the mid-nineties the R&D growth of the German economy has been owed almost exclusively to large-scale enterprises. Thus, the significant deficits compared to Japan and the USA are essentially the result of the lack of commitment of small and medium-sized enterprises. Here, it is necessary to increase and extend the existing incentive structures to include SMEs in R&D.

North-Rhine Westphalia continues to lose ground with respect to the indicator "industrial R&D employees" (indicator 2.3) as compared to the national mean. In this area, Baden-Württemberg, Bavaria and Lower Saxony could increase their values in terms of relative as well as absolute terms. However, this must not blind us to the fact that Germany has lost 3.000 employees in the research and development

---

[18] Ibid. 94 ff.

[19] cf. RUFIS (2007), p. 11.

[20] NIW (2005), 84 ff.

area in absolute terms during the period from 2001 to 2005. Only due to the fact that the loss of R&D employees was relatively low in comparison to the job reduction as a whole the index values of individual Federal Lands were stable or decreased only slightly.[21]

The indicator "Patenting" measures the success of current innovative efforts (indicators 2.4 to 2.13). Here North Rhine-Westphalia (451 patents at the German Patent Office (GPO) per one million population in 2006) fell behind in comparison to the national mean (583 patents). Only Bavaria (1278 patents), Baden-Württemberg (1243 patents) and Hesse (654 patents) are performing above the national mean. Bavaria and Baden-Württemberg could markedly consolidate their lead over time.

In North Rhine-Westphalia, a relatively high number of ideas and innovations continue to be registered at the European Patent Office (267 patents in 2002) (indicator 2.5), despite the downward trend (minus 8 index points) in comparison to 2001. NRW is among the European top performers in this area which is documented by a relative position of 171 index points in the EU_{15} comparison. In comparison with the national performance, NRW comes fifth. Baden-Württemberg's position is excellent (602 patents), since it could improve its relative position, contrary to the EU_{15} and the national trend, by a marked increase of the patent applications filed with the European Patent Office.

Germany, reporting 297 patent applications at the European patent office (EPO) per one million population in 2002, continues to perform well above the EU_{15} mean (159) and Dutch mean (279).

Indicator 2.6 (High-tech Patents at the European Patent Office) does not confirm the picture conveyed by indicator 2.5 for NRW. North Rhine-Westphalia clearly lags behind Bavaria and Baden-Wuerttemberg in the area of highly innovative inventions and innovations within Germany between 1999 and 2002. NRW could just maintain its position.

Germany is not able to keep pace with the dynamics set by the EU_{15} and is slightly losing its lead in this area. In particular the Netherlands made a great leap forward in the area of high-tech patents within a very short time. A detailed analysis of this development could provide interesting learning effects.

Analyses of sector- and technology-specific patenting activities at regional level highlight the existing innovation competence within the meaning of the more recent cluster analyses. For the biotechnological area (indicator 2.7) Germany (12.5 patents per million population filed with the EPO) and the Netherlands (11.2 patents) are clearly leading the field in 2002 within Europe. At the level of the German Federal Lands, in particular the city-lands of Berlin (32 patents per million population) and Hamburg (24 patents per million population) have achieved a very good position, followed by Bavaria (15.4 patents per million population) and Baden-Württemberg (15.3 patents per million population).[22]

---

[21]cf. Iking (2007), p. 99.

[22]Though, in NRW work more people in the future branch biotechnology as for example in Bavaria or Baden-Württemberg, see Wirtschaftswoche (2007), p. 60. The biotech firms are largely

In the field of information and communications technology (indicator 2.8), Germany showed quite a good performance achieving an average of 67.5 patents per million population compared to the EU_{15} performance (41 patents). The regional assessment also clearly shows regional centres of competence in ICT. At the level of the federal lands, Bavaria (152 patents per million population in 2002) and Baden-Württemberg (138 patents) are dominating this important cross-sectional technology. All of the other federal lands are lagging far behind.

In comparison with the European performance, Germany's R&D performance in nano- and microtechnology (indicator 2.10) is very good. This is revealed by the high patent intensity in Germany which is twice as high as the EU_{15} and EU_{25} mean. Five to six patents have been filed per million population in Germany. The Netherlands as well as the EU_{15} mean reported 2.5 to 3 patents per million population between 2000 and 2003. At the level of the federal lands, the analysis exhibits a concentration of technological research in Saarland (10.6 patents in 2002), Rhineland-Palatinate (8.5 patents in 2002) and Hesse (8.1 patents in 2002). Bavaria (6.2 patents), Baden-Württemberg (6.7 patents) and NRW (7 patents) also reported patent applications far above the German and European average.

The health industry is increasingly considered as a strategic growth market having a high innovative and sales potential.[23] In Germany, an average of 21 patent applications per million population (2002) has been filed at EPO by the health sector (indicator 2.11), thus being about 30% above the European average of 14 patents (EU_{15}) or 11.5 patents (EU_{25}), respectively. This corresponded to 150 index points compared to the European performance.

The analysis of the performance of the individual federal lands provides a much more differentiated picture. Above all Hamburg (54.5 patents in 2002) as well as Baden-Württemberg and Schleswig-Holstein (32.5 patents each) are impressive with their high average patent intensity in relation to their residential population.

The logistic competence (indicator 2.12), i.e. extensive services ranging from transport, storage and up to contact logistics, has enormously gained importance due to the globalization and the increasing international movement of goods.[24] After the EU enlargement, Germany had geographical advantages in this promising sector right from the start due to its central position in Europe. Germany has made use of these advantages. Subsequently, logistics grew to take the third place among the economic sectors in Germany, right after trade and car manufacturing.[25] In this

concentrated in Bavaria, Baden-Württemberg, Hesse and NRW; see Statistisches Bundesamt (2003), p. 20.

[23]In autumn 2007, the Federal Government (BmBF) launched a competition "Health Region of the Future" which is to support the five best regions in improving the growth potentials of the health industry by granting € 10 million over a period of four years. The competition aims at establishing networks between research, development and treatment of patients in regional health care. www.bmbf.de (press release 135/2007), compare also FTD (2007).

[24]cf. Axel (2006).

[25]cf. Fraunhofer Institut für Integrierte Schaltungen (2005).

respect, the geographical position is an important location factor.[26] The analysis of the regional patent applications examines this technological aspect in the competition between the logistic locations. Compared to the European performance, the number of patent applications (16 patents per million population in 2002) was nearly twice as high as the EU_{15} mean. This relation has remained almost unchanged in the period between 2000 and 2002. The number of patent applications was especially high in the Federal Lands of Bavaria (25 patents), Baden-Württemberg (29 patents) and Hamburg (21 patents). The Netherlands reached 10.5 patents and performed only slightly better than the European average.

From the German point of view, the vehicle construction industry continues to be the sector having the highest economic importance in respect of sales and employment but also within regard to its innovative potential.

The patent applications filed by the automotive sector (indicator 2.13) document its technology-based innovative power. Here, the German patent applications (19 patent applications per million population) are twice as many as the EU15 mean, with the gap even having slightly widened in the period under observation. After having reached an index value of 263 IP in 2000, the index value of 2003 amounted to more than 270 IP. The Netherlands are lagging clearly behind the German and European mean, reaching about 3.5 applications per million population.

In the comparison of the federal lands, Baden-Württemberg, reaching an average of 59 patents per million population in 2002, is by far in the lead before Bavaria (25 patents), Hesse (23 patents), Rhineland-Palatinate (20 patents) and Lower Saxony (18 patents). In particular Baden-Württemberg has further increased its patent intensity by 27% from a high level in the period from 2000 to 2002.

The transmission of "produced" knowledge (see Fig. 13.6), apart from statutory protective mechanisms to secure innovational profits (patents, utility model protection), is de facto restricted by other circumstances (inter alia by re-engineering costs or by business secrets). In particular, the new models of endogenous growth highlight the special role of the transmission of knowledge ("diffusion").[27] Even under circumstances which do not enable a lasting economic growth (low number of highly qualified workers), a sufficient diffusion rate of new knowledge can induce an economically dynamic development. It can empirically be established that fields of competence and innovation clusters are built in regions characterized by dynamic growth. Such clusters are characterized by so-called "spillovers". In essence, this means “transmission of knowledge”. Thus, dynamic and efficient transmission of knowledge can be considered as a vital factor for developing regional competences and as a crucial determinant of growth.

The sales structure of an enterprise allows drawing conclusions as to the speed of the internal innovation process within the enterprise (indicator 3.1). The higher

[26]The particular importance of geographic location can be shown quite good with the example of Duisburg, which is connected in a perfect way to the German road, rail and water way system. The “Duisport” develops already to one of the leading logistic places within Europe.; cf. Wirtschaftswoche Nr. 25 from 18th of June 2007,66 ff.

[27]Romer (1986, 1990).

the share of sales with new products and services (< 3 years), the better the internal innovation process will work. The assessment[28] concentrates on the structure of the sales made by the businesses and its changes between 2000 and 2003. The European Innovation Scoreboard states that the share of sales attained with new products and new-to-market products has increased from 7.1% in 1996 and 1998 to 7.5% in 2003 in the German economy. The German performance was by about 0.5% better than the European average. Compared to the Netherlands, the EIS 2006 states, the lead even increased to 2.5% in 2005 [7.5% (FRG) compared to 4% (NL)].

However, these results from the evaluation of the survey results conducted by the European Innovation Scoreboard are not confirmed by the data analysis made by IAB-Betriebspanel. Pursuant to the IAB-Betriebspanel, the share of sales attained with new products and new-to-market products in Germany amounted to only 4.2% in 1997 and decreased to 3.4% of the overall sales by 2000. Contrary to the marked improvement stated by EIS, the share of sales achieved with new products or new-to-market products continued to decrease between 2000 and 2003 in Germany pursuant to the IAB data panel.

The assessment of the IAB panel data reveals remarkable structural differences between the Federal Lands:

1. The worsening of the sales share of new products and new-to-market products recorded between 1997 and 2003 at national level does not apply to all of the federal lands. In Berlin and Bremen, the sales structure of the economy was "newer" in 2003 than in 1997.
2. In some federal lands, the sales structure has dramatically worsened between 1997 and 2003: While the sales share of new products and new-to-market products amounted to more than 7% in Bavaria and Saxony-Anhalt in 1997, it fell to only 4.1% or 1.8%, respectively, in 2003! Baden-Württemberg, Brandenburg, Hesse, Mecklenburg-Western Pomerania and Thuringia also reported a decline in the sales share of new products and services by 50% and more.

In 2003, the sales share of new products and new-to-market products was highest in Bavaria (4.1%) and Berlin (4.2%). NRW achieved 2.2% and was slightly below the German national mean (2.6%), Baden-Württemberg (2.7%) was slightly above it.

Innovative activities in the enterprises require sufficient financial means. The development of an efficient debt market (especially equity capital) is therefore of particular importance for the financially less strong SMEs. Indicator 4.1 examines the efficiency of the capital markets. From the point of view of "capital availability", Germany as well as the single federal Lands document a relatively underdeveloped Venture Capital market.

The development of the R&D expenditures in the higher education sector financed by business (indicator 4.2) allows drawing first conclusions as to the

[28]The data are taken from the IAB-Betriebspanel; see also Iking 2007, Table 26, p. 119).

qualitys and the intensity of the transfer of knowledge between higher education and industry (indicator 4.2). In 2005, the businesses commissioned the higher education sector throughout Germany with research services to the amount of almost one billion euros. This was 35% more than in 1999. In 2005, just as in 1999, almost 60% of these third-party funds flowed into the higher education sector in NRW (€ 211.5 million), Bavaria (195 million) and Baden-Württemberg (166 million). NRW which reported an increase in third-party funds of € 45 million or 27.5%. The Federal Lands of Bavaria (+25.6%) and Baden-Württemberg (+33,7%) both reported an increase of about € 40 million.

As regards the structure, the industry-funded share in the R&D expenditures of the higher education sector is highest in Bavaria (35%) and Hesse (34%). The analysis in 2005 showed that the research activities of the higher education sector are practically oriented not only in the whole of Germany (national mean 28%) but also in the federal lands of Baden-Württemberg (27%), Lower Saxony (28.5%), NRW (28.2%) and Rhineland-Palatinate (27%).

Compared to the European performance, these shares document that the higher education sector in Germany is strongly linked with industry.

The expenditures for education are investments in the future. The funds provided per student, per scientific employee and per professor reflect the qualitative conditions for studying (indicator 4.5).

This input-related assessment at the level of the federal lands only allows to draw first tentative conclusions, since the situation may considerably vary from university to university also at the level of the federal lands. This way of assessment also disregards the fact that not every federal land is necessarily educating its "own" graduates due to the mobility of university graduates. It is not exceptional that a student from federal land A studies in federal land B and then works in federal land C after taking his or her degree. The investments made per student, per scientific employee and per professor rather illustrate the preparedness of a federal land to invest in good conditions for higher education. Compared to the international performance, Ger-many's preparedness seems to be quite low. In 2000, the expenditures for educational institutions amounted to 5.3% of the GDP, thus being 0.2% below the average of the OECD countries. Public funds amounted to 4.3% and thus fell below the in-ternational average performance by 0.5%.

It is even more remarkable that the basic funding provided per student was (partially considerably) lower in almost all federal lands in 2004 than in 2000. Only the federal lands of Bremen, Hamburg, Lower Saxony, NRW and Saarland increased their funding in the assessed period.

Gross fixed asset investments comprise the acquisition of reproducible producer durables as well as self-constructed assets and larger repairs increasing the value (indicator 4.4). Investments made in the replacement and extension of an enterprise's capacity are always subject to managerial decisions. Thus, entrepreneurial decisions on investments always reflect the enterprises' confidence in the location.

The gross fixed asset investments made by the entire German industry have fallen from € 81 billion since 1999 to approx. € 70 billion in 2003 (–13%). In relation to the GDP, the propensity to invest decreased from 4% (1999) to 3.3% (2003). Thus, the investment activities in Germany decreased to a lesser degree than in the EU15 (–15%) or EU25 (–13.5%). The Netherlands reported a decrease of the investment activity by 7% in the same period to € 11.3 billion in 2003.

The conclusions as to the confidence of industry as a whole in the location "Europe" are alarming. Significant volumes of investments flow into other parts of the world. However, these figures do not indicate to what extent the diverted funds served to open up new markets (e.g. Eastern Europe and Asia, in particular China), which is very welcome, and to what extent they were the result of entrepreneurial decisions to relocate their business (rather negative).

Compared to the European performance, the German position improved over time, showing a rather declining propensity to invest by the German industry as a whole. After having reached 106 index points in 1999, the relative value amounted to more than 117 index points in 2003.

At the level of the federal lands the gross fixed asset investments decreased in almost all federal lands between 1999 and 2003. However, the investments made in Lower Saxony increased by 4% to € 7.0 billion, in Hamburg by 25% to € 1.4 billion, but in Baden-Württemberg (€ 1.3 billion) as well as in Mecklenburg-Western Pomerania (€ 0.9 billion) the gross fixed asset investments remained almost stable.

All other federal lands reported a nominal decline in the investment volumes ranging between 22% and 38%. Compared to these values, Bavaria's nominal decline in investments by 6% is rather "insignificant".

The development of the trade registrations does not only depend on the cyclical course. Changes over time also signal improved/deteriorated conditions in a region for opening up own business establishments.

In Germany, the number of trade registrations (indicator 4.7) has increased by 10% to 882,000 new trade registrations in the period between 2003 and 2006. In 2006, about 11 trade registrations were filed per 1000 inhabitants. In the federal land of Berlin, the rate of trade registrations was even 13.5, and in Hamburg and Hesse the rate was 12.6 registrations per 1000 inhabitants each.

In 2006, the highest number of trade registrations was filed in NRW. More than 186,000 new business establishments were registered there. Next were Bavaria (146,500) and Baden-Württemberg (109,000).

In most of the federal lands, the number of the trade registrations filed in 2006 was markedly higher than the number filed in 2003, but lower than in 2005. The number of trade registrations has been continuously increasing only in the federal lands of Berlin, Bremen, Hamburg, Rhineland-Palatinate and Schleswig-Holstein.

13.3.1.2 Strengths and Weaknesses Compared to European Average

Key results of the Regional Innovation Scoreboard 2006 are illustrated in Fig. 13.7. In the area "Human Resources" all Federal Lands still can profit from their know-how base which is incorporated in its employees in medium and high tech manufacturing. Here the relative position inside EU_{15} is extremely positive (++).

In addition employment in high tech service sector increased strongly in the last years, especially in Baden-Württemberg and NRW. However in comparison to other

	Indicator	Germany	BY	BW	NRW
Human resources	S&T graduates Trend: 1999–2006	– ↑↑	– ↑	0 ↑↑	– ⇔
	University degree Trend: 1999–2006	0 ↑↑	0 ⬆	+ ⬆	0 ⬆
	Employed in med/hightech manufacturing Trend: 1999–2006	++ ⇔	++ ⇔	++ ↑	+ ⇔
	Employed in hightech services Trend: 1999–2006	0 ⬆	0 ↑↑	0 ⬆	0 ⬆
Knowledge creation	Public R&D expenditures Trend: 1999–2004	0 ↑↑	0 ↑	0 ↑↑	0 ⬆
	Business R&D expenditures Trend: 1999–2005	+ ↑↑	++ ⬆	++ ⬆	0 ↑
	Patent application at EPO Trend: 1999–2003	++ ⇔	++ ⇔	++ ↑↑	++ ↓
	HT-patent application at EPO Trend: 1999–2003	+ ↑	++ ↓	++ ↑↑	0 ⇔
Transmission of Knowledge	Turnover with new projects 1997–2003	0 ⬇	+ ⬇	0 ⬇	0 ⇔
Innovation finance	University projects financed by industry 1999–2003	+ ⬆	0 ⬆	++ ⬆	++ ⬆
	Industrial gross fixed investments1999–2003	0 ⬆	+ ↓	+ ⇔	0 ⬇

Fig. 13.7 Strengths, weaknesses, trends in Germany 1999–2006 in relation to European average (EU_{15})

Current performance relative to EU_{15} average		Trend performance (of absolute data) (static comparison 1999/2000 to 2005/2006)	
(++)	if above 150	+/–(5%)	⇔
(+)	if between 120–150	+ (5–10%)	↑
(0)	if between 80–120	+ (10–20%)	↑↑
(-)	if between 50–80	+ (20–...%)	⬆
(--)	if below 50	- (5–10%)	↓
		(10–20%)	↓↓
		(20–...%)	⬇

EU_{15}-states the high tech service sector in Germany and the Federal Lands in focus is still performing on average.

The low number of graduates in science and technology in the Euroepean context is the biggest threat for the innovation future in Germany and the Federal Lands in focus. Germany faces a slow errosion of its knowledge base even recognizing an increase in absolute numbers by more than 8.000 graduates to 79.000 in this disciplines for Germany between 1999 and 2006. However, this number is much to low to meet demand in industry. Available competences in natural science and engineering, which today are still the most important strength of German innovation systems, are expected to be the core problem in Germany's future. Especially NRW is facing a big challenge.

In general terms the qualification base in Germany and The Federal Lands in focus is still good and a little above EU_{15} average, The positive sign is that the number of people holding an university degree increased strongly in the past years. However, having the strong demand of German firms in mind to employ engineers and scientists, too many young people study the "wrong" subjects.

Business R&D expenditures as well as Patent outcome are still strong points in Germany's innovation system. Although innovation is more complex than the linear innovation model suggests. Financial inputs in research and development still play an important role for the future competitiveness of enterprises. Here German firms, especially in Bavaria and Baden-Württemberg, still make remarkable investments in their future. The slow pace in investment of firms in North-Rhine-Westphalia gives rise to concern. This may be interpreted as a lacking trust by entrepreneurs in the location and urges strong counter measures.

It is also striking that public R&D-expenditures in Bavaria increased slowly in the period from 1999 to 2004 which went hand in hand with a strong decrease of high technology patent outcome at the European patent office.

In general the public R&D efforts in germany and its Federal Lands just have an average level in European terms. Especially NRW increased its efforts between 1999 and 2004. However, in terms of patent application at EPO the trend for NRW is negative.

Quite worrying in the development of the turnover with ne products in the period 1997–2003. Starting from an average position inside EU15, except Baden-Württemberg (+), the overall trend was strongly negative. NRW held its average position. This hints at a certain difficulty of German firms to translate new Know-how and R&D results into new marketable goods and services.

On the other hand the German innovation system releis on a cooperation culture between industry and academia. Being already on a remarkable good level inside EU15 German industry even increased their efforts in the cooperation with universities and R&D institutes.

Very important from a startegic point of view is the development of industrial gross fixed investments. Between 1999 and especially 2003 the development was negatively in Bavaria and especially in NRW whereas Baden-Württemberg could hold its position. Nevertheless, some German regions are still sufficiently attractive in the globalised competition to encourage the enterprises to enforce their investment activities.

13.3.1.3 Conclusions for the Federal Lands North Rhine Westphalia, Bavaria and Baden Württemberg

The innovation capacity and ability of the German Federal Lands Baden-Württemberg, Bavaria and North Rhine-Westphalia show an extremely controversial picture as compared to the national and European performance. The 29 indicators used in the Regional Innovation Scoreboard 2006, taken as a whole, illustrate the innovative profile specific to the regions, thus allowing differentiated analyses of the strengths and the weaknesses. The benchmarking exercise results are most telling if the entire regional picture is interpreted and not just single indicators. Determined profound analyses of individual indicators are hardly helpful, as they imply the risk of over- and misinterpreting the indicators.

There is a high regional diversity in the innovation capacity inside Germany. Especially NRW is lagging behind, compared to Bavaria and Baden-Württemberg.

A positive ray of hope still is the high number of employees in the medium- and high-tech industries. In this area, Germany is at the top compared to the European performance; nationally, Baden-Württemberg clearly leads over Bavaria and outperforms NRW by far. NRW is still above the EU mean, but continues to be clearly below the national mean. This is still mainly due to NRW′s former coal and steel industry, but it is also understood that the structural change has not been completed yet. NRW slightly improved its relative position as compared to the EU_{15} between 1999 and 2006. In comparison to the national performance, however, NRW's relative position has worsened. The results show major weaknesses in med/hightech employment in North Rhine-Westphalia compared to Bavaria and Baden-Württemberg, in production as well as in services. But especially when it comes to the future perspective all German Federal Lands have to rethink their education policy as the development in numbers of graduates in natural sciences and engineering will lead already in a few years to a lack of experts. This result requires immediate action, in particular in NRW, since human capital is the basis for successful innovative activities. A worsening of "today′s human capital" will, with some delay, result in a reduced "innovative ability" in the near future.

It is also meaningful to increase investments in R&D (public and business) even considering the current good perfomance and the positive trend between 1999 and 2005. From NRW's point of view the bad performance in patenting is alarming, especially in the HT-sector.

13.4 Recommendations

In the areas of human resources and knowledge creation, the assessed German Federal Lands have to make considerable efforts to catch up with their European neighbours, which requires an immediate and rigorous paradigmatic change in the German policy regarding university education. "Learning to learn" must become the priority of university teaching under the higher education policy. Hence, efficiency and scientific outcome must be improved significantly in the university sector. It seems also to be important to implement a system that rewards an economically important outcome (patent before scientific publication!). Especially from NRW's point of view it is urgently necessary to attract more high potentials, to improve the patent outcome of firms and to attract more technology-based firms. For the latter aspects it is also necessary to support start up activities at the universities and universities of applied sciences in NRW.

We also have to be cautious in the areas where Germany and the assessed Federal Lands have reached relatively good positions and have a great innovative potential as compared to the national and European mean. In particular, the high employment rates in the technology-intensive industries, but also the business R&D expenditures are still the source of many innovative products and ideas. This is documented by the absolute and relative successes of Germany in the patent area, and in particular the successes of the Federal Lands of Baden-Württemberg and Bavaria. It is alarming, however, that from a German point of view there is a trend of stagnation in the supposed "indicators of success" observed for the whole of Germany, even if it is still at a high level. It is necessary to immediately take adequate counter-measures. Here, public R&D expenditures should be increased. There might be good reasons to concentrate scarce resources (cluster). However, there should be a minimum budget reserved to cofinance risky innovation projects (from all fields)!

As networking between key players in a given sector lead to additional dynamics in innovation performance federal governments should launch initiatives such as technology (transfer) scouts or competence brokers[29] that strengthen networking between players in competence fields and between academia and SMEs. Here is an important job to do since all Federal Lands seem to be lagging behind in the transfer and application of existing know-how. Contests that reward the know how transfer from science to business as well as from business to business are a promising instrument policy decision makers could use.

However, it is generally important to change the people's mind-set and create an innovative climate on regional level (innovation culture). This implies also the necessity to launch an initiative promoting the importance of in-house R&D in industry. Even the public finance systems are in bad conditions. As the future is concerned there should be serious discussions about improved incentive systems to increase private R&D! These activities must go hand in hand with offensive promotion of the location NRW (Talk about it!).

References

BMBF. (2005a). Bericht zur technologischen Leistungsfähigkeit Deutschlands 05. Berlin.

BMBF. (2005b). Studiensituation und studentische Orientierungen, 9. Studierendensurvey an Universitäten und Fachhochschulen, Bonn, Berlin.

European Commission. (2007a). European Innovation Scoreboard 2006 – Strength and Weaknesses report, pp. 34–36,www.proinno-europe.eu/admin/uploaded_documents/EIS2006_strenghts-weaknesses.pdf

European Commission. (2007b). 2006 European Innovation Scoreboard database, www.proinno-europe.eu

European Commission. (2006). European Innovation Scoreboard 2006 – Comparative analysis of innovation performance, Luxembourg, www.proinno-europe.eu/doc/EIS2006_final.pdf

[29] successful technology transfer mechanism used in Norway.

Financial Times Deutschland. (2007). (Grit Beecken und Lukas Heiny), Gesundheitswirtschaft – Wettbewerb der Regionen. FTD.de vom 30.6.2007.
Fraunhofer Institut für Integrierte Schaltungen. (2005). Arbeitsgruppe für Technologien der Logistik-Dienstleistungswirtschaft ATL; Logistikstandort Deutschland: Eine Studie zu Potenzialen aktiver Vermarktung des Logistikstandorts Deutschland im europäischen und globalen Standort – Wettbewerb, im Auftrag der "Invest in Germany GmbH"; Nürnberg.
Mathias, G. Halbwertzeit von Wissen, www.lv1.ifkomhessen.de/halbwert.html
Axel, G. (2006). Logistik ist Deutschlands heimliche Paradedisziplin, in: Handelsblatt vom 08.02.06
Bernhard, I. (2007). Regional Innovation Scoreboard 2006 – Results for North Rhine Westphalia, Bavaria and Baden-Württemberg in national and European comparison, Mülheim an der Ruhr, http://www.zenit.de/d/beratung/innovationsberatung/reg_ref/download/Masterdokument RegIS2006_eng.pdf
Niedersächsisches Institut für Wirtschaftsforschung (NIW) (2005). Forschungs- und Entwicklungsaktivitäten im internationalen Vergleich, No.7-2005. Hannover.
Paul, R. (1986). Increasing returns and long-run growth, *Journal of Political Economy*, *94*(5), 1002–1037.
Paul, R. (1990). Endogenous Technological Change, *Journal of Political Economy*, *98*(5), 71–102.
RUFIS. (2007). Wo steht das Ruhrgebiet im Innovationswettbewerb, ppt.-Präsentation by Rufis on 14th of March 2007 at Sparkasse Bochum.
Statistisches Bundesamt. (Hrsg.). (2003). Unternehmen der Biotechnologie in Deutschland, Ergebnisse der Wiederholungsbefragung 2002, Wiesbaden.
Wirtschaftswoche. (2007). *Logistisches Herz.* WiWo No. 25 vom 18.6.2007, 66ff and *Hand in Hand*, pp. 58–62.
ZEW. (2003). Hochschulabsolventen werden knapp, in: ZEW News 4/2003, S. 1-2, Mannheim, ftp://ftp.zew.de/pub/zew-docs/docus/dokumentation0303.pdf

Chapter 14
Energy Technology Policy in Europe

Jochen Hierl and Peter Palinkas[1]

14.1 Introduction

High fuel prices, the growing concern about global climate change and not least price disputes concerning natural gas supplies from Russia during the years 2005/06 and 2006/07 have highlighted energy policy in European capitals and European Union (EU) decision-making bodies in Brussels. Several political leaders in the EU have called for a common European energy policy – something that national governments have so far resisted in order to preserve their sovereignty over their energy fuel mix, the way in and level at which they choose to tax energy supply and usage, and over their oil and gas stocks (mainly natural gas storage) in order to follow their own energy security, economic and environment/climate change policy targets. Whereas the European Parliament and the European Commission have been calling for a common European energy policy for a long time, powerful national governments have resisted this so far, and even the infamous draft European Constitution does not contain new powers for the EU in this area. But the sentiment in the capitals of Europe has changed and several governments have stressed the need for more common action.

In March 2006, the European Commission issued the Green Paper "Secure, Competitive and Sustainable Energy for Europe" which reflects on the energy future for the EU, taking into account increasing market liberalisation and globalization, climate change and environmental pressures, technological challenges and growing import dependency from politically unstable regions. It follows on from its 2005 communication on climate change "Winning the battle against climate change",

[1] Jochen Hierl is Senior Economist at the European Investment Bank (EIB). Dr. Peter Palinkas is Senior Administrator in the Secretariat General of the European Parliament (EP). The views and opinions expressed in this paper are those of the authors and do not necessarily reflect the positions of the EIB or the EP.
This paper was written in mid-2006 and some elements of the introduction may have been overtaken by the latest developments.

J. Hierl (✉)
European Investment Bank, Luxembourg

P.J.J. Welfens, J.T. Addison (eds.), *Innovation, Employment and Growth Policy Issues in the EU and the US*, DOI 10.1007/978-3-642-00631-9_14,

its Green Paper on Energy Efficiency of 2005 (COM, 2005, 35 final), the Energy Efficiency Action Plan of 2006, the ongoing works of the High Level Group on Competitiveness, Energy and Environment, and the cartel and state aid enquiries into the energy sector by the Commission in 2006.

The EU heads of state and government welcomed the Energy Policy Green Paper and committed itself to adopt a European Energy Action Plan in spring 2007. To this end, in January 2007 the European Commission tabled the Communication

"An Energy Policy for Europe" (COM, 2007, 1 final), accompanied by a Strategic Energy Review over a number of sectoral policies, including renewables, and a ten-point Energy Action Plan. This will be complemented in due time by a Green Paper on climate change policy for the period post-2012 when the Kyoto Protocol expires. Both the EU Strategic Energy Review and the post-2012 Green Paper will outline the EU "energy and climate change vision" and propose a roadmap towards achieving EU energy and climate change objectives, namely, how to ensure the competitiveness of European industries while at the same time combating climate change and ensuring security of energy supply. There will also be an attempt to answer calls for "speaking with one voice" on matters of external relations in the field of energy.

14.2 A Changed Energy and Policy Context

The revival in interest for common and coordinated EU action in the field of energy can be explained by the following developments in the recent years:

- The perceived end of the era of cheap and abundant oil and threat of further drastic global price increases in the medium and long-term ("peaking oil" etc.), and the resulting economic need to diversify the economies' energy reliance on oil and gas;
- The increasing dependency of the EU economies on imported hydrocarbons: In a business-as-usual projection the EU's energy import dependency will jump from 50% of total EU energy consumption today to 65% in 2030. Reliance on imports of gas is expected to increase from 57% to 84% by 2030, of oil from 82% to 93%. In particular, the dependence on natural gas is controversial. There is an increasing penetration of the power markets by gas – with only few major suppliers: Russia (currently representing over 30% of the EU's gas supplies and growing rapidly) and North African and Gulf countries. In this context "speaking with one voice" towards Russia, gas-rich former soviet states and North African and Gulf countries has particular significance.
- The increasingly alarming reports on and signs of climate change and the role the EU took a decade ago in spearheading the global movement in curbing the emission of man-made greenhouse gases;
- The rise in occurrence of inter-regional electricity black-outs due to the growing burden on the European interconnected high-voltage transmission grid (UCTE)

from large commercial transfers of electricity and regional concentrations of wind power, for both of which it was not originally designed;
- The perceived failure in implementing fully liberalised and harmonised electricity and gas markets throughout the EU by the agreed deadline of July 1st 2007 and calls for better coordination and supervision by instituting a European regulator.

In the light of these changes, the EU appears to have shifted its priorities within the triptych of energy policy objectives – consisting of a secure, competitive and sustainable energy supply. Whereas energy policy in the EU was previously impacted from Brussels mainly through the EU objective of completing the internal market and raising the competitiveness of EU economies, now climate change abatement and security of supply have become the new priorities. Of the ten priority action areas identified in the European Commission's latest Energy Policy Communication, nine directly address climate change and security of supply (the tenth being the completion of the liberalisation of the electricity and gas sectors):

1. Effective solidarity between Member States and promoting security of supply
2. A long-term commitment to reducing greenhouse gases and the EU Emissions Trading System (ETS)
3. An ambitious program of energy efficiency measures
4. A longer term target for renewable energy (20% of the EU energy mix by 2020)
5. A European strategic energy technology plan (to be tabled in 2008)
6. Towards a low CO_2 fossil fuel future through carbon capture and storage for fossil based activities (E&P of oil and gas, clean coal power plants)
7. Further development of nuclear technology in the EU
8. Promoting a common external energy policy
9. Creation of an energy observatory office for assessing future investment needs in infrastructure and generation facilities.

There are many references to energy technologies in the policy documents of the EU. For example, the EU's 6-yearly R&D Framework Programmes concerning the allocation of EU support money to R&D and innovation refers to the policy of promotion of new energy technologies, such as in the fields of nuclear and renewable energy. But the EU's action on energy technologies is far wider. The R&D Framework Program is complemented by public and private sector R&D at national level, for which it plays to some extent an integrative and coordination role. The yearly expenses for energy-related R&D at national level can be substantial in the case of large and/or wealthy countries. In addition, the above list of priority action areas contains several areas with important implications for energy technology policies. However, the EU has so far not had a comprehensive energy technology policy, and the proposed European Strategic Energy Technology Plan constitutes a new qualitative step. With this plan the EU wants to lower the cost of clean energy and to put EU industry at the forefront of the rapidly growing low-carbon technology sector. The European Commission sees this plan as a key to achieving the 20% renewable target by permitting a sharp increase in the share of renewables (and by including the roll-out of off-shore wind and the 2nd generation of biofuels).

In the European Commission's vision, technological innovation is needed for

- electricity and heat to be increasingly produced from low carbon sources and extensive near-zero emission fossil fuel power plants with CO_2 capture and storage (by 2030); and
- transport to adapt to using 2nd generation biofuels and hydrogen fuel cells.

The European Commission wants the switch to low carbon in the European energy system to be completed by 2050, with an overall European energy mix that could include large shares of renewables, sustainable coal and gas, sustainable hydrogen, and, for those member states that want, nuclear fission and fusion.

It is clear that such an ambitious plan cannot be achieved solely with R&D support programmes, and the EU is putting great hope into widespread diffusion of new technologies through market-based incentives. This approach was initially started in the 1990s by a few member states in the generation of electricity from renewable energy technologies, and since then has received the support of the EU.

14.3 The EU's Energy Technology Policies

14.3.1 The EU's Double Approach of "Technology Push" and "Market Pull"

Since 2000, the EU has chosen to force the development of new carbon-saving energy sector technologies, particularly renewable energy technologies. Whereas the climate policy debate through technology innovation was often characterised by two polarised views about how to support it, the EU has embraced both options. These can be summarised under the terms *technology push* and *market pull*.

The *technology push* view (for the following see: Grubb, 2004, p. 9) holds "that the primary emphasis should be on development of low-GHG technologies, typically through publicly-funded R&D programmes, rather than regulatory limitations on emissions". According to this concept "it would be preferable to concentrate in the near term on investing in technological innovation, and adopt emissions limitations later when innovation has lowered the costs of limiting GHG emissions, rather than mandating costly reductions now" (ibid).

Technology push implies the identification of a potentially interesting technology and supporting R&D to make a product that would eventually be able to compete in the marketplace.

The opposing *market pull* view (see again: Grubb, 2004, p. 9) holds "that technological change must come primarily from the business sector, and is primarily a product of economic incentives". According to this concept the main emphasis lies on the adoption of regulatory measures (technology-based regulatory limitations, GHG emission caps, or charges). "Profit-seeking businesses will respond by innovating to produce technologies that will reduce emissions at less cost in order to gain competitive advantage over rivals. From this perspective, postponing emissions limitations would simply defer the whole process of innovation required for the private sector to produce these solutions" (ibid). Despite various and frequent market failures in the early stages of innovation, (no adequate incentives for business for companies to invest in basic research because appropriation – through patents etc. – of the knowledge gained and the commercial payoffs may be too uncertain and long-term) *market pull* advocates tend to assume that existing general policies (such as corporate tax breaks for R&D expenditure) are sufficient to overcome these failures" (ibid).

The basic credo behind *market pull* is that it favors industrialization/mass production of the targeted new technology products, and that the scaling-up from prototype manufacturing to industrial manufacturing entails efficiency and cost gains in manufacturing through experience and learning effect and cross-fertilization and spin-offs from other high-tech industry sectors in combination with focused research efforts. New ways of manufacturing the same and possibly improved technology product can also bring new insights and directions for R&D and thus favor further technology development/refinement, leading to a better product etc. The experience and learning effect thus is concerned with cumulative production over time and application of the product over time and in different circumstances and by different users – not the manufacture of a single product/batch or its use at a particular moment in time – and recognizes

a) that it takes less time and cost to assemble a product the more times that product is made, and
b) that lessons can be learned from different applications and users.

It is often argued (see again: Grubb, 2004, p. 10) that divergent perspectives on the process of technology change lead to directly opposing policy prescriptions. This is summarized in Table 14.1.

As far as the stages of the complex process of (energy) technology innovation are concerned, there are at least (in a market economy) six distinct stages in the "innovation chain" (see again: Grubb, 2004, p. 18; see Fig. 14.1a.):

- basic research and development;
- technology-specific research, development and demonstration;
- market demonstration of technologies to show to potential purchasers and users that the technology works in real-world applications, and tests and demonstrates its performance, viability and potential market;
- commercialization – either adoption of the technology by established firms, or the establishment of firms based around the technology;
- market accumulation in which the use of the technology expands in scale, often through accumulation of niche or protected markets;
- diffusion on a large scale.

Table 14.1 The divergent policy implications of different technical change perspectives

	Technology-push: R&D-led technical change	Demand pull: market-led technical change
Process:	Technical change depends mostly on autonomous trends and government R&D	Technical change depends mostly upon corporate investment (R&D, and learning-by-doing) in response to market conditions
Economic/policy implications: Implications for long-run economics of large-scale problems (e.g. climate change)	Atmospheric stabilization likely to be very costly unless big R&D breakthroughs	Atmospheric stabilization may be quite cheap as incremental innovations accumulate
Policy instruments and cost distribution	Efficient instrument is government R&D, complemented if necessary by "externality price" (e.g. Pigouvian tax) phased in.	Efficient response may involve stronger initial action, including emission caps/pricing, plus wide mix of instruments, targeted to reoriented industrial R&D and spur market-based innovation in relevant sectors. Potentially with diverse marginal costs
Timing implications	Defer abatement to await technology cost reductions	Accelerate abatement to induce technology cost reductions
"First mover" economics of emissions control	Costs with little benefits	Up-front investment with potentially large benefits
Nature of international spillover/leakage effects arising from emission constraints in leading countries	Spillovers generally negative (positive leakage) due to economic substitution effects in non- participants	Positive spillovers may dominate (leakage negative over time) due to international diffusion of cleaner technologies

Source: Grubb, 2004, pp. 10–11

During all different stages of technology improvement cost reductions are involved; but (see again: Grubb, 2004, pp. 18–19) "the principal barriers and driving forces change across the different stages. *Technology push* elements dominate early stage research, whilst *market pull* is increasingly important as technologies evolve along the chain".

From a finance and public policy perspective it is useful (see again: Grubb, 2004) to condense the innovation chain into three main components (see again: Grubb, 2004, p. 19; see Fig. 14.1b.): "at one end, the new technology RD&D stages, the main issues concern the funding and management of publicly-financed technology RD&D; at the opposite end, what matters are policies that affect the economic returns to private investors. In the middle, the challenge is the transition from publicly to privately financed operations".

In a similar way (as described above, following Grubb, 2004) the IEA (IEA, 2003) described recently that the "process by which a technology is taken from the stage of exploring a new application of scientific and engineering ideas, through the developments of new equipment, and on to the market is long and complex". The

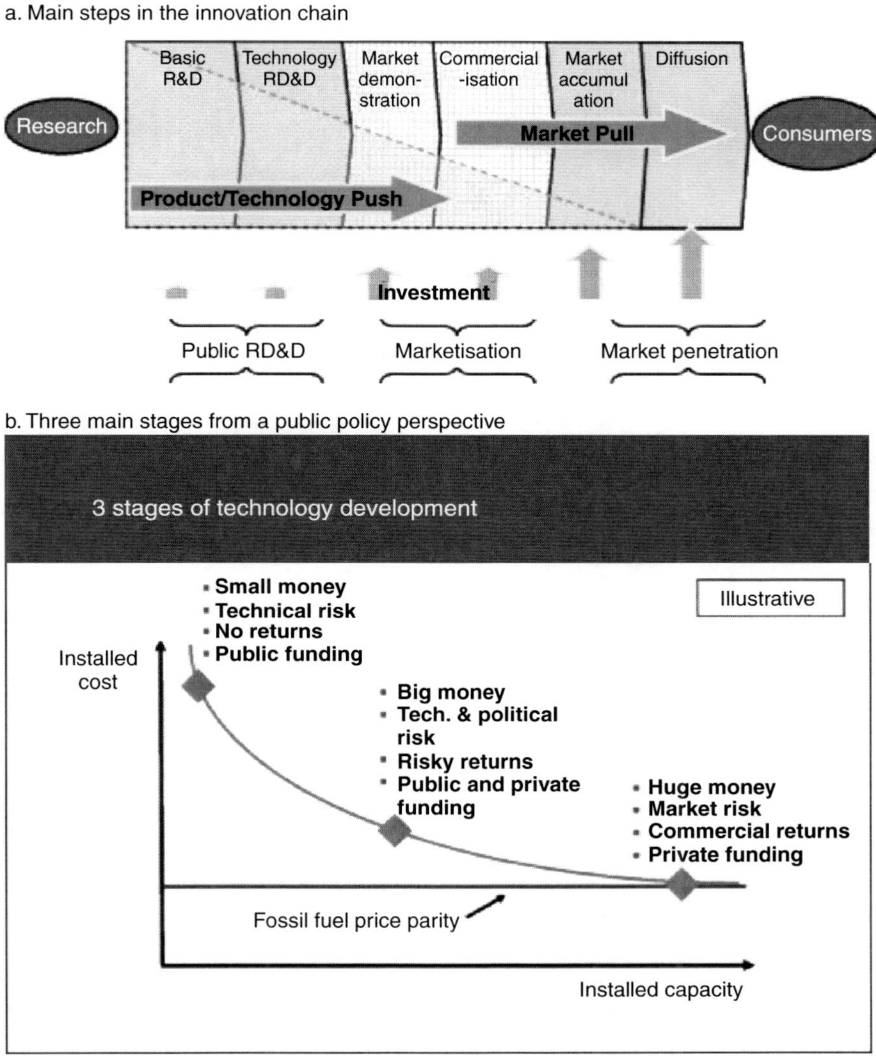

Fig. 14.1 The innovation chain
Source: Grubb, 2004, p. 20

IEA stressed the importance of the link between deployment programmes for new technologies and private sector decisions to invest in the market learning process. "Decision makers in industry may judge the initial costs of market learning for a given technology to be too high and involve too much risk. Though scarce public resources are not sufficient by themselves to bring a new technology through even the early parts of the process, effective government-supported deployment

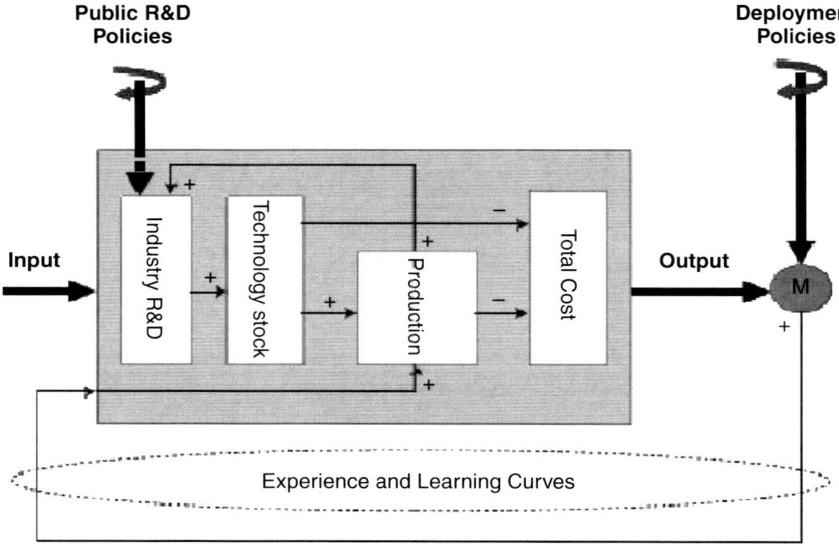

Fig. 14.2 Technology deployment policies and R&D support policies go together
Source: Creating markets for energy technologies, IEA, 2003, p. 43

programmes can play a crucial role in encouraging private investment and activating learning processes among market participants" (IEA, 2003, p. 42).

Figure 14.2 summarizes "how public sector and industry R&D interact to produce a "virtuous cycle" in which government support encourages corporations to try out new technologies in genuine market settings" (IEA, 2003, p. 42).

In an innovation process, as described above, directed towards encouraging innovation and their deployment, the government has a key role across the innovation chain (see again: Grubb, 2004, p. 22), "but its role changes radically along the innovation path and the appropriate extent of involvement may vary greatly between different sectors. At one end, government finances basic and applied technology R&D, and some proof-of concept demonstration, in order to lay a foundation of publicly-available ideas for others to work with. At the opposite end, governments need to define and enforce a basic regulatory structure which can reward innovators" (ibid).

The following sections attempt to describe briefly the EU-policy approach in the field of promoting energy technologies.

14.3.2 Technology Push in the EU

EU countries are providing significant financial support to R&D in new energy technologies – in the field of renewable energy technologies or in carbon emissions saving or capture. Support to R&D is done both at national and supranational level.

National (or even regional) financial support takes many different forms and it would be too exhaustive to go into detail here. Promoters of R&D projects or programmes can solicit financial support, mostly in the form or grants or concessional loans, from both national and EU level (as well as from other international institutions). In addition, national or local governments can grant tax rebates to the institutes and companies in question. At EU level, the financial budget for R&D support is decided and coordinated by the European Commission, the European Parliament and the Council of (national) Ministers. The EU allocates money from its own budget to R&D, innovation and knowledge development in accordance with self-decided targets. Such R&D budgets are laid down for six years at a time in so-called "Framework Programmes". The 7th Framework Program (FP7) has just been decided (OJEU L412 of 30.12.2006) and runs from 01.01.2007 to 31.12.2013 with a total budget of EUR 50.521 billion. FP7 is organised into four specific programmes, corresponding to four major objectives of European research policy:

1. Co-operation (€32.4 bn): refers to gaining leadership in key scientific and technology areas by supporting co-operation between universities, industry, research centres and public authorities across the EU and with the rest of the world. Trans-national cooperation will remain the main instrument for carrying out research activities. This program consists of ten different thematic research areas:
   - Information and Communication technologies (€ 9.1 bn)
   - Health (€6.1 bn)
   - Transport (including Aeronautics) (€ 4.2 bn)
   - Nanoproduction (€ 3.5 bn)
   - Energy (€ 2.4 bn)
   - Food, agriculture and biotechnology (€ 1.9 bn)
   - Environment (including climate change) (€ 1.9 bn)
   - Security (€ 1.4 bn)
   - Space (€ 1.4 bn)
   - Socio-economic sciences and the humanities (€ 0.6 bn)
2. Ideas (€7.5 bn): refers to the establishment of an autonomous European Research Council (ERC), which will support and stimulate basic research carried out by individual teams competing at European level. The ERC has already been established and it has published its draft work program for 2007.
3. People (€4.8 bn): refers to the so called "Marie Curie" actions, which strengthen training, the career prospects and mobility of European researchers.
4. Capacities (€4.1 bn): refers to developing and fully exploiting the EU's research capacities through large-scale research infrastructure, regional cooperation and innovating small and medium-sized enterprises (SMEs) as well as increased international co-operation and bringing science and society closer together.

In addition to the four specific programmes, the FP7 includes specific programmes for the traditional sector of Joint Research on nuclear safety in line with the EU's responsibilities under the Euratom Treaty (€2.7 bn), and the non-nuclear activities of the Joint Research (€1.75 bn).

The energy research area under the Cooperation program will focus on the development of cost-effective technologies for a more sustainable energy economy for Europe (and world-wide) and European industry competitiveness. Activities will cover the whole chain from fundamental and applied research and technological development through to large-scale technology demonstration ("lighthouse projects"), complemented by socio-economic research for policy decisions and market framework development. Specific activities concern:

- Hydrogen and fuel cells
- Renewable electricity generation
- Renewable fuel production
- Renewables for heating and cooling
- CO_2 capture and storage technologies for zero emission fossil-based power generation
- Clean coal technologies
- Smart energy networks
- Energy efficiency and savings
- Knowledge for energy policy making

In addition, the other programmes also cover energy related activities.

14.3.3 Market Pull

In 2001, with the adoption of Directive 2001/77/EC*on the promotion of the electricity produced from renewable energy source in the internal electricity market*, the EU recognised the importance of *market pull* in promoting new energy technologies, rather than *technology push*, based on positive experience in Germany and Denmark with the stimulation of wind power.

The most prominent examples of *market pull* are the stimulation of the production and use of renewable energy by creating new market outlets for the electricity or fuels in question, primarily by paying premium or fixed tariffs for the electricity, and tax rebates on bio-fuels. However, the EU Emissions Trading System should also be included for the incentives it creates for innovation into cost-effective measures in industry and the energy supply sector to reduce carbon emissions.

Based on a strong popular wish to see more use of renewable energies, policy makers in EU states have sought to reconcile the fundamental demand for application of renewable energy with its affordability and competitiveness.

EU states started with innovative regulation of the electricity sector, imposing obligations on market players on one side and on the other providing financial, market-based incentives or compensation for meeting those obligations.

Germany was one of the first to introduce a so-called renewable electricity feed-in tariff system in the early 1990s, under which investors and promoters of new electricity production schemes based on a renewable source (mainly wind) were given a guaranteed fixed tariff by the grid operator for around 10 years for their full electricity output. The tariff (usually per kWh produced) was set at a level to make the enterprise financially attractive and bankable with a minimum of the renewable resource, i.e. wind. The extra costs incurred by the grid company in buying the electricity at non-competitive prices, wasfinanced by cross-subsidization through a fund to which all grid companies had to contribute. Initially, this extra cost was considered low as the share of wind power and other renewable-based electricity was very small, and easily bearable by the cash-rich grid companies that often had the character of regional monopolies. However, in the end, it was (and still is) the captive consumer – i.e. households – who paid for the extra cost burden through the electricity bill. The system was later rendered more transparent and a special grid charge was introduced on electricity bills of low voltage consumers. This regime amounts to a hidden tax and state aid, and was challenged in court in Germany and at EU level. However, the European Court of Justice and the European Commission accepted it because they did not see undue, prohibitive nor discriminatory state-aid, and accepted the promotion of renewable energy generation for environmental reasons as justified. Later, Germany extended this regime to all types of renewable energy technologies applicable for the production of electricity. Basically, an electricity feed-in tariff was set for each renewable energy technology to make its commercial application financially profitable and bankable under minimal normal working conditions. This explains the boom of wind power and photovoltaic solar power in Germany – a country in the middle of Europe, not particularly endowed with a good wind resource nor sunshine. Since then, Spain, Italy, France, Greece and others have followed. Other countries, such as the United Kingdom, have had similar success with a certificates system in which electricity suppliers are obliged to source a minimum share of the electricity from renewable sources or purchase the equivalent certificates from renewable energy producers.

Deployment support is considered legitimate because prices are expected to fall as producers and users gain experience. The key questions are how much support a technology needs to become competitive, and when. Experience curves provide an indication for how much support and time may be required, because they provide a simple, quantitative empirical relationship between price and the cumulative production or use of a technology over the recent past. There is overwhelming empirical support for a price experience relationship from all fields of industrial activities, including the production of equipment that transforms or uses energy. But past experience provides no clear answer as to future cost and price trends. Overall, manufacturing costs can be expected to decline further with the hope of eventually making the technology competitive with the existing energy production and supply means of the market. An indication of future renewable electricity generation costs, based on today's best assessment, is given in Fig. 14.3.

Some of the initial optimistic expectations have been contradicted by the market, however. The regulated *market pull* approach, which proved successful in

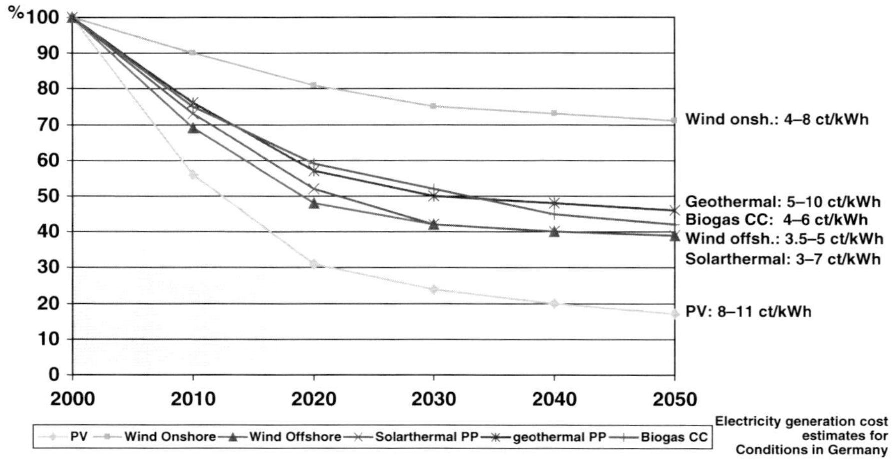

Fig. 14.3 Development of generation costs for RE technologies

developing thriving industries in the few countries to have implemented it in the 1990s, spread quickly throughout the EU after the adoption of *Directive 2001/77/EC on the promotion of the electricity produced from renewable energy sources*, and has spread to large countries outside the EU. For example, China, India and several US states have introduced similar support schemes designed to promote the construction and operation of wind turbines. This has led to a global demand boom for wind turbines in recent years to which the relatively young industry has not yet adapted. The result is bottlenecks in manufacturing capacity and especially in the supply of components, and thus higher prices. In addition, with the global economic growth of recent years, particularly the strong growth of the Asian economies, the prices for steel and copper – some of the main materials used for the construction of wind turbines – have risen significantly. The combined effect is that specific investment costs in wind turbines have increased since 2002 from +/– 1000 €/kW installed to over 1200 €/kW. This 20% price increase is slightly mitigated by the improved wind energy conversion performances of the newer turbines. The solar photovoltaic (PV) sector is showing a similar effect. Since 2004, demand in PV panels has caused a boom in the PV industry, but supply in raw materials has been lagging, particularly the supply in silicon used for the production of crystalline wafers and constituting a major cost element in the production of PV cells. As a result, prices of PV panels that had previously declined at an average annual rate of 5% have increased again since 2004. It is now hoped that the industry will overcome the silicon supply backlog by 2008, and that prices will start to decrease again (prices for most other parts involved in the production and installation of PV modules continued to decline during 2005 and 2006). But this is not assured at this point in time if demand is fuelled by the introduction of attractive feed-in tariffs for grid-connected PV installations in a growing number of countries.

14.3.4 Carbon Capture and Storage

The development of near-zero CO_2 emission fossil-based power plants is done by other means than incentive regulation. Here, member states of the EU choose to grant subsidies, subject to European Commission approval, to the development of pilot plants, as the market does not offer sufficient incentives and the technologies are still in early stages. The full cost of carbon capture and storage (CCS) is not yet clearly established, and the time horizon for CCS to become competitive on a larger scale appears more distant than for many renewable energy technologies.

In parallel, the industry is constantly seeking to increase the energy efficiency of power plants, which brings significant CO_2 emission gains. R&D is directed into new materials that allow new coal-fired plants to run at higher temperatures, thereby achieving higher efficiencies. Today's new coal plants will run at 43–45% efficiency compared to 35% for plants built some 15 years ago, which allows them to emit 25% less CO_2 emissions.

But the power markets, including the EU Emissions Trading System (ETS) offer little stability for deciding on investments of such long-term nature, and only few EU countries envisage new coal-fired power plant projects that would permit significant technology development (mainly Germany and the UK).

The CCS technologies thus still appear to be in a demonstration phase and the policy resembles more *technology push*. But it is imaginable to complement this with *market pull* once the technologies have been proven and show potential for moving down the cost curve.

The pursuit of developing clean coal technologies (or carbon saving technologies) appears justified given that coal represents one of the largest, widespread and most affordable energy resources in the world and is expected to continue to play a major role in the build-up of power generation in the developing and transition economies (as well as in the developed economies due to its security of supply advantages).

14.3.5 Energy Efficiency

As mentioned in the introduction, energy efficiency enhancement and energy saving is another priority action area of the EU. Here the EU targets the markets and regulation of minimum standards, as well as fiscal incentives. This goes for the housing sector, where insulation norms are being raised and fossil-based heating fuels taxed. The most recent prominent example is the obligation on car manufacturers to reduce the CO_2 output of their cars to 130 g/km on average. Heavy investments will be required from the automobile industry and the mineral oil sector to meet the standards and still produce cars that appeal to consumers. *Technology push* is not excluded here, but the onus is on standard regulation.

14.4 Assessment

For the renewable energy sector, the combined approach of *technology push* and *market pull* has already produced impressive results:

- *Wind power* production capacities have increased by 150% since 2001, making the EU the worldwide leader in this field with 60% market share (highest penetrations are found in Denmark, where wind generated electricity amounts to 20% of electricity consumption, Germany and Spain).
- *Solid biomass* gained considerable pace in 2004 and 2005 (electricity generated from solid biomass resulting from the combustion of forestry and agricultural products, as well as biodegradable, non-hazardous municipal solid waste).
- A similar development can be seen for *biogas*, which is not only used for producing energy (nearly two-thirds of biogas is used to produce electricity, the remaining third being used for heat production; it is starting now to be used as a fuel for transport), it also meets environmental considerations of waste management.
- *Solar photovoltaic* capacities in the EU have increased at a record average (annual rate of 70% over the last five years) with Germany in the lead.
- The use of *geothermal* sources for the production of electricity and heat is advanced in Italy (95% of the EU′s installed capacity), as well as in Portugal and France, and is being developed in Germany.
- *Concentrated Solar thermal power* (CSTP) plants are being developed in Spain on the back of long-standing demonstration and development activities made in Spain and the US.
- *Bio-diesel and bio-ethanol* production sites are being developed and regulations provide for their use in the transport sector (e.g. by setting mandatory minimum quantities added to normal fuels).

This apparent success story, however, carries important costs, drawbacks and risks:

14.4.1 Costs

The road to cost-competitiveness of RE and carbon-capture could be long, and over the next decade, billions of Euros will be spent on the promotion of the application and use of the various forms of renewable energy.

The cost-extent of increasing renewable energies can be estimated relatively easily for the power generation sector:

The EU aims to have renewable sources provide 12% of its primary energy supply by 2010 and 21% of the electricity generated in its 25 Member States. This target was established in the EU renewables Directive 2001/77/EC, which sets out differentiated national targets. With current policies and efforts in place, the European Commission expects that a share of 19% will be reached by 2010. In other words, Europe will in all likelihood, come close to its target on renewable electricity by

2010. The new target proposed by the European Commission is 20% of renewable energy in the EU's primary energy supply, supposing roughly 33% in its electricity production by 2020.

Taking the latest statistics available on power generated with RE sources, and multiplying these with standardised discounted generated cost figures established over the years for the different types of RE power generation technologies and applications, the extra cost of the subsidized RE can be estimated as being around €12 billion per year by 2010, increasing to €37 billion in 2020 for the power sector alone (see Table 14.2). After 2020 the experience and learning effect is expected to generate substantial decreases in the cost of renewable based generation as new, cost-competitive and lower cost projects are being commissioned. Thus the extra cost of subsidized RE is expected to decrease rapidly post-2020.

The estimates are based on comparing the estimated discounted cost of electricity generation for the different RE technologies over the coming years with the discounted cost of electricity generation with gas, coal or nuclear driven power plants. For PV, the benchmark used is the cost of distributed electricity.

These cost figures should not be taken too literally as they do not include hidden costs of grid system balancing and regulation, reserve generating capacities and investments in catering for new transmission capacities (particularly relevant for wind power). Also, the average figures can only be very rough estimates, as over time, the generating cost of RE can be expected to decrease (see the cost curve argument made earlier). It may be that in the years up to 2020, some renewable energy technologies will have achieved quasi-competitiveness with fossil and nuclear based power generating options, but the vast investments made in the renewables-based power generation in the meantime and their long pay-back times mean that the effect of cost-competitiveness will really be felt only post-2020.

This exercise shows the dimension of the costs the EU economies are spending and going to spend on *market pull* in the electricity sector for developing the renewable energy technologies. They do not take into account the costs of *market pull* in the transport and heating sectors, so these come in addition. And of course, the *technology push*element has to be added. It shows that the EU targets for RE amount to a mega state subsidy program, whose costs are carried by the consumer in the end. The *market pull* of RE in the power sector will represent over 10% of the EU25 total annual spending on R&D (public and private spending, all economic sectors confounded),[2] or the equivalent of the GDP of a country like Slovenia.

Relating the cost of these subsidies to the CO_2 emissions avoided/saved shows the following: The EU's average generation mix is estimated to have had a carbon dioxide emission factor of around 0,43 tCO_2 per MWh produced in 2004; accordingly the renewables developed so far came at a carbon-related extra cost of 18 EUR/ tCO_2, which is low due to the large share of cost-competitive hydro. If one assumes that the carbon dioxide emission factor would gradually reduce to 0,4 due

[2]In 2004, the EU25 spent nearly EUR 200 billion on R&D – EUROSTAT press release STAT/05/156.

Table 14.2 Contribution of renewables to electricity production and its estimated annual extra cost over fossil and nuclear based generated electricity[3]

	2004 Eurostat TWh	2010 Projections TWh	2020 Projections TWh	Average Discounted Generating Costs for new investments[1] EUR/MWh	2004 Extra Costs Million EUR	2010 Extra Costs Million EUR	2020 Extra Costs Million EUR
Wind	58.5	194	530	65[2]	1170	2522	6890
PV	0.7	7.5	55	300	126	1350	9900
Biomass	67.9	138	300	80	2377	4830	10500
Hydro	303.8	356	384	55	0[3]	522	802
Geothermal	5.5	7	14	70	138	175	350
Solar Thermal	–	2	9	190		290	1305
Ocean	–	3	15	250?		615	3075
Total	435.9	707.5	1313		3811	11662	36532
Total Electricity Generation	3178.6	3300–3500	3250–4000				
Share of RES	14%	20–21%	33–40%				
Average fossil and nuclear				45[4]			
Distributed electricity				120[5]			

[1] very approximate average estimates only; the generating costs of individual projects can differ from these averages depending on the time and location of implementation
[2] for sake of conservativeness this figure underestimates the cost of offshore wind; it also does not take into account the extra cost of frequency and voltage regulation, transmission capacity reservation and supply/demand balancing borne by transmission system operators any electricity supply companies;
[3] hydropower projects implemented up to 2004 are assumed to be generally cost-competitive
[4] very approximate estimate based on a long-term fuel price scenario of 55 USD/bbl crude oil, 59 USD/t coal, and with the cost of nuclear power generation set at 45 EUR/MWh; in order to keep this calculation simple, this figure is kept constant over the future; it is acknowledged that changes in fuel prices and investment costs can alter this figure
[5] rough estimate of electricity distribution cost without taxes and levies for households (4700 kWh/year consumption)

to efficiency gains in fossil-fuel based technologies, the carbon-related extra cost of renewable energy *market pull* in 2010 rises to over 36 EUR/tCO_2 and in 2020 to over 60 EUR/tCO_2 avoided (the new hydro to be developed coming at a higher cost).

[3] Sources: For the TWh figures: European Renewable Energy Council – EREC (2007); For the cost figures: own estimates and calculations

The promotion of renewable energies will undeniably pay off in the long run, and specific renewable energy applications may reach cost competitiveness towards 2020 or thereafter. Thus the carbon-related cost-intensity of the total renewable energy mix is expected to decrease by then. But the huge costs the EU would spend in the run-up to cost-competitiveness of renewables ask for a critical and differentiated approach.

The question should be asked as to whether there are cheaper ways of achieving the EU policy objectives of greenhouse gas reduction and security of supply. For example, is it economically sound to increase renewable energies so fast? And can one optimise the *market pull* in a more targeted and cost-efficient fashion? At costs of over 60 EUR/tCO_2 should one not prioritise the development of clean coal and gas plants with CCS rather than pulling so many renewable energy technologies so fast? Besides, nuclear fission already constitutes a competitive carbon-free technology for power generation. Wouldn't it be more economical to develop its application and long-term storage for spent fuels and contaminated materials?

14.4.2 Limitations to Cost Reductions and Benefits

The *market pull* as currently applied in the EU appears to entail unnecessary costs. This is linked to the fact that the different renewable energy technologies are often pulled, irrespective of

- the cost of the technology and cost-savings potential, or
- the availability of the renewable resource, or
- the effective energy and carbon balance in the harnessing of the renewable energy and its use.

This is illustrated with the following examples:

Capital intensity of the technology and risk of insufficient cost-saving potential:

For example, it is questionable whether fostering the commercialisation via expensive tariffs of large centralised plants running on concentrated solar thermal power will pay-off over the next 15–20 years. These CSTP plants have been under demonstration for decades, and the design of the new commercial applications does not appear to contain significant improvements. Furthermore, for cost-efficiency reasons, these plants are land-intensive and need to be located in flat areas with very high solar irradiation, which are rare in the EU. Mainly Spain, Italy and Greece have suitable solar irradiation, but wide empty level plains are rare in Italy and Greece (unfavorable population density and topography) and often used for agriculture. The north African desert appears to offer better locations for these applications but then this requires long-distance transmission of the electricity which carries high investment costs in transmission lines and costs due to transmission losses. For similar reasons, the development of offshore wind power in deep waters far off the coast seems an expensive way of fostering offshore wind power. The rough climatic conditions, the foundations and the submarine cabling for such projects require extremely high investments in material which exceed the cost-benefits of the stronger wind resource. It would seem better to foster the offshore wind industry by supporting those projects that benefit from shorter distances to the shore, reasonable water depth and ground conditions while still having a good wind resource. This reduces the number of sites in the EU and poses the issue of setting up a selection or prioritisation list of the identified sites at EU level.

14.4.2.1 Availability of the Renewable Resource

Little consideration of availability of the renewable resource has led to the most rapid expansion of solar PV power and wind power in Germany. This country has the highest installed capacity in wind turbines in the world, yet good wind resource locations are concentrated mainly on its short North Sea coastline. The average load factor of wind turbines in Germany is relatively low and many produce electricity at average unitary costs of 60–80 EUR/MWh or higher. By contrast, most wind power in Spain and Denmark is generated at significantly lower cost due to a more constant wind. Some farms there produce already at costs close to the market price. Similarly, tariff regulation has generated strong demand for solar PV panels for grid-connected roof-top applications in Germany. But Germany has less than half the solar irradiation than the south of Europe, and thus the average unitary cost per produced kWh is almost double that in Spain. Another example is the promotion of the use of solid biomass for the generation of electricity and/or heat. The resources of biomass available at local or regional level often set limits to how many biomass-fired generation plants can be developed. With too many projects, the demand for biomass will raise its price, which in turn makes it difficult for biomass-fired generation to become cost-competitive.

14.4.2.2 Energy and Carbon Balance

Another problem can occur with the making of bio-fuels (ethanol or bio-diesel) or solid biomass for electricity/heat production. It is important to keep in mind that the growing of the crops required for energy/fuel production can be intensive in energy (to operate the trucks and machineries for sowing and harvesting) and fertilizer (often an energy-intensive product itself). Not only can the energy balance be negative (i.e. more energy is used to produce the bio-crop than it can generate), but if in such cases the input energy is fossil based, more carbon is released than saved.

In conclusion, the fostering of RE does not automatically achieve cost-competitiveness for each RE technology nor does it automatically bring the environmental and energy security benefits hoped for. It thus appears to be of economic interest to focus on those RE technologies and applications that carry the highest potential for becoming cost-competitive sooner rather than later.

Several renewable energy technologies carry also a number of negative environmental impacts (for example, for bio crops associated with their farming; for wind power associated with sight and noise pollution, bird collisions and shadow casting). But these trade-offs are usually assessed at the permit stage of the individual project, and strict environmental EU legislation generally avoids significant negative impacts for the environment.

14.4.3 Risk of Slowing Cost Reduction

Timely achievement of cost-competitiveness of RE and CCS technologies is far from being guaranteed. Projections of cost-decreases through experience effects are based on theory and past empirical data. But what has occurred in the past does not have to be the same for the future, and what is true for one RE technology does not automatically apply to another. The wind turbine industry is a case-study in disappointment: In 2000, the cost of wind-generated electricity (onshore) was projected to fall to 30 EUR/MWh by 2010. The risk is now that the average cost will hardly decrease from the levels seen in 2001/2002 (50–70 EUR/MWh) in the foreseeable future.

Market pull has a significant drawback in that the EU is not alone in using it, and worldwide countries are following with similar incentive schemes, including China and India. This fuels global demand for RE applications, and causes prices to rise because the industry cannot grow at the same pace and the market turns short:

- the number of manufacturers does not increase in line with demand – oligopolisation (wind turbine sector)
- growth in raw materials does not keep pace with market growth (e.g. solar silicon in the PV sector),
- bottlenecks and concentration in the supply of components (e.g. gears and transformers for wind turbines),
- the locally available renewable resource is limited causing its price to rise (biomass).

At best, these phenomena create periodical stop-and-go or boom-and-bust effects in the RE markets that interrupt the cost decline trend but do not stop it permanently. In the worst case, these interruptions can be for longer periods and cause significant economic cost.

Market pull also offers significant windfall profits to proven technology applications rather than giving absolute priority to the diffusion of innovative technologies, which slows down the cost decline trend. Private investors and financiers tend to prefer a project or business case based on the application of a proven technology rather than on an innovation that may be perceived to bear technical risks (e.g. standard PV panels preferred over concentrated PV panels).

14.4.4 Long-term Potential of Renewable Energy

The EU does support renewable energieswithout seeming to distinguish between RE technologies that could offer a significant contribution to achieving a sustainable energy supply in the long-term and technologies with a limited potential.

For example, PV despite its currently high costs has the potential to make a massive contribution to a sustainable global energy world once it achieves cost-competitiveness. Most other RE technologies do not offer a comparable potential, and given the shortcomings of some, as discussed above, it is questionable whether their sustained growth through (quasi-)taxes and public spending is justified.

14.5 Conclusion

The EU uses massive support to RE and CCS, which are currently expensive means of energy supply but have the potential to become cost competitive over the next 15–30 years. The general rationale of this policy is herewith accepted, but the absence of will to adopt a differentiated and more optimised approach raises criticism.

In view of all the uncertainties involved with bringing the cost of RE down to competitiveness, the EU should prioritise and direct its support to RE technologies depending on

- the marginal generating costs of each technology
- the cost reduction potential and the estimated speed of achieving cost competitiveness
- the availability of the renewable resource and/or of suitable sites
- the potential contribution of the RE technology to the final objective of a sustainable global energy supply.

Support to CCS is of a different nature for the time being and appears justified in view of the prominent role coal will keep in power generation worldwide.

The attention given to climate change and security of supply should not overshadow the need to give preference to cost-efficient means of carbon reduction. This pre-supposes an open mind towards nuclear and large hydropower, and also the courage not to yield to green lobbies that request support for RE with doubtful potential.

A cost-efficient, differentiated approach to the fostering of non-carbon technologies, without compromising the overall objective, requires a coordinated supranational approach. It would require a balanced-score-type listing of the RE technologies and of the regions where their application is most cost-effective. And it would suppose a review of the different instruments used to push and pull the technologies and markets. This should theoretically be feasible at EU level, but it is doubtful whether EU member states would want to transfer the necessary powers to the European Commission. The total cost of non-Europe in this case could well be significant, probably adding up to one year's GDP of the Baltic States.

References

Egenhofer, C. (2005). Technology in a post-2012 transatlantic perspective, *CEPS Policy Brief* No. 86/November 2005, Brussels: Centre for European Policy Studies (CEPS).

European Commission. (2007). An energy policy for Europe, COM(2007) 1 final.

European Commission. (2006a). The state and prospects of European energy research, Luxembourg: Office for Official Publications of the European Communities.

European Commission. (2006b). A secure, competitive and sustainable energy for Europe, COM(2006) 105 final.
European Commission. (2005). Winning the battle against climate change, Green Paper on Energy Efficiency, COM(2005) 35 final.
European Renewable Energy Council – EREC. (2007). Renewable energy technology roadmap up to 2020, Brussels: EREC.
Grubb, M. (2004). Technology innovation and climate change policy. *Keio Economic Studies*, *41*(2), 103–132.
International Energy Agency – IEA. (2003). Creating markets for energy technologies. Paris: OECD/IEA
International Energy Agency– IEA. (2000). Experience curves for energy technology policy. Paris: OECD/IEA
Official Journal of the European Union – OJEU. (2006). 7th framework program of the European community for research, technological development and demonstration activities (2007–2013), L412of 30.12.2006.

Chapter 15
US Climate Change Emissions Mitigation Policy: Energy Technology Push and Other US Policies

Fred Sissine

15.1 Introduction

The efforts to reduce carbon dioxide (CO_2) emissions have been driven mainly by emission goals and targets. This paper tries to describe what has been done in the United States, as a foundation for comparison with actions taken in the European Union (EU). This foundation may help the process of identifying ways to cooperate and jointly innovate in addressing the shared global problem of climate change.

In the effort to reduce CO_2 emissions, the EU has adopted a strategy that emphasizes a "market pull" cap and trade system, while the US has employed a strategy that stresses a "technology push" through focused research and development (R&D) spending. Clearly, the second EU Climate Change Program aims to achieve a "competitive edge" or "prime mover advantage" as an expected side benefit. In contrast, the Bush Administration has expressed concern that an effort to achieve rapid short term CO_2 reductions would be costly and could threaten economic growth.

In the context of concern for economic growth and formulating strategies to curb emissions after 2012, both the EU and US are deepening and broadening their efforts to curb emissions. The EU recently made energy efficiency, a key measure for abating CO_2, a cornerstone of its economic growth policy under its 2005 revisions to the Lisbon Strategy. Also, the EU is planning to increase R&D funding for energy technologies that reduce emissions. Meanwhile, the US has expanded its international technology deployment effort to embrace a new venture with selected Asian nations, and several states have undertaken policies that will reduce emissions, This recent expansion of emission reduction activity may help bring the EU and US strategies a bit closer, and could offer more opportunities for international cooperation.

All opinions and comments that may appear in the content of this paper are solely those of the author, and do not reflect the outlook of the Congressional Research Service or the Library of Congress.

F. Sissine (✉)
Congressional Research Service, Washington, DC, USA

P.J.J. Welfens, J.T. Addison (eds.), *Innovation, Employment and Growth Policy Issues in the EU and the US*, DOI 10.1007/978-3-642-00631-9_15,

US CO_2 emissions grew about 18% from 1990 to 2004 and are projected to grow another 8% by 2010. In contrast to the EU pursuit of a reduction in absolute emissions under Kyoto, the Bush Administration has adopted a goal of reducing the intensity of emissions, indexed to the size of the economy. Measured in terms of CO_2 emissions per $1 million of gross domestic product (GDP), this goal is estimated to cut about 100 million tons of CO_2 from the projected level for 2012. A 2005 draft Climate Change Technology Program (CCTP) Strategic Plan emphasizes R&D and technology for emission reductions in all sectors. The Plan has been criticized for lacking specific objectives and specific links to budgeting and spending. The White House Office of Management and Budget (OMB) estimates that CCTP spending has stood near $2 billion per year in the recent past, but the Government Accountability Office (GAO) says changes in the format and content of OMB data makes it difficult to prepare clear funding trends The US program also includes funding for international assistance, and tax incentives to support low/no carbon technology.

The US is broadening is commitments to international assistance somewhat. In July 2005, it joined the Gleneagles Plan of Action on Climate Change to increase support for broader deployment of energy efficiency, renewable energy, and other measures. In January 2006, the Administration announced that it would seek $50 million for the Asia Pacific Partnership on Clean Development and Climate.

The Energy Policy Act of 2005 was enacted in August. It includes energy efficiency standards for certain equipment, tax incentives for energy efficiency and renewable energy, and a 20% energy efficiency target to "lead by example" in federal buildings. The American Council for an Energy-Efficient Economy (ACEEE) estimates that energy efficiency measures in the law could reduce emissions by 50 million tons in 2010 and by 200 million tons in 2020. Also, the law includes programs to promote technologies that curb GHG The Senate passed a resolution calling for a CO_2 cap and trade system, but it failed to survive conference. In the appropriations process, funding for DOE efficiency and renewables programs stands at $1.1 billion, a slight decline, and funding for efficiency and renewables deployment in climate programs at the Environmental Protection Agency (EPA) and the Agency for International Development held steady at about $100 million each.

In the United States, the states often serve as "policy laboratories" for innovations that can subsequently be applied to the federal level. States, and groups of states, are innovating – taking action on R&D, technology deployment, and other measures to reduce emissions. Some of these state policies were formed to address climate change specifically, and other policies reduce CO_2 emissions even though they were put in place to address other issues, such as air pollution and traffic congestion. These state actions could ultimately influence national policy developments.

As one notable example, California, which ranks 12th in the world in emissions, has a fairly aggressive climate program that includes both technology push (R&D, standards for cars) and market pull (incentives for electricity efficiency) elements. In December 2005, the California Environmental Protection Agency (CalEPA) released a draft Climate Action report that proposed a broad expansion of energy efficiency and renewable energy deployment and options for a cap and trade system.

A new public benefit charge for transportation would fund incentives for more efficient vehicles.

In December 2005, a group of New England and Mid-Atlantic states agreed to establish a CO_2 cap and trade system for electricity, known as the Regional Greenhouse Gas Initiative (RGGI).[1] It aims to stabilize emissions from 2009 through 2014 and then cut emissions 10% by 2019. A wide variety of other states and regional partnerships have established measures that either aim to reduce emissions directly or to promote various energy efficiency, renewable energy, or other measures that would then help reduce emissions.

The American Council for an Energy-Efficient Economy (ACEEE) has studied the potential for how energy efficiency technology can best work with a cap and trade system for controlling greenhouse gas emissions. ACEEE finds that past improvements in energy efficiency have lowered energy intensity in end use sectors. This makes energy a smaller component of economic activity which, in turn, tends to reduce the potential for energy prices to drive further energy efficiency improvements. Also, a cap and trade system tends to pose barriers to energy efficiency because it would depress energy sales and reduce revenue. So even though end-use energy efficiency measures may provide emission reductions at lower costs than measures to sequester CO_2 at fossil-fueled power plants with "smokestack technologies," there may be a need to address these market barriers in order for it to fulfill its potential. ACEEE concludes that policies are needed to encourage energy efficiency both within and outside a cap and trade system. Within the cap, it recommends energy efficiency be treated like a supply resource, with direct allocation to energy efficiency service providers. Outside the cap, efficiency can be used initially to lower the baseline of credits, raise the target, and then later it can be treated like an offset. ACEEE recommends that both strategies (within and outside the cap) be employed.

A recent report by the Pew Center on Climate Change proposes some expansions to strategies for multilateral cooperation to reduce emissions. On one hand, Pew identifies several variations on market pull strategies, including the use of emission intensity instead of absolute targets, technology efficiency standards for a certain product (e.g. cars), and sectoral strategies where cooperation is set up among all nations/companies involved with a certain sector. On the other hand, Pew suggests that new forms of cooperation on breakthrough technologies (e.g. hydrogen, fuel cells, large-scale solar) could evolve that improve coordination, share best practices, and jointly support capacity building.

In conclusion, technology push cooperation could be expanded to include studies of climate-related R&D spending amounts and emission impacts, assessments of unintended impacts of renewable energy technologies, joint efforts to design protection for low-emission supply installations (such as offshore wind and natural gas), sectoral cooperation in setting fuel economy standards for cars and other equipment,

---

[1] In 2006, Maryland joined RGGI, and Pennsylvania may join in 2007.

and shared strategies for assisting developing countries with technology capacity building. On the market pull side, cooperation in technology deployment could help improve economies of scale and expand markets for low/no carbon equipment. For example, Europe has had much success encouraging solar energy development with the "feed laws" and California expects similar results from its new $3 billion Solar Initiative. Cap and trade systems are new and their designs have several soft areas that could be the subject of joint studies, such as how to make the best use of energy efficiency technologies.

A recent report by the Center for European Policy Studies (CEPS) has contrasted the strategies taken by the European Union (EU) and the United States to reduce emissions of greenhouse gases (GHG) (Egenhofer, 2005, p. 8) The EU centerpiece is the Emission Trading Scheme (ETS), a cap and trade market-pull mechanism for carbon dioxide (CO_2) emissions reduction. In contrast, the US government has focused on its portfolio of energy technology research and development (R&D) funding as the technology-push driver to curb emissions. However, the EU is now looking toward a major expansion of R&D funding and other measures to support energy efficiency, renewable energy, and other technologies. Meanwhile, an association of states in New England and the Mid-Atlantic region has established the first cap and trade system in the United States, and California is laying plans to do the same. This paper presents background on these policy developments in the EU and United States, and offers some suggestions for international cooperation that may help promote emission reduction policies in both places.

15.2 Background

15.2.1 EU Lisbon Strategy

In 2000, the leadership of the European Union (EU) met in Lisbon and launched a series of reforms at the national and European levels, including research and innovation, that were aimed at improving economic growth and competitiveness so that the EU could become the most dynamic and competitive knowledge-based economy in the world by 2010. However, at the half-way point in 2005, the European Commission found that the results to that point were not very satisfactory, and announced new emphases on productivity and employment.[2]

15.2.2 EU Climate Policy

The first European Climate Change Program (ECCP) began in 2000, with a primary goal of developing a policy response to the Kyoto Protocol that had been proposed

---

[2]EU Commission. A New Start for the Lisbon Strategy. EUROPA Growth and Jobs. January 20, 2006. p. 1.

in 1997. The main result of ECCP was the establishment of the European Emissions Trading Scheme (ETS), a cap and trade system for carbon dioxide (CO_2) emissions. ETS is the first international trading system for CO_2 in the world, and it is the centerpiece of the EU policy on climate change and its commitment to the Kyoto Protocol. The purpose of ETS is to help EU member nations achieve compliance with Kyoto Protocol targets at the lowest cost. It was established in 2003 and began operation in January of 2005.[3] In December 2005, ETS covered only large energy-intensive industries, including the electric power industry, which accounts for nearly half of Europe's CO_2 emissions.[4] At that time, CO_2 allowances varied in cost from 23 to 27 dollars per metric ton.

Within the backdrop of the Lisbon Strategy, the EU has pursued its commitment to reduce GHG and support the Kyoto Protocol. In February 2005, the Commission adopted a communication to the European Council and other EU institutions entitled, Winning the Battle Against Global Climate Change.[5] In March 2005, the European Council endorsed the Commission's communication.[6] The Commission's policy document noted that the Kyoto Protocol was only a first step, and emphasized that the Protocol called for planning to begin in 2005 for further GHG reductions after the Protocol's first commitment period ends in 2012. The document found that the transition to a "climate-friendly" society offered economic opportunities that support the Lisbon Strategy, and it proposed new initiatives to increase energy efficiency, enhance energy supply security, and reduce emissions.

For the international policy arena, Winning the Battle recommends that the EU seek to broaden multilateral participation, with a focus on involving all large GHG emitters through projects or programs to improve energy efficiency, low-carbon supply technologies, and climate policies. Further, it calls for expanding the scope of emissions reduction to cover all GHG and sectors. Also, the document argues for innovation to transform energy and transport infrastructure with a technology policy that employs a mix of "technology-push" and "demand-pull" policy instruments. Additionally, it calls for continued use of emissions trading, and increased resources for adaptation efforts.

For the policy arena within the EU, the Commission's policy document notes that the EU has reduced GHG emissions to 3% below the 1990 level, but needs to engage several new measures (including a market assessment for green certificates) to reach the 8% reduction target for 2012. A key element is a recommendation that the EU

---

[3] ETS was established by Directive #2003/87/EC of the European Parliament and European Council on October 13, 2003.

[4] Europa. European Commission (Environment). Questions and Answers on Emissions Trading and National Allocation Plans. (Memo/05/84) March 8, 2005. p. 1.

[5] EU Commission. Communication from the Commission to the Council, the European Parliament, the European Economic and Social Committee, and the Committee of the Regions. Winning the Battle Against Global Climate Change. [COM (2005) 35 final] February 9, 2005. p. 17.

[6] Council of the European Union. Presidency Conclusions. March 23, 2005. (Matter #IV, Climate Change, paragraphs 43–46, pp. 15–16).

budget for the period 2007 through 2013 include a major new initiative, a European-wide Energy Efficiency Initiative. Additional measures include increased public information about climate change, increased spending under the seventh Framework Program (FP7) for climate-friendly technology R&D, especially in the energy production and transportation sectors. Enhanced technology transfer and R&D cooperation with third countries (pre-EU accession countries and other EU neighbors) is another recommended policy measure. Also, within the context of the Lisbon Strategy, the Commission indicated that it will explore new initiatives in energy efficiency, renewable energy, air and sea transport, and carbon sequestration.

As part of the 2005 mid-term review of the Lisbon Strategy, the EU has made energy efficiency a cornerstone of its economic growth policy. The narrative in the Commission's document finds that action on energy efficiency

> ... complements the Lisbon Strategy, strengthens the security of energy supply, and creates a significant number of new jobs in Europe and a more competitive industry consuming less energy. Estimates show that in the EU-15 it would be economically feasible to realize energy savings of up to 15% over the coming decade...[7]

Further, in the European Council's 2005 mid-term review of the Lisbon Strategy, the Council

> ... emphasizes the importance of energy efficiency as a factor in competitiveness and sustainable development and welcomes the Commission's intention of producing a European initiative on energy efficiency...[8]

Thus, both the Commission and the Council have directly endorsed energy efficiency as a key element of economic growth.

Along this same theme of contribution to economic growth, the Commission narrative finds that innovation in climate-friendly technology could give Europe a competitive edge in a low carbon future. Specifically, it asserts that the EU could gain a "first mover" advantage and develop a competitive edge through a focus on resource-efficient climate-friendly technologies that "other countries will eventually need to adopt." As an example, it notes that countries that led in promoting wind energy had captured 95% of the fast-growing wind turbine industry. This, the Commission finds, could also happen with other technologies and sectors, such as cars and aviation.[9]

In October 2005, the EU Commission launched the second European Climate Change Program (ECCP II). Several working groups were established, and tasked with preparing policy recommendations for the Commission in 2006.[10] The aims of

---

[7]EU Commission, Winning the Battle, pp. 6–7.

[8]EU European Council, Presidency Conclusions. (Matter #II, Mid-term review of the Lisbon strategy, paragraph 19, p. 6).

[9]EU Commission, Winning the Battle, p. 7.

[10]EU Commission (Environment). Climate Change: Start of the Second European Climate Change Program. October 24, 2005. p. 1.

the second Program are to assess results under the first Program, provide a strong push for innovation in climate-friendly technologies, and to expand mitigation activities to all emitting sectors. In particular, it focused on aviation, shipping, and road transport sectors and on carbon sequestration (capture and storage). Additionally, it aimed to address issues involving adaptation to effects of climate change that are unavoidable.[11]

The EU Commission estimated that it would have have spent a total of 440 million euros on renewable energy technology research during the Sixth R&D Framework Program (FP6) from 2002 through 2006, accounting for nearly half of all EU energy research spending during that period.[12] The goal is for renewables to produce at least 12% of the EU's energy supply by 2010. EU spending constitutes about one quarter of total public spending on renewables research.[13] In November 2005, the Commission proposed that the same level of funding for renewable energy be maintained throughout the Seventh R&D Framework Program (FP7) that runs from 2007 through 2013. FP7 will have a major focus on Joint Technology Initiatives (JTIs) for public-private partnerships, including JTIs for renewables, hydrogen, and fuel cells. Also, FP7 is intended to be a key element to achieve the Lisbon Strategy for growth and competitiveness.[14]

In a report to the European Parliament on the Montreal Climate Change Conference and its implications for the EU, Mr. Dimas of the European Commission stated his belief that the EU's response to the climate challenge can be cost-effective and that it can even "give European companies a competitive edge." Further, he noted that the "...strong interest in our trading scheme made it clear that we are setting an example for the world to follow." As for next steps, Mr. Dimas said the EU should "... work with all parties, in particular major emitters such as the United States and emerging economies ...." He further noted that the Group of Eight Industrialized Nations (G8) has concluded its Plan of Action, the EU has formed climate partnerships with China and India, and the EU is the major contributor to a $410 million per year climate change fund for developing countries. Mr. Dimas stressed that the Bush Administration's initiative to create the Asia Pacific Partnership on Clean Development and Climate Change will complement other technology initiatives to help address climate change.[15]

[11]EurActiv (Online EU News). EU's Second Climate Change Program to Put Onus on Technology. October 24, 2005. p. 2.

[12]In January 2006, the Euro was valued at about $1.21.

[13]EU Commission (Research). Green Energy Sector Comes Together to Examine its Research Agenda. November 21, 2005. p. 2.

[14]Refocus (Elsevier). Research Investment for Renewables will Promote Sustainable Energy in Europe. November 30, 2005. p. 1.

[15]European Commission. Dimas, Stavros. Climate Change: Montreal and Beyond. (Speech/06/9) Presented at European Parliament Debate. January 16, 2006. p. 4.

15.3 US National Policy

15.3.1 US CO_2 Emission Trends

The Table 15.1 shows CO_2 emission trend estimates prepared by the US Department of Energy's (DOE's) Energy Information Administration (EIA). The EIA estimated that the United States' energy-related CO_2 emissions stood at 5002 million metric tons (MMT) in 1990,[16] and grew to 5900 million metric tons (MMT) of CO_2 in 2004, an increase of 18%.[17] Further, relative to 1990, EIA projects that current policies would lead US energy-related CO_2 emissions to reach 6365 MMT (27% higher) in 2010, 6536 MMT (31% higher) in 2012, 7119 MMT (42% higher) in 2020, and 8114 MMT (62% higher) in 2030.[18] Also, EIA reports that from 1990 through 2004, total GHG intensity fell by 23.5%, at an average rate of 1.9% per year.[19]

15.3.2 US Goals for Emission Reductions

In 2001, President Bush launched his National Climate Change Technology Initiative. It continued several policies and programs that had been developed previously and proposed a modest new initiative which did not receive much funding from Congress.

In 2002, the Bush Administration put forth its target for curbing CO_2 emissions.

It contended that a rapid short-term reduction in emissions would be costly and threaten economic growth[20] (p. 5). Thus, the Administration described its approach as one in which

> "...economic growth will make possible the needed investment in research, development, and deployment of advanced technologies. This strategy is one that should offer developing countries the incentive and means to join with us in tackling this challenge together."[21]

---

[16]US Department of Energy (DOE). Energy Information Administration (EIA). Emissions of Greenhouse Gases in the United States 2004. December 2005. Table ES2. p. x. [ftp://ftp.eia.doe.gov/pub/oiaf/1605/cdrom/pdf/ggrpt/057304.pdf].

[17]In 1990, energy-related CO_2 accounted for 81% of total US GHG emissions in CO_2 equivalence terms. In 2004, energy-related CO_2 share grew to 82% of total GHG, p. x.

[18]DOE. EIA. Annual Energy Outlook 2006 (Early Release). Table 18, Carbon Dioxide Emissions by Sector and Source. December 12, 2005.[http://www.eia.doe.gov/oiaf/aeo/aeoref_tab.html].

[19]DOE. EIA. Emissions of Greenhouse Gases, p. 14.

[20]The White House. Global Climate Change Policy Book. February 2002. p. 5.

[21]The White House. Climate Policy Book, p. 1.

Table 15.1 US CO_2 (energy-based) emissions (in millions of metric tons of CO_2)

	1990	2000	2001	2002	2004	2010	2012	2014	2019	2020	2030
US											
Baseline/BAU [1,2]	5.002	5.802	5.728	5.746	5.900	6.365	6.536	6.654	7.030	7.119	8.114
Intensity (CO_2/GDP)Improvement [4]											
(metric tons/$1 million GDP)	—	—	—	183	—	—	151	—	—	—	—
Resultant CO2 Reduction [4]	—	—	—	0	−20	−80	−100	—	—	—	—
Net Emissions Projected	5.002	5.802	5.728	5.746	5.880	6.285	6.436	—	—	—	—
California [3,5]											
Baseline/BAU	358	376	380	360	—	435	—	—	—	503	—
Existing Programs Only	—	—	—	—	—	−22	—	—	—	−67	—
Total CAT Plan	—	—	—	—	—	−59	—	—	—	−145	—
Net Emissions Projectged (CAT)	358	376	380	360	—	376	—	—	—	358	—
RGGL [3,6]											
Total Baseline/BAU	510	534	524	—	—	—	—	—	—	—	—
Electric Baseline/BAU	127	123	122	—	—	—	—	—	—	—	—
Electric Cap & Trade	—	—	—	—	—	—	—	—	−13	—	—
Net Emissions Projected	127	123	122	—	—	134	134	134	121	—	—

Source 1: US DOE. EIA. Emissions of Greenhouse Gases in the United States 2004. Table ES3 (CO_2 from Energy). December 2005.
Source 2: US DOE. EIA. Annual Energy Outlook 2006 (Early Release). Table 18 (CO_2 Emissions by Sector and Source). December 2005.
Source 3: US EPA. Table, Carbon Dioxide Emissions from Fossil Fuel Combustion (by State and Sector). [http://Yosemite.epa.gov/oar/globalwarming.nsf/uniqueKeyLookup/RAMR6E9KSV/$file/CO2FFC_2001.pdf?OpenElement].
Source 4: The White House. Climate Policy Book. February 2002. The figures shown assume all GHG reduction is CO_2.
Source 5: California Environmental Protection Agency (CalEPA). Climate Action Team (CAT) Report. December 2005.
Source 6: Regional Greenhouse Gas Initiative. Revised Staff Working Group Proposal. August 24, 2005.

In particular, the Administration set a target to reduce the GHG intensity (the ratio of annual CO_2-equivalent[22] emissions relative to the gross domestic product [GDP]) by 18% by the year 2012.[23] It said that in 2002, the GHG intensity was 183 metric tons (MT) of CO_2 emissions per $1 million of GDP. Under the 18% target, this would fall to 151 metric tons per $1 million of GDP in 2012.[24] Also, under this target, CO_2 emissions are projected to be 100 million MT lower than that projected for the business-as-usual policy in the year 2012.

Further, the Administration stated that

> ... When the annual decline in intensity equals the economic growth rate (currently, about 3% per year), emission growth will have stopped. When the annual decline in intensity exceeds the economic growth rate, emission growth will reverse. Reversing emission growth will eventually stabilize atmospheric concentrations as emissions decline.[25]

The Administration has also stated that a key aspect of climate policy is that it is one of several goals that must be addressed simultaneously. This includes reducing pollution, improving energy security, and mitigating climate change, while promoting economic growth at home and abroad.[26]

15.3.3 US Draft Strategic Plan for the Climate Change Technology Program (CCTP)

In September 2005, the Bush Administration issued a draft Climate Change Technology Program (CCTP) Strategic Plan.[27] It provides guidance to federal agencies for coordinating climate-change-related technology research, development, demonstration, and deployment. These CCTP activities form the core technology portion of the US approach to climate change that includes short-term actions to reduce GHG emissions intensity, advances in climate science, and promotion of international cooperation. The activities guided by this Draft Plan are expected to stimulate innovation.[28]

---

[22] CO_2-equivalent emissions is used as a universal unit measure, representing the effect of methane, nitrous oxides, and other non-CO_2 greenhouse gases in terms of an equivalent amount of CO_2 emissions.

[23] The Administration explains that the 18% goal represents a 4% acceleration above the previous trend.

[24] This change is illustrated in a chart prepared by Mr. James Connaughton, Chair of the US Council on Environmental Quality. Presentation at the Center for European Studies. November 21, 2005. Available on the web (p. 5) at [http://www.ceps.be/files/USClimateChange211105.ppt#497,5, Climate Policy Components].

[25] The White House. Climate Policy Book, p. 5.

[26] US Department of State. Effective US Actions to Combat Climate Change. Op-Ed for Les Echos, by Craig Roberts Stapleton, Ambassador to France. August 25, 2005. p. 1.

[27] The Draft Plan is on the web at [http://www.climatetechnology.gov/stratplan/draft/index.htm].

[28] Bodman, Samuel W. et al. Committee on Climate Change Science and Technology Integration. US Climate Change Technology Program (CCTP): Strategic Plan. September 2005. p. iii.

The Draft Plan is defined by six goals. They are (1) reduce emissions from energy end-use and infrastructure, (2) reduce emissions from energy supply, (3) capture, store, and sequester carbon dioxide (CO_2), (4) reduce emissions of non-CO_2 greenhouse gases, (5) improve capabilities to measure and monitor GHG emissions, and (6) bolster basic science contributions to technology development.[29] There are a number of near-term, mid-term, and long-term programs and initiatives that support the emission mitigation strategies proposed in goals one through four.[30]

To achieve these goals, the Draft Plan proposes seven approaches, or strategies. These approaches are (1) strengthen climate change technology R&D; (2) strengthen basic research contributions; (3) enhance opportunities for partnerships; (4) increase international cooperation; (5) support cutting-edge technology demonstrations; (6) ensure a viable technology workforce of the future, and (7) provide supporting technology policy.

For the first goal to curb energy end-use, the Draft Plan stresses that, except for electricity generation, the transportation sector my have the highest growth in global CO_2 emissions over the next 25 years. Thus, the CCTP emphasizes low-carbon fuels, hydrogen (fuel cell vehicles), and efficiency measures for trucks. Regarding the second goal to reduce emissions from energy supply, one highlight is the DOE FutureGEN program, which aims to show the viability of zero-emissions, high-efficiency coal-based power plants with CO_2 capture and storage. There are also initiatives to support hydrogen fuels and nuclear power plants. Activity under the third goal is focused on participation in the international Carbon Sequestration Leadership Forum (CSLF), seven Regional Carbon Sequestration Partnerships, and advanced management systems for forests and crops. The fourth goal, for other greenhouse gases, is focused on mitigation options and research needs to curb targeted gases that include methane emissions from energy production, waste, and agriculture, and nitrous oxides emitted from combustion and industrial sources.

Goal five aims to develop measurement systems that could establish baselines and guide opportunities for technology development. These activities include a focus on changes in soil carbon and the carbon exchange between the land and the atmosphere. The sixth goal involves basic science (biology, nanoscience, computer science) to support climate-related technology, including carbon sequestration and hydrogen research needs.

Other, more general, proposed activities include efforts to assess long-term technology potentials, encourage more public-private partnerships, increase international cooperation, foster capacity building in developing countries, consider cutting-edge technology demonstrations, and study policy options for stimulating private investment.[31]

---

[29]CCTP Strategic Plan, Chap. 10.

[30]These programs and initiatives are illustrated in Fig. 10-3 (p. 211) of the CCTP Strategic Plan.

[31]CCTP Strategic Plan, Chap. 10.

15.3.4 Critique of the Draft CCTP Strategic Plan

On December 21, 2005, the Administration posted public comments that had been submitted on the draft CCTP Strategic Plan. Several key points surfaced from the comments regarding a lack of key strategic elements. They included observations that the Plan: (1) does not specify a concrete goal in terms of a GHG concentration level or annual GHG emissions target, (2) lacks a time frame, (3) does not have objectives prioritized and sufficiently specific to be observable or measurable, (4) does not indicate the scale of deployment needed, (5) does not provide a link to spending and budgeting, and (6) does not incorporate the results of the Energy Policy Act of 2005, which was enacted in early August 2005.[32]

15.3.5 CCTP Programs and Funding History

The cabinet-level Committee on Climate Change Science and Technology Integration is responsible for overseeing the implementation of climate science and technology initiatives and programs within the Administration. The US Climate Change Technology Program (CCTP) was established under this Committee to coordinate and focus climate change technology research, development, and deployment programs and initiatives. A variety of voluntary partnership and grant activities are also part of the program portfolio. The Department of Energy (DOE) is responsible for coordinating the CCTP.[33]

In March 2005, the President's Office of Management and Budget (OMB) reported on federal climate change expenditures.[34] There are three categories of expenditures that directly address GHG mitigation: climate change technology, international assistance, and energy tax incentives.[35] The climate change technology category comprises the US Climate Change Technology Program (CCTP). CCTP is a multi-agency effort coordinated by DOE that incorporates a variety of technology research, development, and deployment activities, as well as voluntary partnerships and grant programs, that reduce GHG emissions. This category includes funding for the Hydrogen Fuel Initiative, FutureGen (zero emissions coal-fired powerplant),

---

[32]The comments are posted on the web at [http://www.climatetechnology.gov/]. See for example, comments number 4, 14, 22, 25, and 26.

[33]Executive Office of the President. Office of Management and Budget (OMB). Federal Climate Change Expenditures, Report to Congress. March 2005. p. 8. [http://www.whitehouse.gov/omb/legislative/fy06_climate_change_rpt.pdf].

[34]Executive Office of the President. Office of Management and Budget (OMB). Federal Climate Change Expenditures, Report to Congress. March 2005. p. 8. [http://www.whitehouse.gov/omb/legislative/fy06_climate_change_rpt.pdf].

[35]The report also includes funding figures for a fourth category, climate change science activities.

International Thermonuclear Experimental Reactor (ITER, fusion energy), Generation IV (nuclear power), Climate Leaders (EPA partnerships with businesses), Climate VISION (DOE partnerships with industrial companies), and Methane to Markets (EPA international initiative).

Programs under the international assistance category provide funding and technical assistance to developing countries to support their efforts to address climate change through improved energy efficiency, renewable energy, and through land use and forestry practices. Energy tax incentives include the Administration's proposals for tax measures to encourage renewable energy, hybrid and fuel cell vehicles, cogeneration, and landfill gas conversion. Many of these incentives were, in some form, adopted or extended by the Energy Policy Act of 2005.

In August 2005, the US Government Accountability Office (GAO) released a review of OMB's report on climate change funding.[36] GAO cites OMB's data showing that federal funding for climate change (including science) increased from $2.4 billion in 1993 to $5.1 billion in 2004, a 116% increase. However, after adjusting for inflation, GAO says the increase in constant dollar terms was from $3.3 billion to $5.1 billion, or 55%. During this period, the inflation-adjusted funding share for technology increased from 36 to 56%, and the share for international assistance decreased from 9 to 5%. However, GAO cautions that "...it is unclear whether funding changed as much as reported because modifications in the format and content of OMB reports limit the comparability of funding data over time." Also, GAO said it was unable to compare climate-related tax expenditures over time because OMB reported data on proposed, not actual, tax expenditures.[37] The table in Appendix shows the funding for CO_2 mitigation (excluding climate science) as reported by OMB, the Internal Revenue Service (IRS) of the Department of the Treasury, and GAO.

15.3.6 US International Programs and Activities

Although the United States is a signatory to the 1992 United Nations Framework Convention on Climate Change (UNFCCC), it is not a signatory to the 1997 Kyoto Protocol accords. However, the United States does conduct a variety of activities that support efforts to reduce climate change. These activities include several programs of international cooperation and assistance. Some examples are the Carbon Sequestration Leadership Forum, International Partnership for the Hydrogen Economy, Generation IV International Forum, ITER fusion project, and Methane to Markets Partnership. Further, the United States has several foreign assistance programs geared toward the environment. Though these programs are not in place

[36] US Government Accountability Office (GAO). Climate Change: Federal Reports on Climate Change Funding Should be Clearer and More Complete. (GAO-05-461) August 2005. p. 34.

[37] GAO. Federal Reports on Climate Change Funding, summary page.

solely for climate change purposes, they do provide climate change benefits. The primary federal agencies administering these programs are the US Agency for International Development (funding and technical assistance to developing countries), the Department of State (funding for UNFCCC and Intergovernmental Panel on Climate Change [IPCC]), and the Department of the Treasury (funding for the Tropical Forestry Conservation Act and the Global Environment Facility). OMB estimates that about 23% of current GEF funding supports climate-related projects. Energy efficiency and clean energy account for about one-third of GEF projects through 2005.[38]

15.3.7 G8 Gleneagles Plan of Action

In July 2005, the Group of Eight (G8) industrialized nations issued The Gleneagles Communique, which included a Plan of Action on Climate Change, Clean Energy, and Sustainable Development. The White House reported that President Bush and the G8 Leaders agreed to speed the development and deployment of clean energy technologies to address climate change, air pollution, and energy security in the United States and throughout the world. The agreement covers promotion of nuclear power, clean coal technologies, clean diesel and methane, renewable energy, bioenergy, more efficient electric power grids, and R&D on hydrogen vehicles. The G8-proposed transition to cleaner energy would be financed through strengthened World Bank and national policies that support markets, remove barriers to investment, leverage private capital, and promote investment.[39]

In the Communique, the G8 Leaders reaffirmed a commitment to the UNFCCC and its objective to stabilize GHG concentrations. They also agreed to take further action to "... promote innovation, energy efficiency ... and accelerate deployment of cleaner technologies, particularly lower-emitting technologies." Further, the G8 Leaders agreed to take forward a Dialogue on Climate Change, Clean Energy, and Sustainable Development to transform energy systems, monitor Gleneagles commitments, and share best practices.[40] The Plan of Action covers a broad range of topics including energy efficiency, buildings, appliances, surface transportation, aviation, industry, nuclear energy, cleaner fossil fuels, CO_2 sequestration, renewable energy, electricity, R&D international cooperation, finance of the transition, management of climate impacts, and risk management.[41]

---

[38]OMB. Federal Climate Change Expenditures: Report to Congress. March 2005. pp. 12–13.

[39]The White House. Office of the Press Secretary. Fact Sheet: Action on Climate Change, Energy and Sustainable Development. July 8, 2005. (Cited on [USINFO.STATE.GOV], "G8 Climate Change Plan Serves Energy, Environment, Development.")

[40]The Gleneagles Communique. [Agreement on] Climate Change, Energy, and Sustainable Development, pp. 1–3.

[41]Gleneagles Plan of Action: Climate Change, Energy, and Sustainable Development, p. 10.

15.3.8 Asia Pacific Partnership on Clean Development and Climate (APP)

In July 2005, this six-nation partnership was announced at the South East Asian Nations (ASEAN) forum. The Partnership includes the United States, Australia, Japan, China, India, and South Korea. President Bush said the Partnership will "... develop and accelerate deployment of cleaner, more efficient energy technologies to meet national pollution reduction, energy security, and climate change concerns in ways that reduce poverty and promote economic development."[42] At its January 2006 meeting, the Partnership agreed to form six public/private task groups to study options for various energy technologies and other measures. Australia indicated that it would contribute $75 million over 5 years and Bush Administration representatives announced that the President would request about $50 million in the fiscal year 2007 budget request to Congress.[43] Critics of the Partnership say proposed funding amounts were too small to make a real difference in promoting climate-friendly technologies. Secretary Bodman of DOE said that one of the best ways to assist developing economies "... is to help them invest in cleaner, more efficient energy technologies....[44] Further, he said, "In order to achieve meaningful results, we must engage growing and emerging economies from the outset and encourage the implementation of technologies that have demonstrated success."[45]

15.3.9 Energy Policy Act of 2005 (EPACT, P.L. 109-58)

EPACT was the first major energy law to be enacted since 1992. It has several provisions for low/no carbon energy technologies, mainly energy efficiency and renewable energy measures.

For energy efficiency, the law set 19 new efficiency standards for consumer appliances and commercial equipment. Most of these standards are set by law, but some are at the discretion of a future DOE rulemaking. The American Council for an Energy-Efficient Economy (ACEEE) estimates that these new standards will save more energy than any other efficiency provisions in the law. Further, the law (§141) requires that DOE report to Congress when the rulemakings are behind schedule, and include steps to respond to missed deadlines. Also, the law §102) sets a goal to reduce energy use in federal facilities by 20% (relative to 2003) over the 10-year period from 2006 to 2015. Further, the law (§101) calls for congressional buildings

---

[42]The White House. Office of the Press Secretary. Statement by the President. July 27, 2005.

[43]More information about the US role in the partnership is available at [http://www.state.gov/g/oes/climate/c16054.htm].

[44]US DOE. Asia-Pacific Partnership Ministerial Statement; Remarks of Energy Secretary Bodman. (Energy/Gov/News) January 12, 2006. p. 2.

[45]US DOE. Energy Secretary Announces $52 Million Request for Asia Pacific Partnership on Clean Development and Climate. (Press release) January 12, 2005. p. 2.

to meet the energy efficiency goal for federal agencies. Provisions to improve energy efficiency in public housing are also included. Additionally, the law includes a total of about $1.26 billion in extended and new energy efficiency tax credits for commercial property, new home construction, existing home improvements, appliances, residential fuel cells, and business fuel cells. For all energy efficiency provisions in the law, ACEEE estimates a total CO_2 emission reduction of about 50 million metric tons in 2010 and 200 million metric tons in 2020.

For renewable energy, the law (§1301) extends the production tax credit (PTC) for 2 years at a total value of $2.7 billion, and creates new tax-related incentives for solar, geothermal, and biodiesel at a total value of about $700 million. There is also a renewable fuel standard (RFS) that will increase the use of ethanol and biodiesel blended into gasoline from 3 billion to 7.5 billion gallons by 2012. Also, the law (§1253) terminated the mandatory purchase and sale requirements for a new renewable power facility, provided that the Federal Energy Regulatory Commission (FERC) finds that the new facility has access to wholesale power markets and transmission services.

The law has some provisions addressing climate change, and several other proposals were debated during consideration of the bill (H.R. 6). Title XVI establishes programs to promote the adoption of technologies to reduce greenhouse gas intensity, similar to those adopted on the Senate floor in S.Amdt. 817. The House version of H.R. 6 did not contain climate change or greenhouse gas provisions. Two other climate change amendments included in the Senate bill as passed were not included in the final version of the bill: S.Amdt. 866 expressed the sense of the Senate that climate change is a serious concern and that Congress should establish mandatory, market-based limits on greenhouse gas emissions; and S.Amdt. 826, based largely on S. 1151, would have required mandatory emission reduction and a cap-and-trade system.[46]

15.3.10 Funding in 2006 for Climate-Related Energy Programs

For the 2006 federal fiscal year, technology-related appropriations included about $1.2 billion for DOE energy efficiency and renewable energy programs, of which about $240 million was for renewable (biomass, solar, wind, geothermal, small hydro) energy technologies and $470 million was for energy efficiency (hydrogen, fuel cells, vehicles, buildings, industry) technologies. The EPA climate protection (energy efficiency) programs received nearly $100 million, and the Department of State's climate assistance programs for developing countries received about $100 million targeted for energy efficiency and renewables.

[46]US Congress. Congressional Research Service. Global Climate Change. (Issue Brief IB89005) Updated. September 7, 2005. p. 14.

15.3.11 Climate Legislation and Policy Debates in the 109th Congress

In July 2005, as part of the debate over the Energy Policy Act, the US Senate approved (approved 53-47) a non-binding resolution on climate change, which marked a major shift from a 1997 Senate resolution (approved 99-1) that had opposed US participation in the Kyoto Protocol. The first part of the 2005 resolution affirms findings that GHG emissions are causing global warming with consequent climate disruption and that human activity is a substantial cause of the GHG accumulation. The key remaining parts of the resolution say

> ... mandatory steps will be required to slow or stop the growth of greenhouse gas emissions into the atmosphere ... It is the sense of the Senate that Congress should enact a comprehensive and effective national program of mandatory, market-based limits and incentives on emissions of greenhouse gases that slow, stop, and reverse the growth of such emissions at a rate and in a manner that (1) will not significantly harm the economy, and (2) will encourage comparable action by other nations that are major trading partners and key contributors to global emissions.[47]

In January 2006, the Senate Energy Committee Minority Leader stated his intent to submit legislation that would seek a "middle road" between the President's reliance on R&D with voluntary measures and the 2005 McCain-Lieberman bill that had proposed a hard cap on emissions and a target of reducing 2010 emissions to the year 2000 level. The new bill would adopt a measure of GHG emissions "intensity" relative to economic growth, instead of absolute emissions levels.[48]

Also, the Senate Energy Committee Chairman and Minority Leader plan to jointly introduce a bill that would create a new energy R&D development agency within DOE. This "Advanced Research Projects Authority – Energy (ARPA-E) would be similar to the Pentagon's Advanced Research Projects Agency (DARPA), and the proposal is based on one of a package of proposals in the "Augustine Report" prepared by the National Academy of Sciences (NAS) to address the use of science to improve American competitiveness. In the NAS Report, the ARPA-E portion would cost $300 million in the first year and grow to $1 billion in its fifth year. In designing legislation, one concern is that the Department of Defense, with a $300 billion budget, serves as the customer for DARPA projects, but there is no parallel customer yet for ARPA-E. Also, there is a concern about such a large infusion

---

[47] US Senate. Committee on Energy and Natural Resources. Bingaman's Climate Speech in Canada. December 7, 2005. p. 7.

[48] Congressional Quarterly Green Sheets (Global Environment). Senate May Act to Slow Greenhouse Gas Emissions. January 16, 2005. p. 2.

of money to ARPA-E could have a negative impact on DOE's national laboratory system.[49]

15.4 US Regional and State Policies

In US energy and environmental policy making, states sometimes serve as "policy design" laboratories where "trial and error" approaches in multiple states can help generate information about the relative merits of different policy approaches.[50] Elements from the best models may then be used in the design of a federal policy. In regard to climate change policy, many states believe that climate change may pose significant risks to their population and economy. Further, many individual states emit high levels of CO_2 and other greenhouse gases.[51] In a recent review of state climate policies, the Pew Center on Global Climate Change found that state and regional action on climate change emissions mitigation may not be as effective as a coordinated national response, but it could provide a solid basis on which to design a federal policy.[52]

Many states embrace GHG emissions mitigation measures, especially those involving energy efficiency and renewable energy, as part of broader efforts to address other problems, such as traffic congestion and air pollution. They see the potential for multiple benefits and, even further, some states see climate policies as an economic opportunity to produce and sell alternative fuels, become renewable energy exporters, or to sell carbon emission reduction credits. The Pew Center finds that regional (multi-state) climate programs can be more efficient than a single-state program. Regional programs can cover a broader range of clean energy sources, reduce the administrative burden, and create a more uniform regulatory environment.[53]

15.4.1 California's Climate Policies

California is the 12th largest source of GHG in the world, emitting 360 million metric tons of CO_2 in 2002, which was 7% of the US total.[54] Several energy technology and other policies are in place that help reduce CO_2 and other GHG emissions.

---

[49]Energy Washington Week. Senate Energy Committee Re-Thinking Proceeding with DARPA-Like DOE Program. January 18, 2005. p. 2.

[50]For example, in the early 1980s, several states enacted a variety of efficiency standards for consumer products and appliances, which ultimately led the respective manufacturing industries to seek a more uniform federal standard.

[51]For example, Texas emits more GHG than France and California emits more GHG than Brazil.

[52]Pew Center on Global Climate Change. Learning from State Action on Climate Change: November 2005. Update. p. 9.

[53]Pew Center, Learning from State Action, p. 3.

[54]California Environmental Protection Agency (CalEPA). Climate Action Team (CAT) Report to the Governor and Legislature. (Draft) December 2005. p. 9.

15.4.2 California's Current Programs and Emission Reduction Targets

California has an energy R&D program that supports technology development to help curb CO_2 emissions. In 1996, California Assembly Bill AB1890 was enacted into law to restructure the electric power industry. In order to address energy R&D efforts not adequately provided by the new regime of competitive and regulated electricity markets, the law set a charge on electricity ratepayers to generate funding of about $62 million per year for a Public Interest Energy Research (PIER) Program administered by the California Energy Commission (CEC). In 1997, Senate Bill SB90 was enacted, requiring that PIER include the five priority areas of renewable energy technologies, environmentally preferred advanced generation, energy-related environmental enhancements, end-use energy efficiency, and strategic energy research. Two of the PIER's specific goals are to reduce the environmental impacts of electricity generation, distribution, and use, and to enhance California's economy.[55]

Further, California state agencies have undertaken several other actions to reduce emissions, which will remain in place or be expanded. Prominent among these actions was a regulatory policy adopted in 2004 by the California Air Resources Board (CARB) that will control emissions for passenger vehicles, starting in 2009.[56] Other key past policies that support emissions reduction include a renewable energy portfolio standard (RPS),[57] new building and appliance efficiency standards (including green buildings), waste reduction, and the Hydrogen Highway Initiative.

Also, California has several regulatory and deployment programs for energy efficiency and renewable energy technology and for other measures that will reduce CO_2 emissions. One example is electric utility energy efficiency programs which have an estimated budget of $2 billion for 2006 through 2008, with projected emission savings of nine (9) million metric tons by 2020. Another example is a $3 billion solar photovoltaic program, with projected savings of three (3) million metric tons by 2020. The California Environmental Protection Agency (CalEPA) estimates that the entire group of existing programs will cut annual emissions by 22 million tons in 2010 and by 67 million tons in 2020.[58] In 2005, California launched a new climate planning effort that would augment many existing programs, and add several

[55] California Energy Commission. PIER Program. [http://energy.ca.gov/pier/about.html].

[56] However, in December 2004, the Alliance of Automobile Manufacturers joined some California car dealers to file suit in federal district court against the regulation. They argue that the new law actually regulates fuel economy, not CO_2, and federal law grants sole authority to the Department of Transportation to regulate fuel economy.

[57] For retail electricity suppliers, an RPS sets a minimum requirement (often a percentage) for electricity production from renewable energy resources or for the purchase of tradable credits that represent an equivalent amount of production. In 2002, California Senate Bill 1078 was enacted, setting a 20% RPS target for 2017. In 2005, due to electric utility progress toward that target, the California Public Utility Commission (CPUC) and the California Energy Commission (CEC) accelerated the 20% target to 2010, and set a new goal of 33% by 2020.

[58] CalEPA, CAT Report, pp. 36–37.

new ones, in order to attain certain targets that would require additional emission reductions of up to 40 million tons in 2010 and 100 million tons in 2020.[59]

15.4.3 California's Draft Strategic Plan: Climate Action Team (CAT) Report

California has planned a major consolidation and expansion of its effort to combat climate change. In June 2005, the Governor issued an executive order[60] which calls for GHG to be reduced to the 2000 level by 2010,[61] to the 1990 level by 2020,[62] and to 80% below the 1990 level by 2050. The executive order directs CalEPA to lead a multi-agency Climate Action Team to implement emission reduction programs and report on progress toward the statewide targets. In December 2005, the Team released a draft report that made four major proposals for state action during 2006 and 2007. The proposals were intended to encourage investment in technologies that reduce emissions, create jobs, and stimulate economic growth.[63] Overall, the draft Report estimated that the proposed strategy to meet the targets would reduce emissions by 59 million tons in 2010 and by 145 million tons in 2020. By 2020, the climate policies were estimated to make a net contribution to economic growth of $4 billion (0.2%) and increase employment by 83,000 jobs (0.4%).[64]

The first proposal is to use the existing climate action registry to build a system of mandatory emissions reporting that enables CalEPA to assess progress toward the targets. This data would allow the targets to be translated into a statewide emissions cap for 2010 and 2020, and create a foundation for a cap and trade program. Under the second proposal, a public goods charge (PBC) would be set for the transportation sector.[65] The resulting funds would be used to curb emissions and reduce dependence on oil. Third, a coordinated investment strategy would be crafted to provide incentives to industry for developing emission reduction technologies. This

---

[59]CalEPA, CAT Report, pp. 41–42.

[60]Executive Order (EO) S-3-05. June 1, 2005.

[61]This would be a 59 million ton reduction, or 11% below the business-as-usual projection.

[62]This would be a 145 million ton reduction, or 25% below the business-as-usual projection.

[63]California Environmental Protection Agency (CalEPA). Climate Action Team Report to the Governor and Legislature. December 8, 2005. Executive Summary.

[64]CalEPA. CAT Report. Revised Draft Chap. 8 on Economic Assessment. Posted January 12, 2006. p. 10.

[65]The Climate Action Team reports that it considered and rejected several tax proposals, including a "carbon" tax. As part of their electric utility restructuring process, many states have established a public benefit fund (PBF) program. This type of fund has been generally used to support renewable energy resources, energy efficiency initiatives, and low-income electricity support programs. Money for the fund is often obtained through a "system benefits charge," or SBC. This is a small (for example, 0.2 cents/kWh) incremental cost that is charged to all electricity consumers. Existing state PBF funds are used in various ways that include rebates, public education, and R&D for renewable energy, energy efficiency, and other measures.

strategy may include use of the state pension system, the Public Interest Energy Research Fund (PIER), and other state investment programs. The fourth proposal is to establish an "early action credit" for California businesses that could support the transition to federal and international emission reduction schemes, including a cap and trade program. The credit would ensure that companies that had been proactive in advance of such schemes would not be penalized.[66]

The draft CAT Report finds that the transportation sector accounts for nearly 40% of California's emissions. As one option, it describes a public goods charge that, if implemented in parity with the existing public goods charge on electricity production, would be equivalent to 2.57 cents per gallon of gasoline or diesel at the wholesale level. Given annual consumption of about 18 billion gallons per year of these two fuels, this charge would generate about $426 million annually for the California Energy Commission,[67] which would be available to fund incentives for clean and efficient vehicle and fuel technologies.[68]

15.4.4 California's Options for a Cap and Trade System

The draft CAT Report outlines two key options for a state cap and trade system: a sector-based emissions cap and a fuels-based carbon cap. A sector-based emissions cap would cover up to 30% of California's emissions by focusing on five key industries: electric power (including imports), oil refining, oil and natural gas extraction, landfills, and cement production. A specific strategy for mobile (transportation) sources was not proposed. Alternatively, there could be a cap on the total carbon content of oil, natural gas, and coal consumed in the state. This would encompass all sectors using fossil fuels, and would cover 75% of the emissions. A hybrid approach, for example, could cap emissions from the electric power industry (including imports) and cap the content of transportation fuels. Further, emission offsets could be used to stimulate reductions from sources outside the cap and, through substitution, be used to lower the cost of complying with the cap. The Report notes that the main weakness of a cap and trade program in California would be its vulnerability to emission "leakage."[69] Without having neighboring states involved in the program, there would be an incentive for energy production to

---

[66]CalEPA, Climate Action Team Report, p. iii.

[67]CalEPA, CAT Report, p. 75.

[68]Energy Washington Week. New Fuel, Vehicle Programs May Enjoy California Funding Windfall. January 12, 2006. p. 2. This newsletter reported that one state official said that the level of charge was too high, and it would be more likely that a charge of about one cent per gallon could gain approval, which would generate about $180 million.

[69]For example, about half of California's electricity, accounting for nearly 10% of the state's total emissions, is imported from neighboring states. CalEPA. CAT Report. Chap. 6, Cap and Trade Options. p. 61.

shift to neighboring states to avoid the cap. In this case, emissions would decline in California, but rise in those neighboring states.[70]

Additional cap and trade policy design issues are discussed in the CAT Report. Setting a maximum value on carbon allowances is described as one possible "safety valve" to control the impact of a carbon cap on fuel prices. This may be needed to ensure that unacceptable price increases are avoided during transition periods. Also, if California is the only state to implement a fuel approach to cap and trade, a "black market" for fuels could develop, especially for liquid transportation fuels that might be brought in by tanker trucks from neighboring states. Also, forest management (tree planting) was identified as a potential source of emission offsets. The Report strongly urged that emissions coverage be limited in the initial phase to gases and sources that could be measured reliably. Additionally, enforcement was noted as a key to achieving targets, and the two main flexibility options for enforcement were identified as (1) acquiring allowances and offsets to make up a shortfall, or (2) paying a fee per ton. The Report concluded that new legislative authority was needed, and should be sought, for a cap and trade system. Further, it recommended that the budget requirements to administer a cap and trade system should be designed.[71]

15.4.5 Regional and Multi-State Climate Policies

15.4.5.1 New England (RGGI) Cap and Trade Initiative

The Regional Greenhouse Gas Initiative (RGGI) was established on December 20, 2005. Seven Northeastern and Mid-Atlantic states signed a memorandum of understanding (MOU) that sets a cap on each state's power plant CO_2 emissions from 2009 through 2014 and then calls for CO_2 to be cut 10% below 2014 levels by 2019.[72] Collectively, these states currently emit about 121 million tons of CO_2 per year.[73] The MOU outlines the program in detail, including the framework for a Model Rule to be developed by March 2006. RGGI will allow trading of emissions credits to help achieve the caps, establish credits for projects that capture and burn landfill methane, and encourage energy efficiency measures for natural gas, propane, and home heating oil. This is the first "CO_2 cap-and-trade" program in the United States. The MOU says the participating states hope to become "world leaders" in technology for carbon emission control, renewable energy, energy efficiency, and demand-side management.[74]

---

[70]CalEPA. Climate Report, pp. x–xi.

[71]CalEPA. CAT Report, pp. 65–72.

[72]The seven states are Connecticut, Delaware, Maine, New Hampshire, New Jersey, New York, and Vermont.

[73]EnergyBiz Insider. Carbon Constraints are Coming. January 18, 2006. p. 2.

[74]Regional Greenhouse Gas Initiative (RGGI). Northeastern States Sign Pact to Limit Greenhouse Gases. [http://www.rggi.org/agreement.htm].

The American Council for an Energy-Efficient Economy (ACEEE) has studied the potential for RGGI's operation, and expects that energy efficiency measures will play a major role. ACEEE's models showed that the states that invest the most in energy efficiency would have the lowest energy prices, allowance prices, and "leakage" rates. CO_2 "Leakage" is defined as taking place, for example, when a neighboring state exports coal-based electricity, thereby increasing that state's CO_2 and thereby offsetting the CO_2 reductions from states within RGGI. For sufficient investment in efficiency to take place, ACEEE suggests that the states will need to adopt energy efficiency targets, increase public benefit fund spending, or take other measures.[75]

15.4.5.2 Additional State Climate Plans and Policies

The Pew Center reported that, by the end of 2005, 28 states had a climate action plan, of which nine also had state-wide emission targets.[76] For achieving targeted reductions, the Center stresses the importance that the targets be enforceable and supported by strong incentives.

In 2004, the West Coast Governors' Global Warming Initiative (WCGGWI) was formed as a cooperative effort involving the states of Washington, Oregon, and California. Taken together, the three states would be the 7th largest emitter of CO_2 in the world. They agreed to increase energy efficiency measures in state building energy codes, adopt energy efficiency standards for several products not regulated by the federal government, create incentives to increase renewable energy sales, and agreed to acquire more efficient vehicles in their fleets.[77]

For transportation emissions, CalEPA estimates that California's proposed regulation of CO_2 from new cars and light trucks would cut 30 million tons of CO_2 equivalent by 2020. Further, if the regulation is upheld in court, another 10 states had announced their intent to follow with similar vehicle standards.[78] For electric power emissions, Massachusetts and New Hampshire require CO_2 reductions from existing plants, and Oregon and Washington require CO_2 offsets from new power plants. Also, West Virginia and Ohio have a pilot project to sequester CO_2 underground, along their mutual border. Further, Illinois, Nebraska, North Dakota, Oklahoma, and Wyoming have set up advisory committees to investigate agricultural (soil tillage) CO_2 sequestration options.[79]

---

[75] ACEEE Grapevine Online. The Magnificent Seven: States Take the Lead on Global Warming. January 17, 2006. p. 2.

[76] The nine states are California, Connecticut, Maine, Masschusetts, New Jersey, New Mexico, New York, Oregon, and Washington. Further, Pew reports that North Carolina and Arizona have begun a process to formulate a plan.

[77] WCGGWI. [http://www.ef.org/westcoastclimate/].

[78] The states are Connecticut, Maine, Massachusetts, New Jersey, New York, Oregon, Pennsylvania, Rhode Island, Vermont, and Washington.

[79] Pew Center, Learning from States, pp. 6–8.

15.4.5.3 Other State (and Canadian Province) Programs that Mitigate Emissions

About 23 states have established a public benefit fund dedicated to supporting energy efficiency and renewable energy. Twelve of these 23 states have formed a Clean Energy States Alliance to coordinate public benefit fund investments. Also, some 20 states have established a renewable energy portfolio standard (RPS) to increase electricity production from renewables. Similarly, at least 3 states have set up energy efficiency targets: New York, at 25% below 1990 primary energy use per unit of gross state product by 2010; Texas, at 10% reduction in electricity load relative to 2004; and Iowa, at 10% reduction in total energy end-use.[80] Also, electric utilities in 41 states have set up a net metering arrangement that allows customers to sell renewable-generated power back to the grid, and 36 states have utilities that offer green pricing, where customers can pay a small premium to purchase power from renewables.[81]

Three other regional partnerships with an energy focus have also addressed GHG emission reductions. In 2001, the Conference of New England Governors and Eastern Canadian Premiers (NEG/ECP) adopted a regional action plan for climate change, the first in North America. The plan set a target of bringing 2010 emissions down to the 1990 level, bringing 2020 emissions to 10% below the 1990 level, and bringing long-term emissions to 80% below the 2001 level.[82] One task in the plan was to explore a cap and trade system. Many of the US states in the NEG/ECP plan are now covered by the RGGI cap and trade system. Powering the Plains (PTP) is another regional partnership, that involves cooperation among Upper Midwest states and the Canadian province of Manitoba.[83] The partnership emphasizes renewable energy and agriculture issues, but has a focus on reducing the risk of climate change. It is studying carbon sequestration options.[84] In 2004, the Western Governors' Association (WGA) agreed to study the potential to develop 30,000 megawatts of clean energy by 2015 and to increase energy efficiency 20% by 2020.[85] The policy proposals were due in March 2006, with a decision to follow in June 2006.[86] Also, WGA has prepared a study of CO_2 sequestration options involving injection into depleted oil and

[80] Personal communication with Ms. Sylvia Bender, California Energy Commission. January 18, 2006.

[81] Pew Center, Learning from States.

[82] Conference of New England Governors and Eastern Canadian Premiers. NEG/ECP Climate Program. [http://www.neg-ecp-environment.org/page.asp?pg=46].

[83] The US states are Iowa, Minnesota, North Dakota, South Dakota, and Wisconsin.

[84] In 2003, a delegation of representatives from the PTP partnership visited countries of Northern Europe to gather ideas for climate-related initiatives.

[85] The WGA membership includes 18 states and three US-Flag Pacific Islands.

[86] Western Governors' Association. Clean and Diversified Energy Initiative.[http://www.westgov.org/].

natural gas reservoirs, unmineable coal seams, afforestation, and various farming techniques.[87]

15.5 Policy Design Issues

15.5.1 Energy Efficiency Technology in a Cap and Trade System

The American Council for an Energy-Efficient Economy (ACEEE) prepared a study of the potential for energy efficient technology and equipment in a cap and trade system for emissions. ACEEE concluded that policies should be designed to encourage energy efficiency "both within and outside the cap and trade framework."

ACEEE says the results of its study shows that energy efficiency is a cost-effective approach to meeting emissions targets. One reason is that efficiency achieves reductions through a single technology at end use. In contrast, achieving multi-pollutant emissions reductions at the smokestack may require multiple technologies, some of which may have limiting or counter-productive effects on reductions of other pollutants. As an example, ACEEE cites flue-gas scrubbers, which usually consume energy and thus increase energy use and emissions at the power plant. As another example, ACEEE notes the energy efficiency savings at end use also eliminate power plant combustion heat losses and electricity distribution losses.[88]

15.5.2 Energy Efficiency as a Resource Inside the Cap

One main finding was that energy efficiency requires explicit policy treatment to be an effective emission reduction method within a cap and trade framework. A key factor is that past improvements in energy efficiency have led to lower energy intensities in the end use sectors which, in turn, reduced the effect of energy price signals as a market force because energy is an increasingly small fraction of the cost of most economic activities. Further, there are market barriers that impede market forces that would otherwise drive energy efficiency. Moreover, the basic design of a cap and trade system limits its ability to encourage the most cost-effective levels of efficiency investment. Some key barriers are: power generators

[87]WGA. Carbon Management Working Group. Advanced Coal Task Force. Report. p. 43. [http://www.westgov.org/wga/initiatives/cdeac/Coal-carbon.pdf].

[88]ACEEE. Cleaner Air Through Energy Efficiency: Analysis and Recommendations for Multi-Pollutant Cap-and-Trade Policies. (By Bill Prindle et al.) April 2005. p. vi.

tend to oppose end use efficiency because it lowers sales and reduces revenue,[89] the need to aggregate small end use resources raises transaction costs, and "indirect" end use emission reductions could be double counted as "direct" reductions at the smokestack.

ACEEE recommends the direct allocation option as the most effective way to obtain energy efficiency benefits within a cap and trade system. This option would award allowances to providers of energy efficiency resources. This could occur within an output-based, updating allocation system at the outset of the cap program, in order to avoid double counting. The providers could sell allowances and use the proceeds to invest in more efficiency measures. This approach could overcome the major barriers of high transaction costs and limited allowance values.[90]

15.5.3 Energy Efficiency as an Offset Outside the Cap

Alternatively, ACEEE says that energy efficiency policies could be pursued in parallel with (outside) the cap, but not within it. As an example, ACEEE says that member nations implementing the EU Emissions Trading System (ETS) are generally taking this approach. In this approach, energy efficiency is used to set a reduced emissions baseline and a more aggressive target, while also reducing the overall cost of meeting the cap. After serving targets inside the cap, energy efficiency is used primarily as an offset mechanism, from sources outside the cap. In this way, energy efficiency investments in developing countries could be employed as emission offsets for the ETS.[91]

ACEEE concludes that energy efficiency can, and should, be pursued outside the cap. This approach focuses on using energy efficiency to reduce baseline emissions and to reduce cap levels. Overall, ACEEE recommends that policy makers pursue energy efficiency policies on their own merits and use energy efficiency within the cap and trade system. Further, ACEEE recommends that the initial allocations of allowances be made to public entities or their designees, as they have demonstrated to be the most reliable as cost-effective providers of energy efficiency in the past. Also, public entities are more likely than private entities to channel proceeds from allowance sales into additional delivery of energy efficiency resources.[92]

---

[89] This issue of "lost revenue" was a key barrier to US electricity demand-side management programs in the 1990s. This issue involved much study of the potential to "decouple" electric utility profits from sales volume, as a means for ensuring that profits would flow to cost-effective energy savings resulting from energy efficiency measures.

[90] ACEEE. Cleaner Air, pp. vii–xi.

[91] ACEEE. Cleaner Air, p. viii.

[92] ACEEE. Cleaner Air, p. x.

15.5.4 Strategies for Multilateral Action

In November 2005, the Pew Center issued a report on workshops held to discuss ways of expanding the international effort to combat climate change and reduce GHG emissions.[93] The report identifies six "elements" that can be combined and/or elaborated in various fashions to produce options for strengthening multilateral action. The first element is the long-term climate objective. Pew notes that the objective has taken different forms (temperature increase limit, CO_2 concentration, emissions reduction by fixed date) with various entities that seek to make a contribution. The existence of such an objective is seen as providing a motivational signal, and the variations could ultimately be unified into a common view to guide international efforts. The second element is adaptation. Here, the report says the top priority is to meet the needs of the most vulnerable and, thus, adaptation should be worked into the development process, the funding commitments by donor countries, and multilateral development banks should incorporate climate risk assessments and adaptation measures into lending guidelines.[94]

The third element is targets and trading, which are the centerpiece of the Kyoto Protocol. Cost-effectiveness is portrayed as the principal value of the third element. For the future, the report suggests that other possible types of acceptable targets could be "intensity" targets, that limit emissions relative to an indicator such as GDP; "no-lose" targets, that allow a county to market emission reductions below its target; and conditional targets, that remain binding unless a criterion, such as cost, is exceeded. Also, parallel to the existing Clean Development Mechanism (CDM) crediting for discrete projects that reduce emissions, a new "programmatic" credit mechanism could be established for reductions across a given sector driven by policies such as energy efficiency standards.[95]

The fourth element, sectoral approaches, would structure international efforts around economic sectors. In this element, governments or companies would agree to reduce emissions in a sector such as transportation, power, land use, steel, or cement. For example, in the automobile sector, an agreement among 15 countries (and fewer companies) would cover most of the global market. Further, sectors with significant international trade could address competitiveness concerns by ensuring some comparability of effort across a given sector. Sectoral efforts could take the form of emission targets, performance/technology-based standards, or best practice agreements. Some examples are vehicle fuel economy or GHG standards, intensity targets for energy-intensive manufacturing, and standards that require phase out of old technologies or the phase-in of new equipment.[96]

---

[93] Pew Center on Global Climate Change. International Climate Efforts Beyond 2012: Report of the Climate Dialogue at Pocantico. November 25, 2005. p. 25.

[94] Pew Center. International Climate Efforts, pp. 13–14.

[95] Pew Center. International Climate Efforts, pp. 14–15.

[96] Pew Center. International Climate Efforts, p. 15.

The fifth element of policy-based approaches would take policies formed for other purposes, such as energy service expansion or strengthening of energy security, and expand them to include measures to improve energy efficiency or increase the use of renewable energy.[97]

While most of the preceding approaches employ a market-pull approach, the sixth element is focused on technology cooperation as a way to push technology, especially long-term technology. New technologies and "breakthrough" technologies are needed to help attain the steeper reductions needed over the long-term. Further, groups of countries could form agreements to increase, and better coordinate, support for options such as hydrogen, fuel cells, next generation nuclear reactors, large-scale solar generation, biomass fuels, and carbon sequestration. Governments could form partnerships, share best practices, and support capacity building. One example of a way to frame an R&D effort could be for governments to agree on a long-term goal such as zero net emissions in a key sector, such as for automobiles in 2080.[98]

15.6 Conclusion: Some Possible Areas for Transatlantic Cooperation

There are several areas for possible international cooperation. Some may involve an expansion of an existing area of cooperation, while others could involve an entirely new area. The suggested areas are categorized under the three topics shown below.

15.6.1 Technology Push and Competitiveness

Several types of cooperative studies could be undertaken. The GAO, as noted above, has found significant uncertainties in the reporting of climate-related R&D spending in the United States. Similarly, one observer finds that the EU Commission is unable "... to tell accurately how much R&D spending (both at EU and member states level) is earmarked for climate-friendly technologies.[99] One reason for this uncertainty is likely that many of these programs were initially, or primarily, created to serve other goals, such as energy security, other environmental objectives (such as reducing air pollution), or economic competitiveness. Thus, one possible study could involve an inventory of spending on R&D and deployment related to reduction of emissions. A further area of cooperation could involve assessing the impacts of

---

[97] Pew Center. International Climate Efforts, p. 16.

[98] Pew Center. International Climate Efforts, pp. 16–17.

[99] EurActiv. Transatlantic Divide Remains Ahead of UN Climate Summit. November 25, 2005. p. 2.

these various technology development and deployment programs in terms of cost-effectiveness and emission reduction estimates.

Some installations of low-emission energy supply facilities may be more physically vulnerable to climate disruption than other facilities. For example, offshore wind and natural gas facilities may be particularly vulnerable to an increasing threat from hurricanes. This topic might be a fruitful area for international cooperative study on ways to strengthen the climate "resilience" of such installations. Also, perhaps energy security studies of facilities for liquified natural gas (LNG) tankers could serve the additional goal of ensuring low-emission energy supply.

Following the Pew Center's concept of sectoral approaches, there may be several areas that could be the basis for fruitful international study and cooperation. One area might focus on international fuel economy standards for passenger cars and light trucks. With involvement from all companies and nations that have a stake in the economic benefits of this industry, a level playing field might be created that overcomes companies concerns about competitiveness and national standards. Similarly, cooperation could be undertaken on international energy efficiency standards for a variety of consumer products and commercial equipment. Several energy-intensive industries may benefit from cooperation on standards for industrial process energy efficiency technology. This could include, for example, steel, chemical, and cement industries. Perhaps cooperation could take place for certain types of electric power plant equipment as well.

Capacity building has been shown to be an effective way to transfer technology to developing nations. Maybe initiatives like the G8 Gleneagles Plan of Action could be expanded to embrace greater international cooperation in a way that accelerates the push of climate-friendly technologies into broader use.

International cooperation might also be employed to address emerging problems with large scale deployment of renewable energy technology. For example, the emergence of giant windfarms, both onshore and offshore, has created a number of environmental and other concerns that might be fruitfully addressed by cooperation among the nations and companies that are involved in the production of wind energy equipment.

15.6.2 Market Pull Through Technology Deployment and Trading Systems

International cooperation in technology deployment could possibly be employed to expand markets for low/no carbon equipment. For example, California has launched a major program to stimulate the installation of solar photovoltaic (PV) equipment. If the other nations with PV manufacturing were to create similar programs, it could multiply economies of scale, thereby increasing market potential and providing greater incentive to manufacturers. Further, the nations and companies involved could cooperate in setting standards for energy performance or other aspects.

Emission trading systems could be a focus of international cooperation. The European ETS is in operation, but it may face significant challenges. One report says that the ETS has stimulated electric power producers to raise electricity prices which, in turn, has made large electricity users unhappy.[100] Further, the 8-year planned duration of ETS, combined with uncertainty about what would follow Phase II may not be sufficient to pay back utility investments in CO_2-reducing capital equipment, which typically have a 30-year lifetime.[101] In the United States, the RGGI is poised to start up, and California is taking steps toward an ETS. Studies related to RGGI and the strategy for California have raised concerns about "leakage" and the need for a "safety valve." California is also concerned about the potential black market for petroleum fuels. These challenges and concerns provide a rationale to consider international cooperation in studies of emission trading systems.

15.6.3 Technology Push Interaction with Market Pull

Given that both the EU and the United States have technology-push and market-pull types of policies, there may be an opportunity for cooperation to study the most productive way to coordinate these policies. In particular, ACEEE has raised important questions about how to integrate deployment of energy efficiency technologies where a cap and trade ETS is operating.

Appendix: US Federal Climate Change Funding for CO_2 Mitigation

[100]Enviros Consulting. Briefing Report on EU ETS. Cited in Energy Washington Week. EU Climate Change Emissions Trading Scheme Future Still Shaky. November 15, 2005.p. 2.

[101]The report says allowance prices stood at 20–25 euros per ton in November 2005, but that prices would have to rise to nearly 200 euros per ton for a new combined cycle gas turbine to be repaid over a period of 8 years.

Table 15.2 US federal climate change funding for CO_2 mitigation

Fiscal year	In current dollars						In constant 2004 dollars					
	Technology (CCTP)	Intern'l assist	Tax incentives	Total	Deflator 2000	Deflator 2004	Technology (CCTP)	Intern'l assist	Tax incentives	Total	Annual change	% Annual Change (%)
93	845.0	201.0	–	0.0	0.8838	0.8164	1,035.0	246.2	0.0	1,281.2		
94	1,038.0	186.0	–	0.0	0.9028	0.8340	1,244.6	223.0	0.0	1,467.6	186.5	15
95	1,283.0	228.0	38.1	1,549.1	0.9395	0.8515	1,506.7	267.7	44.7	1,819.1	351.5	24
96	1,106.0	192.0	159.7	1,457.7	0.9559	0.8679	1,274.3	221.2	184.0	1,679.6	−139.6	−8
97	1,056.0	164.0	40.5	1,260.5	0.9675	0.8830	1,195.9	185.7	45.8	1,427.4	−252.1	−15
98	1,251.0	186.0	53.7	1,490.7	0.9802	0.8938	1,399.7	208.1	60.1	1,667.9	240.5	17
99	1,694.0	325.0	48.0	2,067.0	1.000	0.9055	1,870.8	358.9	53.0	2,282.8	614.9	37
00	1,793.0	177.0	227.0	2,197.0	1.0236	0.9238	1,940.9	191.6	245.7	2,378.2	95.5	4
01	1,675.0	218.0	142.9	2,035.9	1.0426	0.9456	1,771.4	230.5	151.2	2,153.1	−225.1	−9
02	1,637.0	224.0	215.1	2,076.1	1.0614	0.9631	1,699.6	232.6	223.4	2,155.6	2.5	0
03	2,555.0	270.0	205.5	3,030.5	1.0825	0.9805	2,605.8	275.4	209.6	3,090.8	935.5	43
04	2,868.0	252.0	194.0	3,314.0	1.1045	1.0000	2,868.0	252.0	194.0	3,314.0	223.2	7
05	–	–	83.0	3,244.5	1.1045	1.0203	–	–	81.3	3,261.2	−52.8	−2
*6	–	–	524.0	3,535.5	1.1261	1.0403	–	–	503.7	3,902.3	641.1	20

*For FY2006, the figures shown are Administration-requested amounts.

[1]The year 2000 Deflator (GDP Chained Price Index) was obtained from the Administration's Budget for FY2006, Historical Tables, Tables 10.1, Gross Domestic Product and Deflators Used in the Historical Tables: 1940–2010. p. 184.

[2]The tax incentive figure for 2005 fell, due to an 8-month termination of the renewable energy production tax credit.

Source (Technology and International Assistance): GAO Analysis of OMB Reports, August 2005, Table 6, p. 34.
Source (Tax Incentives, 1995–2003): IRS, Office of Tax Analysis. Personal communication with Mr. Curtis Carlson. January 24, 2006.
Source (Tax Incentives, 2004–2006): OMB, Federal Climate Change Expenditures, March 2005, Table 6, p. 17.
Source (Total, 2005–2006): US Climate Change Technology Program, Draft Strategic Plan, Table A-1, pp. A-4–A-5.

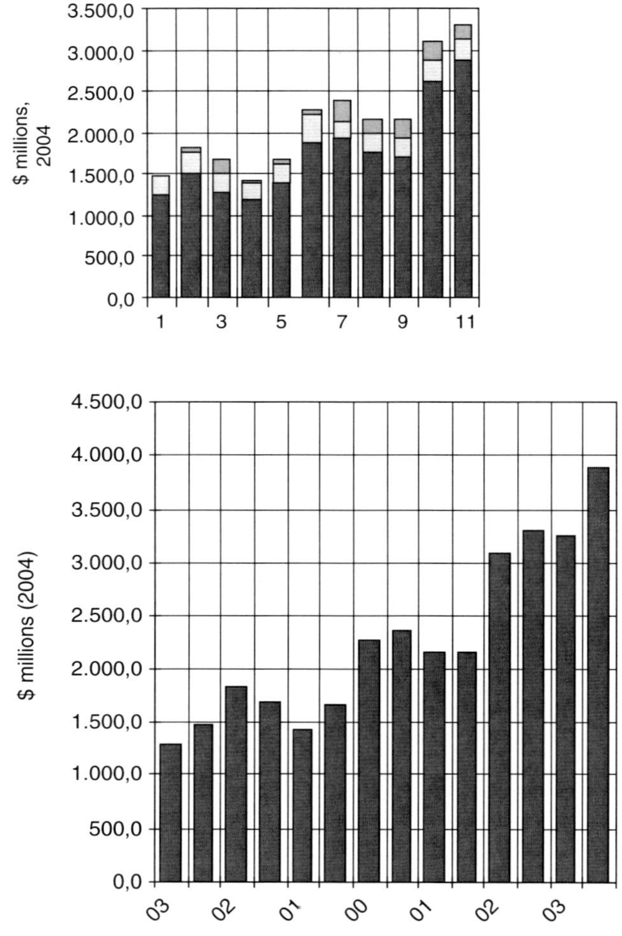

Fig. 15.1 US climate change funding

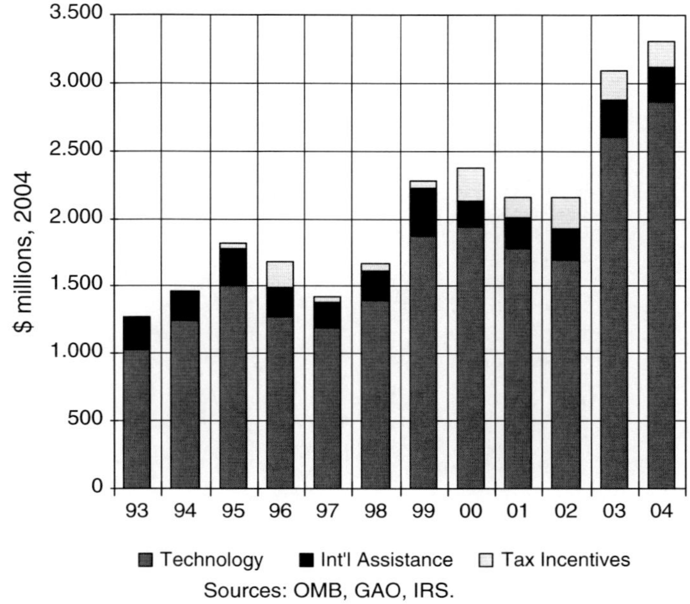

Fig. 15.2 US climate change funding

Reference

Egenhofer, C. (2005, November). *Technology in a post-2012 perspective* (CEPS Policy Brief No. 86, p. 8). http://shop.ceps.be/BookDetail.php?item_id=1281

Chapter 16
EU-US International Relations: A Political Science Perspective

Martin List

16.1 Research Programs in Political Science Analysis of International Relations (IR)

Political science analysis of transatlantic relations – which in this contribution will be used as shorthand for the relations between the United States of America, on the one hand, and EU-Europe, on the other-sees these relations as a wide-ranging complex set of nested games, including domestic, transatlantic and global games. It is conscious of the complex nature of the political governance problems involved. Any political coordination of these relations requires optimization of multi-criteria goal catalogues (with internal trade-offs) among multiple independent actors rather than maximization of a short list of goals by one unitary actor. In other words, conflicts in these relations are to be expected, and traditional experience of transatlantic relations has born out this expectation. However, the level and scope of transatlantic differences[1] seems to have risen since the end of the Cold War and in particular in the context of the war against Iraq under Saddam Hussein undertaken by the administration of President G.W. Bush. Political science analysis of international relations analytically follows differing perspectives, also called paradigms or research programs. They will be introduced shortly in a first section and then used mainly as a presentational device in the context of this contribution, which is to give an overview of the political science perspective rather than a rigorous test of hypotheses. Also, comprehensiveness is out of the question, and exemplary points are made to give a broad overview.

Political science research on international relations is proceeding within what is now conventionally grouped as four research programs, paradigms or basic approaches (or combinations thereof). Each is characterized by certain basic

[1] Such is the title given to a collection of contributions of one of the many conferences held on the subject: Zacharasiewicz, 2004.

M. List (✉)
FernUniversität at Hagen, Germany

P.J.J. Welfens, J.T. Addison (eds.), *Innovation, Employment and Growth Policy Issues in the EU and the US*, DOI 10.1007/978-3-642-00631-9_16,

assumptions about the object of research, i.e. international relations, and tries to get maximum explanatory mileage out of its respective explanatory strategy.[2]

Realism regards IR as a game of unitary (state) actors pursuing self-interest (power maximization) under conditions of anarchy, i.e. in the absence of superior authority.

Institutionalism highlights the fact that under conditions of interdependence state and non-state actors pursue their interests within the framework of shared norms, institutions and organisations whose structure co-determines political outcomes.

Constructivism emphasizes the fact that IR is founded in collectively shared perceptions of situations and understandings of self and foe (self- and enemy images).

Finally, *critical approaches* share realism's emphasis on power which, however, is interpreted as resulting from transnational structures of inequality among societal groups (classes).

In what follows, we shall take a look at transatlantic relations through these four theoretical lenses.

16.2 A Realist Perspective

The power-oriented realist view understands well the – relatively – harmonious nature of transatlantic relations during the East-West conflict. Given the Soviet threat which was seen as serious and immediate, at least in the European theatre, alliance building (formation of NATO) seems a natural reaction of balancing Soviet power (or, seen from Western Europe, as bandwagoning with the US against the Soviet threat). This may not be (I would say: certainly is not) the whole story to be told about East-West-conflict era transatlantic relations, but realists do have a point in emphasizing the cohesiveness building effect of a commonly percieved threat. However, from a realist point of view, it is somewhat remarkable and bears emphasizing that the US side in this constellation, while clearly pursuing its self-interest (keeping Soviet Russia at bay and leadership in the alliance), exerted its power (which after World War II was overwhelming power relative to its European partners) in a rather moderate, if you will enlightened, way. Not only did the US transcend its traditional aversion against "entangling alliances",[3] it also opted for a rather multilateral institutional design of the alliance. NATO thus became not only a case of leadership by the primus inter pares the US certainly was; it also was an effective forum for debates about diverging views especially in the field of security policy.[4]

---

[2]For a didactical introduction into these research programs see List, 2006, Chaps. 2 and 13. A broader choice of IR approaches is excellently presented in Sterling-Folkert, 2006.

[3]Though not completely: The famously loose assistance obligation contained in article 5 of NATO's founding treaty was the price to be paid for acceptance of it by the US Congress.

[4]As shown in Risse-Kappen, 1995.

There was, however, from a traditional realist point of view, a clear expectation for the post-Cold War era of a declining role of the alliance or declining alliance cohesion due to the removal of threat which leads to increased transatlantic conflicts. Add to this the economic (re)emergence not only of European countries individually (Germany in particular), but their – US-sponsored, one might say[5] – economic integration into what by now is a 450 million consumer market acting rather cohesively in international economic policy fora such as GATT/WTO and again increased economic competition; and hence transatlantic conflicts in international economic policy making is what realists would expect.

A final point to be made here from a realist perspective concerns what might be called "superpower psychology". The current global power structure is, in realist terminology, unipolar (there currently is only one superpower). This alone leads realists to expect more independent (unilateral[6]) use of power by the US who, under conditions of anarchy, must constantly be on the watch for potential power challengers and threats. Among the challengers, it is probably true that it is the People's Republic of China rather than EU Europe that comes to the realist US mind. EU Europe is seen mostly as an economic competitor (hence trade conflicts) but not as a security challenge – and be it only because it lacks the status of unified actor in this field. There is, however, the lingering problem of just how much autonomous capacity for action in the field of foreign policy in general and security policy in particular the US would like the EU to have. The US has always demanded a strengthening of the European defence contribution *within* NATO; it has reacted rather squeamishly to even minor steps toward an institutionalization of a European defence and security policy.[7] When it comes to new threats, the divergence between the US, who after the terrorist attacks on New York and Washington in September 2001 see themselves under immediate attack and in president Bush's words are involved in a "war on terror", and the somewhat less dramatic immediate reaction on the European side have brought this US tendency towards unilateral action into full relief: unilateral war against Iraq (supported by the ad hoc coalition of the willing rather than NATO which, however, declared its defence condition for the first time in its history after 9/11) and rejection of the European-sponsored project of an International Criminal Court (for fear of political abuse against the US playing their new role as intervener in the periphery in the war on terror as well as in the context of humanitarian interventions). Add to this several highly publicized unilateral US steps in, from a realist point of view, low politics areas such as non-ratification of the Kyoto climate-protection protocol and the realist expectation of a differing psychology of the lone

[5]On the US role in fostering European integration see Neuss, 2000 and Lundestad, 1998; also see Katzenstein, 2005 for a comparative perspective on regionalism under conditions of US hegemony.

[6]On the sources, forms and fields of US unilateralism see Malone and Khong, 2003.

[7]See Heise and Schmidt, 2005 for recent developments and Hamilton, 2004 on future NATO perspectives.

superpower is well borne out. Less powerful states neither would be inclined to such unilateralism – nor would they get away with it – so (seemingly[8]) easily.

16.3 An Institutionalist Perspective

From the point of view of the institutionalist research program which also is the inheritor of the theoretical concept of interdependence developed originally in the 1970s,[9] the real basis of transatlantic relations is precisely that: tight factual interdependence, especially in the economic field. In fact, transatlantic trade has more than doubled (in value) between 1990 and 2000.[10] Even more important is the foreign direct investment link between the US and the EU (adding up to 1.5 trillion USD in 2002 and resulting in foreign affiliate sales, i.e. sales by US and/or EU companies immediately on the other side of the Atlantic and hence not showing up as transatlantic trade five times the value of that trade in 2001[11]). This kind of deep economic interdependence also means that a lot of (also politically) powerful economic actors on both sides of the Atlantic have literally invested a lot in smooth transatlantic relations and would have a lot to lose from any politically induced disturbance in their relations. At the aggregate level, then, anarchy alone is no longer an adequate characterization of the transatlantic situation. While this does not exclude transatlantic conflicts in economic policy (and certainly not transatlantic economic competition, as it should not), institutionalists emphasize that existing institutions, especially GATT/WTO, are precisely the fora in which these kinds of conflicts can be managed in a regulated way, i.e. following established procedures and hence avoiding classical trade wars.

A similar view is taken of the field of international security relations. From an institutionalist point of view, the alliance (NATO) continues to be a valuable asset. Its comprehensive institutional setting, built over a fifty-year – not to forget by no means conflict-free – history constitutes international social capital[12] which can be put to good future use, and in fact already has been so. It is in this light that the not always smooth process of NATO's reinventing itself must be seen, in a changed landscape of security policy which includes new threats (civil wars in the context of state failure such as former Yugoslavia, but also Afghanistan, and transnational

[8]The negative reaction provoked by this unilateralism has not gone unnoticed, even (and especially) among classical realist analysts who reckon with counter-balancing to US power anyway; see Walt, 2005.

[9]See the classic work by Keohane and Nye, 1977.

[10]In figures: from 273 to 557 billion USD according to Gartzke, 2005 who concludes his overview of transatlantic economic relations by stating that the deep economic interdependence "leads to the conclusion that the process of geo-strategic estrangement will not have significant negative impacts on transatlantic economic relations" (184, my translation).

[11]Ibid., 175.

[12]As was argued early on after the end of the Cold war, and convincingly, by Chernoff, 1995.

terrorism), new members (through NATO's eastward expansion which, among others, served the vital function of socializing East European militaries – like the German military before – into Western attitudes of democratic civilian control of the military[13]), and new "partners" in security like Russia concerning which institutional tying in was, a somewhat realist tack, balanced by NATO's eastward expansion into the former Soviet sphere of influence.

A final point made by institutionalists again concerns economic transatlantic relations, especially the changed societal context within which existing institutions have to work. International economic relations after 1945 can be characterized as embedded liberalism,[14] i.e. transnational economic integration used to be accompanied by domestic buffering of negative impacts of this liberalization by social welfare measures. Up to the 1980s, arguably, a sort of de facto Keynesian consensus had emerged in the transatlantic, interdependent societies concerning social and economic policy. With the challenge of this socio-economic policy consensus since the 1980s[15] and the "reformatting of domestic state-economy-relations" through deregulation, privatization and – as we have come to term it – resulting economic globalization, the fact that capitalism has always been played in transatlantically distinct varieties[16] has become more prominent. Arguably, a dis-embedding of liberalism (denounced by critics as "neo"-liberalism) has taken place. What is more, not only is the leading role of the US in these changes in economic policy seen, and probably rightfully so. The whole process is also regarded, by critics of differing political kinds, as a process of "Americanization", carrying negative connotations and often leading to an overlap of the critique of (concrete forms of) economic globalization with anti-American sentiments. The extent of Americanization of European business, both factual and the perception thereof, is a fascinating subject in itself.[17] From the late 19th century on the US was perceived in Europe as a model of modernity, also – though not only – in economic terms,[18] not always, however, as universally a positive one; Americanization (or Westernization) of European societies after 1945 always had its critics.[19] Even today, there is both empirical evidence for regionalized business practice – hence transatlantic differences – among MNEs,[20] and there are pleas for a European business model.[21] However correct it may be to see globalization as Americanization: the association of both phenomena lets WTO-provided

[13] See Forster, 2006, 152ff and Gheciu, 2005.

[14] The term was introduced by Ruggie, 1983.

[15] This global process is vividly described in Yergin and Stanislaw, 1998.

[16] A good overview of the varieties-of-capitalism literature and a central contribution to it is the volume by Hall and Soskice, 2001.

[17] See Schröter, 2005 for a general overview; Costigliola, 1984 for Europe in the interwar period and Nolan, 1994 for Germany in particular.

[18] As is vividly demonstrated in Schildt, 1997.

[19] See Pells, 1997; for France: Kuisel, 1993; on Germany Doering-Manteuffel, 1999.

[20] Rugman, 2005.

[21] Kalff, 2006.

mechanisms for dispute settlement appear as part of the problem rather than as a solution. This is also a point made, if for different reasons, from a realist point of view, which points to the ambivalent effect of these institutions for (economic) conflict management. Under the domestic pressure of neo-mercantilist economic interest groups,[22] governments are inclined to make tactical use of these mechanisms, which not so much settle the differences but rather broaden the keyboard on which they can be played out.

16.4 A Constructivist Perspective

In talking about changes not only in the economic "philosophy" and the way capitalism (in its differing varieties) has come to be played since the 1980s but also how these processes have been *perceived*, we have already entered the field theoretically best covered by constructivist approaches. Their key point, to remember, is that perceptions matter in international relations, including self-perceptions and enemy images. And it is, arguably, indeed in this field of "soft factors" that transatlantic relations in recent times have suffered most.

Probably the largest transatlantic perceptual differences have come out in the time after 9/11 and in the context of the Iraq war.[23] As noted, "9/11" was widely perceived in the US not only as an act of transnational terrorism, which it certainly was. The Bush administration came to construe (construct) this event as the opening of a new kind of *war* which the US, in this view, has been waging ever since: the war on terror. It deserves emphasizing that this construal was not the only one possible, is contested in the US as well[24] and has contributed to transatlantic differences. The problem of fighting a war on terrorism rather than against concrete terrorists or state sponsors (such as Taliban Afghanistan, a case that was accepted in Europe) is the open-ended nature of that enterprise: when would a war on terrorism (as a method of furthering political goals) be ended? Isn't it just as open-ended as the – metaphorical – war against stupidity? And, what is more, through concrete military action like the wars against Taliban Afghanistan and Saddam's Iraq, the war on terrorism has lost all metaphorical character; it has become the reason – critics would say, the pretext – for real wars. And this is where another transatlantic difference has become relevant. While it is perhaps mere polemics to paint Europeans (all of them) as being "from Venus" while the US is "from Mars" (as done in the strikingly effective polemic of Kagan [2003]), there is a marked difference in war experience between Europe (victim of two self-destructive world wars) and the US (victorious in both wars plus the Cold War).[25] This difference is even greater between certain

[22]These domestic political games of international trade policy are well researched for the US case, see List, 2004 for an overview.

[23]The controversy has been reconstructed in Gordon and Shapiro, 2004.

[24]See Record, 2003 for a critical assessment of the concept by a knowledgeable expert.

[25]On the US war experience cf. Snow and Drew, 1994; Meernik, 2004.

parts within the US (there is a marked Texan attitude towards the use of violence, even international violence[26]) and *parts within* Europe (e.g., Germany who – to the relief of its neighbours – has drawn wide-ranging anti-war, though not completely pacifist, consequences out of its recent history). It was not helpful for the quality of transatlantic relations to find these real, and explicable, differences in attitudes stylized into images of a "trigger-happy Texan cowboy" on the one side and "peace-loving but duped Old Europeans" (a term coined by US Secretary of Defence Rumsfeld) on the other side. Leadership rhetoric, partly for domestic purposes (such as, in the case of Germany, an election campaign to be won by a chancellor who not only distanced himself from the war on Iraq, but from American "adventures") does make a difference here – and in this case to the worse.

This stylizing of images is a second important point to be made from a constructivist perspective. Out of differences in concrete questions (such as: Are there good reasons legitimating the use of military violence in any given case?) grow self-images characterized by a heavy dose of self-*righteousness*. A whole group of US authors and political commentators suddenly became advocates of a new US role: it was no longer only "bound to lead" (as had long ago been argued, from a liberal-institutionalist point of view by Joseph Nye [1990]) but had to shoulder the burden of new imperial responsibility. US imperialism, long time a bogeyman of US critics, suddenly seemed to turn into a US self-image – something that fits rather uneasily with the republican and anti-imperialist tradition of the political culture of the USA, after all itself the "first independent nation".[27] This was (and continues to some extent to be) matched by EU-European self-stylizing as a "post-imperialist", peace-building and countervailing force in the new global order. At its maximum of self-stylizing, this went along with an attitude of "moral superiority" for having drawn the "right lessons of history" (and hence, once more, downplaying the fact that there may well be good reasons for diverging views as to what the right lessons might be). Changes, however, are going on here, too. The new European security strategy[28] does recognize the threat of transnational terrorism, and European action on Iran and its suspected search for nuclear weapons is markedly more firm in rhetoric than was the case before the war on Iraq. However, conflict prevention is stressed in all these contexts rather than the propagation of a preventive war (as aired in President Bush's national security strategy of 2002[29]).

Stylized self-images easily lead to mutual perceptions as not only differing in view and attitude, but differing in the moral quality of these attitudes. To this one,

---

[26]See Lind, 2003 (the author is a Texan himself).

[27]An overview of this new US-internal discussion on a – potential – imperial role can be found in the contributions in Bacevich, 2003; a major non-US contribution to the debate arguing for a dominant US role, but also that the US would be unfit to really fulfil it, is from British historian Ferguson, 2004. In the light of the Iraq experience US willingness for imperial action – especially if that means staying on the ground and getting involved in nation building - indeed seems to be more limited than ever.

[28]http://ue.eu.int/uedocs/cmsUpload/78367.pdf (13.03.06).

[29]http://www.whitehouse.gov/nsc/nss.pdf (13.03.06).

we should add as a last point from a constructivist perspective: the *blowback of stylized images*.[30] Escalating rhetoric by political leaders ("Old Europe", "no adventures") impacts public perception and attitudes. Among the more educated parts of the publics it may lead to the (re)discovery of attitudinal gaps in many other fields, e.g. concerning death penalty in particular[31] or the penal system in general.[32] This is supported by severe transatlantic policy differences[33] between the current US administration and its European counterparts when it comes to questions like the treatment of inmates in prisons in Iraq, in Guantanamo Bay – or allegedly, in secret CIA prisons on European soil. The fact that even a leader as supportive of transatlantic relations as German Chancellor Merkel came only to agree with President Bush to disagree on the rightfulness of the Guantanamo policy is sign enough for persistent differences, but also for the attempt to keep resulting rhetoric in check. Not only does this correspond to the alleged quality of transatlantic relations which indeed should not only withstand, but encourage "criticism among friends"; it also shows the condition for that to work: that the way this criticism is articulated makes it credible that criticism among friends is actually what is going on – instead of unilateral self-stylizing for domestic or transatlantically intimidating purposes. Otherwise, there is a not to be ignored danger that escalating leadership rhetoric seriously contributes to estrangement on a societal level by seemingly legitimizing in Europe the use of traditional anti-American stereotypes (there seems to be a larger stock of that than of anti-European stereotypes easily invoked in the US public, an asymmetry itself worth pondering). Elements of this turning of self-images based on political differences into stylized and normatively charged attitudes towards the other with a tendency to degenerate into negative stereotyping are not limited to the field of security policy alone. It can also be seen in the field of socio-economic policy[34] where the difference between analytically debateable varieties of capitalism is turned into the stylized image pitting the "European social model" against socially cold US (or Anglo-Saxon) capitalism which becomes stereotypically the protest against the sell-out to globalization seen as Americanization. Interestingly, there is a reverse stylizing in the US which sees the Anglo-Saxon model as more competitive and innovative (which might be argued[35]), turning into self-congratulatory images as

---

[30] The pernicious impact of leader rhetoric on public views – and the feebback of that on leaders room for political manoeuvre has been analyzed by Snyder, 1991.

[31] Martschukat, 2002.

[32] Whitman, 2003.

[33] Not all transatlantic policy differences, however, are due to underlying differences in values, even in value-related policy areas, as is argued for the field of data protection policy by Heisenberg, 2005.

[34] On these transatlantic differences in social and economic policies see, respectively, Streissler, 2006; Blum and Ludwig, 2006.

[35] Aiginger, 2004 shows the US to be leading in 16 future growth indicators in comparison to the EU; he also points out, however, that comparative evaluation of the success of national economic models is bound to involve normative questions.

that of “the United States, the most sophisticated and successful nation on Earth”.[36] The latter, of course, is a typical example of what is called US exceptionalism – in the normative rather than analytical version of that phenomenon.[37]

16.5 A Critical Perspective

The transatlantic escalation of self-congratulation is an apt point at which to turn to our forth theoretical perspective – that of critical approaches. Indeed, the attribute “critical” refers mainly to the attitude towards the socio-economic model that, from this perspective, is *shared* by both the US and its European partners. They are, rightfully, seen as sitting in one boat when it comes to the basic economic – capitalist – and political – liberal-democratic – structure of their respective systems. While the traditional sense of superiority, based on the supposed knowledge of the laws of history and its necessary end[38] which inspired dogmatic representatives of Marxist criticism has waned with the downfall of communism in Eastern Europe, non-dogmatic critical views are still a healthy corrective mechanism also for putting transatlantic relations into perspective.

From this point of view, the transatlantic community traditionally (after, 1945) constitutes the core of the capitalist world system, made up of the dominant states of the North and tied together by a transatlantic ruling class.[39] One does not have to share the – in both the literal and figurative sense – re-ligious (literally, backward linking) connection of class analysis to the philosophy of history of classical Marxism to see the double point made here. From an external perspective, the societies woven together by transatlantic ties are certainly the major beneficiaries of the existing international or rather transnational economic system which indeed has been propagated after 1945 by the US in its role as hegemon, benign hegemon from a (West) European point of view, to be sure, but rather less so from many a “Third World” point of view (though, with the rise of Newly Industrializing Countries, not only has the Third world become rather heterogeneous; so have the southern views on global capitalism: more often than not it is now demanded by southern elites

---

[36]The – not untypical – quote is deliberately taken from a non-suspicious source: the US-authored entry (by David Shenk) on “information overload” in an internationally produced and marketed major reference work (Johnston, D. H. (Ed.). (2003). *Encyclopedia of International Media and Communication* (Vol. II, p. 395). Amsterdam.). Contrast this with the recent headline in the British *Guardian* (as quoted in Newsweek 09.01.2006, p. 20) which referred to *Sweden* as “the most successful society the world has ever known”.

[37]Analytically, macro-social comparison makes it perfectly sensible to speak of American exceptionalism. See Lipset, 1996. The normative version sees the – indeed special – historical experience of the US as a model for the rest of the world, a view widely shared in the US (see Junker, 2003) – and, if not at all times and in all places, often outside the US as well.

[38]In socialism or communism; there is an interesting parallel here between overly self-congratulatory views on the Western side after the end of the Cold War as expressed in Fukuyama’s (1992) declaration of the end of history (in liberal capitalism).

[39]The classical analysis is by Van Der Pijl, 1984.

that they be allowed to play along in that game; calls for a different game altogether have become rare). From this critical point of view, the transatlantic "haves" not only have a lot in common; they also have a lot to lose – their privileged position. We shall come back to this point in our conclusions. The other point made about the transatlantic ruling class is that it is under its direction that the process of transatlantic coherence building is conducted. Translate this, if you will, into a dense elite network of actors from business live, politics and academia sharing the view that good transatlantic relations are in their own (as well as their societies) interest and you, arguably, have indeed one of the firmest sociological bases of traditional transatlantic relations.[40]

Two trends have come to undermine this basis of transatlantic elite harmony recently. On the one hand, there has been a south western shift in the composition of the US foreign policy elite. Already under President Reagan, a former governor of California, the direction of view of this US elite started to turn eastward, across the Pacific. To this must be added under President Bush, Jr., a notable Texan tinge in the foreign policy outlook. From a European point of view, a prioritization of transatlantic relations by US policy makers can no longer be taken as the standard assumption.

The other accompanying trend undermining transatlantic elite cohesion is the gradual dissolution of the *transatlantic* ruling class into what has been termed a *global* ruling class.[41] New players have emerged on the international level such as China and India, and they not only require strategic policy attention, they also have their own ruling elites which have come to play the game of capitalism, and quite well at that. Often, members of this elite (or their children at least) are trained at US universities. Personal networking across the Pacific may soon rival that across the Atlantic. While this new global elite certainly is not anti-European, sharing with Europe's elites the pro-capitalist if not completely the pro-liberal-democratic outlook, the extent to which this is a transatlantic, Europe-focused elite is certainly on the decline. The overall effect on transatlantic relations is not easily assessed: in part, as just mentioned, the classical transatlantic agenda of promoting a liberal international order is supported – even if the capitalist game continues to be played according to *nationally* determined development strategies (as is clearly the case in China); the European voice in that global concert may be heard less loudly,[42]

---

[40]This elite community has been open to the extent that, e.g., exchange programs like Fulbright or Rhodes scholarships have allowed for young students, tomorrow's elites, to make transatlantic experiences ever since the 1950s. Other reach-out activity into the European societies has been done by institutions such as America Houses. Finally, on the most popular level, transatlantic TV program exchange, though rather a one-way street, has popularized a certain image of the US in Europe which overall has attractive rather than repelling qualities (see Pells, 1997 however on the mixed effects). The resulting popular familiarity with things American is somewhat treacherous, however, since it is rather superficial.

[41]See Sklair, 2001; Robinson, 2004; Mazlish and Morss, 2005; also a liberal thinker like Dahrendorf has come to see the phenomenon (2003).

[42]The extent to which Europe has, e.g. not only a common foreign policy strategy, but also an economic development strategy (as distinguished from continuing economic nationalisms) may be

however, and even a transatlantically orchestrated voice may no longer have the weight it used to have – again a point to return to in our conclusions.

The final useful point made from a critical perspective is the inequality increasing effect of the new capitalist game now being played globally. This certainly is a hotly debated subject not easily empirically assessed, especially when inequality is not measured by (weighted) national averages,[43] but truly globally, i.e. not between nations aggregated, but across borders. Milanovic (2005, Chap. 10) sums up his carefully argued view on the issue in the phrase: a world without a middle class. To the extent that this finding is either valid, or even only easily perceived as such, it is likely to contribute to the rise of anti-globalization forces. Both, increasing inequality, if that is what we get globally, and the perception that economic globalization is one of its main reasons, should be on the mind of any sensible person in what is still a privileged transatlantic West. An increase in the feeling of being excluded from the healthy – as opposed to the negative – effects of a globalized system of Western origin may easily turn back on that West itself – and the transatlantic societies still form the core of that West. The 9/11 has certainly shown that political grievances, certainly not by the poor of the non western world but by its most angry offspring trained well in Western technology, may hit hard, and an unsustainable global petro-economy[44] is certainly one of the underlying reasons for these political grievances.[45] The other effect of trans-nationally increasing inequality which, as said, affects the developed regions of the world as well and hence the transatlantic societies themselves may also well contribute to deteriorating transatlantic relations to the extent that, as mentioned, this particular form of economic globalization is seen as Americanization. It furthers a European positioning as anti-American while at the same time undermining integration projects such as the EU constitutional treaty. To put it in a nutshell, transatlantic relations may well fall victim to the success of their own liberal-economic agenda if the unintended effects of the globalization of that model increase global inequality and if the global elite turns a cold neo-liberal shoulder to that outcome.

---

questioned. The absence of both certainly does not increase Europe's impact in either the transatlantic or the global games.

[43]These measures tend to be skewed towards a positive view by the weight given to the spectacular southern Chinese economic development over the last years: that development has happened here is hardly to be denied. The problem lies in increased poverty in other regions of the world, including increased relative poverty in developed areas.

[44]On which see Klare, 2005; Seifert and Werner, 2005; notwithstanding catchy titles both publications are to be taken seriously. Note also that emphasizing the geo-strategic relevance of the factor oil in the process leading to the Iraq war is not tantamount to subscribing to a crude "blood-for-oil"-interpretation.

[45]Try the following argument: If there was either no oil in the Middle East or no western dependence on it, wouldn't the West (represented mainly by the US) have acceded to Bin Laden's demand of drawing back from "holy territory"? Indeed: Would the West (the US) have built this presence in the region in the first place?

16.6 Conclusions

A short contribution like the present one can only provide glimpses at what – this should have become clear by our theoretically guided tour d'horizon of transatlantic relations – is a complex set of nested games. Developments both within the societies linked by transatlantic relations as well as developments at the global level beyond those relations are likely to have a decisive impact on their future development. In conclusion, let us try to distil three lessons from what has been said so far.

Transatlantic differences in longer-term attitudes as well as in concrete questions are to be expected. The fact that they *may* not only concern (material) interests but also basic value questions should not come as a complete surprise.[46] Add to this that there are multiple independent actors involved on both sides: president and congress, not to speak of the wider public, on the US side, by now 27 still rather independently inclined EU member states on the European side (with marked differences even among the "big three": France, Britain, and Germany) and transatlantic conflicts are certainly to be expected. Much hinges, therefore, on how they are dealt with. In the light of recent experience, the following can (and should) be said on (and to) both sides. As the only superpower, the US is not only more able to act unilaterally than others. It is also tempted to do so. This tendency should be resisted. International cooperation is necessary in many fields where even the only (lonely) superpower cannot fix it alone. Examples range from fighting viruses to fighting terrorists. Such cooperation may not always be easy to solicit, even from European allies. The way forward, however, lies in leadership, not in the arrogance of power.[47] For some members of the current US administration that seems hard to see. They either think that, given enough US pressure, others will follow anyway, or even seem to mistake unilateral showing of strength for leadership. This is a mistake, as is seen by many US analysts as well. They understand (and recommend) that US leadership be exerted through the use of soft power which among others means a willingness to argue – and listen to others' arguments.[48] On the European side, the advice is not to essentialize differences, even where values are concerned. After all, it probably is true that the US and European societies are closest to each other even on fundamental value questions, closer than with any other player in world politics. There is, however, an even harder to swallow lesson for Europe which often remains unspoken – and nevertheless seems hardly deniable.[49] US leadership and potential for action are practically indispensable in tackling almost all major problems of global policy. The difficulty, from a European point of view, will consist in finding ways of accepting this sheer fact without therefore renouncing own views

---

[46]Such surprise is, however, precisely what seems to have happened recently. This probably bespeaks transatlantic illusions left form Cold War times; we have to learn to live without these illusions.

[47]There was a time when leading members of he US foreign policy elite had recognized the dangers, to US interests, too, of this arrogance; see Fulbright, 1966.

[48]Nye (2003, 2004) in particular has emphasized this point; it is also made by Haass, 2005.

[49]Again, there is at least one plain statement of this point from an US analyst: Mandelbaum, 2005.

and an own voice. More thinking than recently should then go into how this voice should be brought to hearing in Washington, provided there is one European voice to be heard (again, not a naturally given condition but one to be worked on).

A second lesson to be drawn concerns the joint interests (and value commitment) the US and Europe share in the present global architecture. It was heavily shaped by US inputs, both in the form of the United Nations as the global institution, and in the economic field (through GATT/WTO and IMF and World Bank). Not only have the transatlantic partners invested in building this architecture (which is, just like NATO, an international form of social capital). They also stand to lose most from a disintegration of this global governance architecture. Short-sighted hardliners in Washington may not easily come to admit this, again seeing imperial imposition as a workable alternative. This, however, is a view refuted by reality even today – and more enlightened US thinkers recognize that the time of American preponderance may come to a close.[50] As the present security problems with the nuclear ambitions of both North Korea and Iran make clear, we are already past the point where Western concern and opposition alone are enough to determine outcomes.[51] There is, thus, a common transatlantic interest in preserving global governance architecture precisely because the times when the West alone was able to dominate outcomes are running out. This certainly is a tough selling message in today's Washington, but as pointed out, leading US thinkers in fact see the point.

Finally, for the time being, it is still true that the US and Europe have not only most to gain by acting in concert; they also have most to give, to the rest of the world: lived openness in the widest sense of that term, including a model of open argument as basis for concerted action; openness in thinking (still an important value commitment shared transatlantically – and to be imitated in many regions of the world); openness for migration (even and especially in times of terror) and finally openness in terms of economic market access. In all of these fields, transatlantic finger pointing is inappropriate. In fact, there is a shared transatlantic, Western, need to practice what has been preached: Do not sacrifice basic human rights when – jointly – fighting terrorism; do not disregard the need for social embedding of economic liberalism to (re)gain acceptability (and avoid anti-globalization backlash) without, however, letting this degenerate into rent-seeking protectionism or innovation blocking institutional sclerosis. That way, a transatlantic community and transatlantically shared interests and values will gain a chance to survive in a world beyond Western dominance.

---

[50]Brzezinski, 1993 (298ff) is clear on that point, reflecting on the time when relative US dominance will have waned; in terms of economics, Wilson and Purushothaman, 2003 have reflected on the rise of BRIC (Brazil, Russia, India and China) by 2050.

[51]See Kerr, 2005 who argues that the governance structure in the North Korean case is actually a regional concert similar to the 19th century's concert of Europe; on the Iran case see List (2007).

References

Aiginger, K. (2004). Competition between the US and Europe: which model is economically more successful? In W. Zacharasiewicz (Ed.), *Transatlantische differenzen – transatlantic differences* (pp. 197–221). Wien/Köln/Weimar: Böhlau.

Bacevich, A. J. (Ed.) (2003). *The imperial tense: Prospects and problems of American empire*. Chicago: Ivan R. Dee.

Blum, D., & Ludwig, A. (2006). Divergence Between the European Union and the USA. The Economic Perspective. In G. E. Gustenau, O. Höll, & Th. Nowotny (Eds.), *Europe – USA: Diverging Partners* (pp. 121–169). Baden-Baden: Nomos.

Brzezinski, Z. (1993). *Die einzige Weltmacht. Amerikas Strategie der Vorherrschaft*. Munich: dtv.

Chernoff, F. (1995). *After Bipolarity: The vanishing threat, theories of cooperation, and the future of the Atlantic Alliance*. Ann Arbor: University of Michigan Press.

Costigliola, F. (1984). Awkward Dominion. *American political, economic, and cultural relations with Europe, 1919–1933*. Ithaca/London: Cornell University Press.

Dahrendorf, R. (2003). Klassen ohne Kampf, Kampf ohne Klassen: Der moderne soziale Konflikt. In id. *Auf der Suche nach einer neuen Ordnung* (pp. 81–130). Munich: Beck.

Doering-Mannteuffel, A. (1999). *Wie westlich sind die Deutschen? Amerikanisierung und Westernisierung im 20. Jahrhundert*. Göttingen: Vandenhoeck & Ruprecht.

Ferguson, N. (2004). *Colossus: The price of America's empire*. London: Penguin.

Forster, A. (2006). *Armed forces and society in Europe*. Basingstoke/New York: PalgraveMacmillan.

Fukuyama, F. (1992). *The end of history and the last man*. New York: The Free Press.

Fulbright, J. W. (1966). *The arrogance of power*. New York: Random House.

Gartzke, U. (2005). Die transatlantischen Wirtschaftsbeziehungen: Ein Pfeiler in der Krise? In Th. Jäger, A. Höse, & K. Oppermann (Eds.), *Transatlantische Beziehungen. Sicherheit – Wirtschaft – Öffentlichkeit* (pp. 173–185). Wiesbaden: vs Verlag.

Gheciu, A. I. (2005). *NATO in the "NEW Europe": the politics of international socialization After the cold war*. Stanford: Stanford University Press.

Gordon, Ph. H., & J. Shapiro (2004). *Allies at war: America, Europe, and the crisis over Iraq*. New York: McGrwa-Hill.

Haass, R. N. (2005). *The opportunity: America's moment to alter history's course*. New York: Public Affairs.

Hall, P. A., & D. Soskice (Eds.) (2001). *Varieties of capitalism: The institutional foundations of comparative advantage*. Oxford: Oxford University Press.

Hamilton, D. S. (Ed.) (2004). *Transatlantic transformations: Equipping NATO for the 21st century*. Washington, DC: Center for Transatlantic Relations.

Heise, V., & P. Schmidt (2005). NATO und EU: Auf dem Weg zu einer strategischen Partnerschaft? In Th. Jäger, A. Höse, & K. Oppermann, (Eds.), *Transatlantische Beziehungen. Sicherheit – Wirtschaft – Öffentlichkeit* (pp. 65–86). Wiesbaden: vs Verlag.

Heisenberg, D. (2005). *Negotiating privacy: The European Union, the United States, and personal data protection*. Boulder/London: Lynne Rienner.

Junker, D. (2003). *Power and Mission: Was Amerika antreibt*. Freiburg i.Br.: Herder.

Kagan, R. (2003). *Of power and paradise: America and Europe in the new world order*. New York: Alfred A. Knopf.

Kalff, D. (2006). *An unamerican business: The rise of the new European enterprise*. London/Philadelphia: Kogan Page.

Katzenstein, P. J. (2005). *A world of regions: Asia and Europe in the American imperium*. Ithaca/London: Cornell University Press.

Keohane, R. O., & J. S. Nye (1977) (^{3}2000). *Power and interdependence*. Boston: Little, Brown.

Kerr, D. (2005). The Sino-Russian Partnership and US. Policy Toward North Korea: From Hegemony to Concert in Northeast Asia. *International Studies Quarterly*, *49*, 411–437.

Klare, M. (2005). *Blood and oil: How America's thirst for petrol is killing us*. London: Penguin.

Kuisel, R. (1993). *Seducing the french: The dilemma of Americanization*. Berkeley/Los Angeles/London: University of California Press.
Lind, M. (2003). *Made in Texas: George W. Bush and the southern takeover of American politics*. New York: Basic Books.
Lipset, S. M. (1996). *American Exceptionalism. A Double-Edged Sword*. New York/London: Norton.
List, M. (2004) Vereinigte Staaten von Amerika. In M. Neu, W. Gieler, & J. Bellers (Eds.), *Handbuch der Außenwirtschaftspolitiken* (Vol. 1, pp. 359–368). Münster/Hamburg: Lit-Verlag.
List, M. (2006) *Internationale Politik studieren. Eine Einführung*. Wiesbaden: vs Verlag.
List, M. (2007). Im Kern gespalten: Zur Lage des internationalen Nichtverbreitungsregimes für Kernwaffen. In A. Hasenclever (Ed.), *Macht und Ohnmacht internationaler Institutionen*. (pp. 252–282). Frankfurt a.M./New York: Campus.
Lundestad, G. (1998). *"Empire" by integration: The United States and European integration, 1945–1997*. Oxford: Oxford University Press.
Malone, D. M., & Y. F. Khong (Eds.). (2003). *Unilateralism and US foreign policy: International perspectives*. Boulder/London: Lynne Rienner.
Mandelbaum, M. (2005). *The case for Goliath: How America acts as the world's Government in the twenty-first century*. New York: Public Affairs.
Martschukat, J. (2002). *Die Geschichte der Todesstrafe in Nordamerika*. Munich: Beck.
Mazlish, B., & E. R. Morss (2005). A global elite? In A. D. Chandler, & B. Mazlish (Eds.), *Leviathans: Multinational corporations and the new global history* (pp. 167–186). Cambridge: Cambridge University Press.
Meernik, J. D. (2004). *The Political use of military force in US foreign policy*. Aldershot: Ashgate.
Milanovic, B. (2005). *Worlds apart: Measuring international inequality*. Princeton/Oxford: Princeton University Press.
Neuss, B. (2000). *Geburtshelfer Europas? Die Rolle der Vereinigten Staaten im europäischen Integrationsprozess 1945–1958*. Baden-Baden: Nomos.
Nolan, M. (1994). *Visions of modernity: American business and the modernization of Germany*. Oxford/New York: Oxford University Press.
Nye, J. S. (1990). *Bound to lead: The Changing Nature of American Power*. New York: Basic Books.
Nye, J. S. (2003). *The paradox of American power: Why the world's only superpower can't go it alone*. Oxford: Oxford University Press.
Nye, J. S. (2004). *Soft power: The means to success in world politics*. New York: Public Affairs.
Pells, R. (1997). *Not like US: How Europeans have loved, hated, and transformed American Culture Since World War II*. New York: Basic Books.
Record, J. (2003). Bounding the global war on terrorism. Strategic Studies Institute. Available at: www.carlisle.army.mil/sci/.
Risse-Kappen, Th. (1995). *Cooperation among democracies: The European influence on US foreign policy*. Princeton: Princeton University Press.
Robinson, W. (2004). Global class formation and the rise of a transnational capitalist class. In id. *A theory of global capitalism* (pp. 33–84). Baltimore/London: Johns Hopkins University Press.
Ruggie, J. G. (1983). International regimes, transactions, and change: embedded liberalism in the postwar economic order. In St. D. Krasner (Ed.), *International regimes* (pp. 195–232), Ithaca: Cornell University Press.
Rugman, A. M. (2005). *The Regional multinationals: MNEs and 'Global' strategic management*. Cambridge: Cambridge University Press.
Schildt, A. (1997). *Reisen in die Moderne: Der Amerika-Diskurs des deutschen Bürgertums vor dem Ersten Weltkrieg im europäischen Vergleich*. Berlin: Akademie Verlag.
Schröter, H. G. (2005). *Americanization of the European economy: A compact survey of American economic influence in Europe since the 1880s*. Dordrecht: Springer.
Seifert, Th., & K. Werner (2005). *Schwarzbuch Öl: Eine Geschichte von Gier, Krieg, Macht und Öl*. Wien: Deuticke.

Sklair, L. (2001). *The transnational capitalist class*. Oxford: Blackwell.

Snyder, J. (1991). *Myths of empire: Domestic politics and international ambition*. Ithaca/London: Cornell University Press.

Sterling-Folkert, J. (Ed.) (2006). *Making sense of international relations theory*. Boulder/London: Lynne Rienner.

Streissler, A. (2006). Europe: USA. A Comparison of the Social Policy Systems. In G. E. Gustenau, O. Höll, & Th. Nowotny (Eds.), *Europe – USA: Diverging partners* (pp. 47–120). Baden-Baden: Nomos.

Snow, D.M., & Drew, D.M. (1994). *From lexington to desert storm: War and politics in the American experience*. Armonk/London: M.E. Sharpe.

Van Der Pijl, K. (1984). *The making of an Atlantic ruling class*. London: Verso.

Walt, St. M. (2005). *Taming American power: The global response to US primacy*. New York/London: Norton.

Whitman, J.Q. (2003). *Harsh Justie: Criminal Punishment and the widening divide between America and Europe*. Oxford: Oxford University Press.

Wilson, D., & Purushothanam, R. (2003). Dreaming with BRICs: The Path to 2050. Global Economics Paper No.99. Available at: www.gs.com/insight/resaerch/reports/99.pdf .

Yergin, D., & Stanislaw, J. (1998). *Commanding hights: The battle between government and the market place that is remaking the modern world*. New York: Simon&Schuster.

Zacharasiewicz, W. (Ed.). (2004). *Transatlantische Differenzen – Transatlantic differences*. Wien/Köln/Weimar: Böhlau.

Chapter 17
Policy Issues and Fields of Cooperation

Jackson Janes

17.1 Introduction

When I began my graduate studies in international relations thirty years ago, I was immediately immersed in the work of Thomas Kuhn and his book "*The Structure of Scientific Revolutions*". The intent was to begin our approach to the study of global affairs with some questions about the role of what Kuhn called paradigms: Paradigms understood as maps of the theory, methods and standards used to understand problems and challenges. Changes in those paradigms, Kuhn argued, caused shifts in the criteria determining the legitimacy of problems and proposed solutions. What caused those changes constituted the revolutionary process when one paradigm could no longer explain the world and another was needed.

Yet it need not take a scientist to recognize that need. Nothing is more revolutionary in world politics than the unexpected event and the unanticipated individual and their impact on history. British Prime Minister Harald Macmillan was asked by a reporter to identify his biggest problem in office. He responded "Events, dear boy, events."

In looking at the last three decades, we can see continuous examples of the revolutions unfolding in the form of determined Polish workers in a shipyard in Gdansk, a committed prisoner on Robben Island in South Africa, a questioning Soviet leader in Moscow, and millions of brave people from Leipzig to Kiev willing to change their paradigms.

I understand that you have been discussing the nuts and bolts of major economic policy issues of concern on both sides of the Atlantic. And I am somewhat intimidated to be standing here trying to offer insights to such a prestigious cast of experts on policy issues. The agenda you have set for yourselves covers a wide range of challenges all in need of solutions, and perhaps a change in paradigms.

J. Janes (✉)
AICGS/The Johns Hopkins University, Washington, USA

P.J.J. Welfens, J.T. Addison (eds.), *Innovation, Employment and Growth Policy Issues in the EU and the US*, DOI 10.1007/978-3-642-00631-9_17,

What I would like to do is to try and provide flank support to all of you by underlining why we need to be thinking about how our paradigms or our maps of these individual issues need to add up to a common agenda shared not only by Germany and the US but in a global framework.

Let me start with the premise that the last sixty years of transatlantic relations and German-American relations in particular, is a remarkable and unique success story in the annals of world history. Out of the ashes of World War II, we have seen Germany and Europe emerge to embody many of the visions of those who were present at the creation of the Federal Republic and the European communities. In essence, leaders on both sides of the Atlantic shared a map which was once a point of shared orientation about the world, but, on the basis of a shared sense of legitimacy about our goals, also helped us plot a common course into the future. That map evolved along with the changes and challenges we confronted as did our methods of dealing with them. On occasion, we would argue about those methods and standards, but we had general agreement on the theory: where our interests and our goals overlapped. In the arena of German-American relations, nothing captured that more dramatically than on the steps of the Reichstag on October 3, 1990 when the world welcomed a unified Germany.

But much more changed in the following years in Eastern Europe, Russia, and the Balkans that our Cold War paradigm was in urgent need of replacement. But what were the new ones to be? What were our problems, our methods for dealing with them, our standards for measuring success? How were we to explain and stop the genocide happening on the doorsteps of Europe in the Balkans, or in Africa? What was changing in the Middle East in the wake of Iraq's invasion of Kuwait? How should we deal with China after Tiananmen Square? What were we to make of the Taliban in Afghanistan? The raging Intifada in the Middle East? What was the meaning of the attack on the World Trade Towers in 1993? And who has what responsibility for what burdens within the transatlantic alliance, the most successful alliance in history?

And those were just the challenges of the Nineties.

After 9/11, the US and the world were shaken by threats no longer coming from states but from non-state actors, indeed individuals equipped with mass killing capabilities. Responding to them meant the need for new methods, standards and indeed theories to explain where they emerged and how to combat them. Once defined as Mutual Assured Destruction during the Cold War, the paradigm of MAD now meant Mutual Assured Interdependence but along with that came not only mutual interests but also a reaction to a new sense of vulnerability.

In the 20th century we had fought states that were too threatening; in this century we are facing states that are too weak – failing states as we call them – that become havens for those seeking to take advantage of these weaknesses and who shun the interdependence brought on by globalization.

In a world full of imbalances and asymmetries, the centripetal forces of globalization are pulling countries and populations ever closer together, yet those same forces are generating centrifugal forces within and among nations as they try to adapt. In the past ten years, we have seen many painful illustrations of the dark side of a globalizing world in the killing fields of Srebenica, Ruanda and Dafur, the extreme poverty in half the world, in the ethnic and religious tensions arguing over what the purpose of life is and the codes by which it should be lived.

Conflicts over energy supplies, nationalist tensions, and separatist movements spawning terrorism continue to evolve, creating more vulnerability and demands for protection and security measures among, and within, states and regions. The need to understand our problems demands a new understanding of mutual security in the 21st century. It was not longer a paradigm of territorial security. It had to include achieving stability with justice and inclusion in a globalizing world.

17.2 EU-US Perspectives

Amidst all these developments, the European Union represents the promise and purpose of shared sovereignty and the ability of states to find common ground institutionally. This is one of the most important experiments in world history when it comes to generating a new paradigm of national self-interest and methods and standards to pursue them jointly. It has never been accomplished on this scale anywhere else. And the process, despite its setbacks, is continuing. The United States supported that process as a method of achieving stability and democracy in a region which had given birth to two world wars during the first half of the 20th century.

In the wake of the debate over the constitution, the EU is challenged to assess how that process can best be nurtured without losing the support and commitment of the citizens of each member state, who are coping with their own sense of dislocation, anxiety, frictions and searching for a sense of legitimacy in making adjustments and changes to a global environment. The forces of European politics, particularly those that converge in Brussels, are still driven by national leaders with their individual priorities, all vulnerable to elections.

All of these developments demand a rethinking of our definition of security at multiple levels and well beyond that of strictly military security. Just as we no longer speak of the East-West divide in Europe, our approach to concepts such as the Third World, the North-South Divide, and other alignments from the Cold War period are now becoming obsolete. Concepts such as national sovereignty are also in transition due to the increasing bonds of globalization, sometimes freely chosen such as in the EU and other times forced by the economic and technological developments reshaping our choices, and by new forms of vulnerabilities. Non-state actors, such as global companies and NGOs or organized crime, and of course groups like Al-Quaeda become increasingly important in this new paradigm, for better and too often for worse.

The tools with which we have managed our foreign policies heretofore, including the international institutions created after World War II such as the UN, the World Bank, and the IMF, will all be challenged to adapt to the new parameters of security, which will require enormous investments in economic, political and social development in regions of the world serving as incubators for terrorism and resentment.

The twenty-first century will rearrange the cast of political actors on the world stage and we will need to know how to negotiate with all of them, one way or the other. Yet the question is: in the name of what theory, what vision, what shared goal do we enter into that discussion?

For the foreseeable future, the United States will remain the most powerful player given its predominance in so many areas. But American leadership will be increasingly challenged from many sides. With an economy that is vulnerable to fluctuations in a globalized world, with a high level of dependence on foreign oil and demand for energy and with challenges on a range of issues dealing with environmental policies, trade, and social concerns, the US will need to pursue its security interests by making sure that it can manage conflict as well as cooperation and deal with threats within a complicated web of interdependence. That is a challenge to both, the US foreign policy as well as the domestic debate about American priorities. With one fifth of the US economy financed by foreign sources, the choices facing policy-makers in Washington are going to be driven by external and domestic concerns, requiring compromise and cooperation within the world wide web, not only confrontation. And we will need a strong link with the core of those who share the paradigm.

At the moment, the transatlantic links are the strongest within this web. They make up more than half of worldwide trade and investments to the tune of 2.5 trillion dollars. With a total of almost 800 million consumers shared across the Atlantic, there is literally much at stake in maintaining a secure relationship, even allowing for the inevitable conflicts and competition we will see within this web. In this transatlantic arena we share profits, investments, jobs and markets. But more importantly, we share methods and standards – or we should – to create a common paradigm in dealing with adjusting to this intimate relationship. German-American tensions over the war in Iraq did not dampen the appetite for profits on either side of the Atlantic with sixty five billion dollars channeled in foreign direct investment from the US into Germany and more than double that from Germany aimed at the US in that tension filled year of 2002.

Economic interdependence, however does not necessarily translate directly into a sense of political interdependence. That was clearly evident in the clash over Iraq. This is in part due to the fact that it is increasingly difficult to separate domestic from foreign policy when it comes to the security issues we face today. For that reason, if the perceptions of security, social, political, military, differs between the respective domestic debates on either side of the Atlantic, there is fertile ground for friction. That could involve everything from homeland security to food safety, from data protection to social security policy, as you have been discussing here today. As each society attempts to cope with the domestic economic challenges and structural weaknesses, none of which can be solved in isolation from the global forces at work, finding common ground will be a dual challenge at home and abroad. If Americans believe that having access to a job with lower wages is better than a higher level of unemployment insurance, does that constitute a value clash or an opportunity to measure best practices against common standards? If the EU has enormous economic capability but cannot generate the political or military capacities to match it, does that mean we cannot have a security dialogue or does it mean that we need to find better ways to pool our resources.

Germany's economic stagnation during the last decade has been a burden not only on Germany but also on Europe and the global economy as a whole. On the other hand, increasing American deficits raise fears for long-term financial and economic stability throughout the world. But the debates in Berlin and Washington also reflect the perception of different choices acceptable based on the paradigm of how a market economy can best function. In Germany, those who fear importing American-style conditions point at the imbalances and asymmetries in American society in terms of social security or health care, whereas American critics of the German conditions point at the record of static economic growth and question German choices as a viable path toward reforms.

What we are seeing in these debates is an increasing interweaving of our domestic arguments about common concerns, without having the structures or institutions to engage them together. The trade offs within our respective debates are no longer strictly domestic but they become part of our shared economic and political space. Creating opportunities to explore that space together with the objective of finding acceptable compromises or best solutions has been a challenge at the level of EU-US relations in dealing with everything from regulatory issues to international financial markets to global poverty. Yet it seems that the institutions have not evolved sufficiently to address this need and we have been sometimes overly dependent on the luck of the draw in individual leaders taking that initiative within the current channels.

Our transatlantic arguments are always a mixture of economic interests, domestic lobby groups and contrasting frameworks of popular debates. Arguments over how to deal with regulatory issues, be it corporate governance, product standards or privacy, have been challenging national and international frameworks to come up with ways of dealing with coordinated approaches to policies in order to avoid protracted conflicts. The US reaction to its corporate scandals produced the Sarbanes Oxley legislation to improve corporate oversight but was also the source of enormous tensions between transatlantic regulators as they tried to find common ground between German and American systems of corporate regulation. Indeed, recent criticism has also come from American courts. Charges of US exertion of extraterritorial authority may not be new but they have only grown since 9/11. One thinks here of the debate over the exchange of airline passenger information, or dealing with the provisions of seaport security.

We also find ourselves engaged in the future of scientific frontiers, be it in stem cell research, environmental developments or genetic discoveries, areas in which we ought to be able to learn from each other but often find ourselves all too wrapped up in our domestic debates about ethics and legal frameworks that we are unable to secure common ground to explore what we know and what we don't on either side of the Atlantic.

After September 11, the public debate among Americans quickly accepted the notion of being in a state of war, while the Europeans on the whole perceived the terrorist threats as dangers but not as an object toward which you could declare war but rather as means chosen by groups to achieve their goals. While there was an agreement on the dangers, there was less of a consensus on how to proceed against them. Parts of our paradigms were out of sync. And when that happens, as I said at the outset, there are gaps in the legitimacy with which we see a paradigm able to address the changes and challenges we face.

The Bush administration's understanding of what security means was most clearly outlined in the National Security Strategy (NSS) in the fall of 2002. While the language, which most Europeans focused on, spoke of American power beyond challenge, and preemptive and preventive actions against threats, the main thrust of the document was to present a vision of the world shaped by free markets, human rights, the rule of law and democracy as the formula to rid the world of the roots of terrorism with the added dimension of a military response to terrorist organizations. A year later, the European Security Strategy document was presented as a response to the American version. While there was agreement on challenges such as non-proliferation or homeland security goals, how to achieve those goals remained controversial.

The conflict with Europe that emerged over Washington's decision to remove Saddam Hussein was at one level less about whether to get rid of Hussein but rather about the best means to do it. The European preference for diplomacy, sanctions, engagement and dialogue contrasted with the Bush approach to cutting through the problem with a sharp knife and opening up the possibilities of change, risking both better and worse scenarios. While many of the transatlantic arguments over the past decades have been about means rather than ends, the serious clash over the means to deal with Saddam Hussein seem to represent a breakdown in consensus on what constitutes agreed principles to respond to threats. Whether this was an exception – there had been a transatlantic consensus on dealing with the conflict in the Balkans even without a UN mandate for example – depends on whether we can find a process to respond to the many future threats we need to expect.

While we need to address the need to increase our harmonization within the transatlantic arena, our mutual security is threatened primarily by factors outside the framework of US-EU relations. And it is the particular region of the Middle East that holds the most dangers and also causes the most arguments across the Atlantic as to how to deal with that danger.

Parallel to the sharper focus on the Middle East, there is also the loss of the centrality of Europe as a geopolitical concern for the US. The East-West conflict was symbolized most clearly in Berlin where the exposure to threats was most visible at the Berlin Wall. Today's conflicts are elsewhere and it is the United States that now feels the same vulnerability that was felt in Germany for so many years during the Cold War. Since September 11, the US sees itself as the primary target of terrorist threats, which lies at the heart of the Bush White House strategy to respond.

The organizing principles or paradigms we used for the Cold War are not going to transfer to the situations we face today, where we are not talking primarily about standing armies on either side of a very visible line, but rather a very tangled web of contradictions and conflicts constantly in flux.

Yet again we have to balance the fact that the nuclear threat runs parallel to concerns about human rights abuse, support for terrorists groups, the despair of the increasing numbers of young people without jobs, and the dangers of militant Islamic groups. We cannot see Middle East region simply through the single lens of Iraq, we need to see the entire region through multiple lenses of concerns we share.

There are many other test cases for European-US cooperation: Ukraine, the Balkans, Afghanistan where there are clearly common sets of goals to maintain stability and security. Furthermore we have to continue to explore how we can deal with issues such as the pandemic of HIV, access to clean water not available to millions of people in Africa, and adequate food. As Timothy Garton Ash tells us, it is sobering to remember that "on September 11, 2001, the [same] day just over 3,000 people were killed in the al-Qaeda attacks on New York and Washington, some 30,000 children died around the world from preventable disease. And the next day. And the day after that. And every day of the year."

Looking at how the European and American publics reacted in unison to the tragic disaster in Southeast Asia during the past winter, it is striking how much synergy and common goals were pursued in such a short period of time when the focus was so dramatic. How do we channel that same capacity on a more regular basis to deal with the permanent problems we face around the globe?

During the last three years, the White House has done a very poor job of explaining how and why we need to change our way of thinking about security in the 21st century. Bush may have changed the status quo as we knew it in the 20th century, but the White House did not sufficiently explain how and why we need to change our way of thinking about security in the 21st century. At the same time, Europeans have been slow in generating their own security dialogue and policies and will continue to struggle with the forming of a European Security and Defense policy,

In the next few years, events we can anticipate, and those we cannot, will offer a number of opportunities to find practical steps to deal with the dangers we face. However, just as we need to revise our thinking shaped by the experiences in the Cold War, we also need to revise our understanding of our goals and our self-perceptions. What will be the content of the West as a basis for that understanding? Where do we draw our borders in defining who we are? That will require a new definition of what we are or should be, rather than of what we are not. It will require a paradigm which can help explain our goals and our methods for reaching them.

The challenge we face now is sorting out our questions, let alone our answers. Germany and the EU make up the most important partner for the United States in securing global stability, which is indeed a necessary requirement for both partners. Indeed, the US is in need of a partner, which can help it avoid an isolationistic attitude and demonstrate the value of harmonization of perspective and policies. In turn, European developments today underscore the need for a partner which consistently points toward the global agenda and helps to avoid self-absorption in a huge internal arena.

Fore those of us, who know the numbers, appreciate the levels of integration, and see the values of expanding our ties, there is a working paradigm. But for all too many people on the globe, there is none, indeed not even a glimmer. Or there are those whose paradigm is not to create a future but to recreate a past. And we cannot see solutions to the gaps in our world by building walls around them, as every German must know so well.

In the end, my message to you who make up the world of economic expertise is that security and economics are intimately connected in our efforts to generate a new paradigm for security in the 21rst century. There needs to a greater and more comprehensive dialogue between those who wish to confront violence with defense and those who seek to build an enlarged market connecting the six billion people on the globe with a stake in a common future. We need both sides of the dialogue to speak and listen to each other. And the experience of the last half century across the Atlantic should serve as a reminder and a benchmark to see how successful that dialogue can be if it is based on a shared map of the present and the future.